JOURNAL OF APPLIED LOGICS - IFCOLOG
JOURNAL OF LOGICS AND THEIR APPLICATIONS

Volume 6, Number 5

August 2019

Disclaimer

Statements of fact and opinion in the articles in Journal of Applied Logics - IfCoLog Journal of Logics and their Applications (JALs-FLAP) are those of the respective authors and contributors and not of the JALs-FLAP. Neither College Publications nor the JALs-FLAP make any representation, express or implied, in respect of the accuracy of the material in this journal and cannot accept any legal responsibility or liability for any errors or omissions that may be made. The reader should make his/her own evaluation as to the appropriateness or otherwise of any experimental technique described.

ISBN 978-1-84890-311-1
ISSN (E) 2631-9829
ISSN (P) 2631-9810

College Publications
Scientific Director: Dov Gabbay
Managing Director: Jane Spurr

http://www.collegepublications.co.uk

Editorial Board

SCOPE AND SUBMISSIONS

This journal considers submission in all areas of pure and applied logic, including:

pure logical systems
proof theory
constructive logic
categorical logic
modal and temporal logic
model theory
recursion theory
type theory
nominal theory
nonclassical logics
nonmonotonic logic
numerical and uncertainty reasoning
logic and AI
foundations of logic programming
belief change/revision
systems of knowledge and belief
logics and semantics of programming
specification and verification
agent theory
databases

dynamic logic
quantum logic
algebraic logic
logic and cognition
probabilistic logic
logic and networks
neuro-logical systems
complexity
argumentation theory
logic and computation
logic and language
logic engineering
knowledge-based systems
automated reasoning
knowledge representation
logic in hardware and VLSI
natural language
concurrent computation
planning

This journal will also consider papers on the application of logic in other subject areas: philosophy, cognitive science, physics etc. provided they have some formal content.

Submissions should be sent to Jane Spurr (jane.spurr@kcl.ac.uk) as a pdf file, preferably compiled in LaTeX using the IFCoLog class file.

CONTENTS

Introduction to Legal AI

Livio Robaldo
University of Luxembourg
`livio.robaldo@uni.lu`

Leendert van der Torre
University of Luxembourg
`leon.vandertorre@uni.lu`

Legal AI is the research area concerned with the AI-driven processing of norms occurring in legislation and related documents (jurisprudence, international standards, doctrine, etc.), in order to achieve compliance of the systems with the regulations in force.

Compliance checking in computer systems is the process of ensuring that the specification requirements of such systems are in accordance with prescribed and/or agreed set of norms, a.k.a. compliance requirements. Compliance requirements may stem from legislation and regulatory bodies (e.g., Sarbanes-Oxley, Basel II, HIPAA), standards and codes of practice (e.g., SCOR, ISO9000), and business partner contracts.

Two fundamental strategies are identified in the literature to characterize norm enforcement and the concept of compliance in computer systems.

First, norms may be hard constraints and the system compliance is achieved by design. This option is usually implemented by adopting the so-called norm regimentation strategy, which can amount to designing the system in such a way as illegal states are ruled out and made impossible in it, or by imposing that the occurrence of any illegal states is, in theory, possible but leads to signaling a system failure.

Second, norms are soft constraints and so do not limit in advance the system's behavior. Compliance is then ensured by system mechanisms stating that violations should result in sanctions or other normative effects to recover from violations. In general, certain situations must be avoided by design, for example any serious failure affecting the system's overall functionality: norms can be modeled here as

This special issue is supported by the European Union's Horizon 2020 research and innovation programme under the Marie Sklodowska-Curie grant agreement No 690974 for the project "MIREL: MIning and REasoning with Legal texts".

hard constraints. In other cases, where it is of paramount importance to design flexible and adaptive systems, coordination and social models are used to set up self-organizing systems: whenever the overall functionality of the system is not directly in jeopardy, then norms (as soft constraints) can play in this second case a decisive role to guide and control the desired system behavior.

Legal AI has recently received a lot of investments from industry and institutions, due to the well-known rise of RegTech and FinTech, which is in turn due to the 2008 global financial crisis. Legal scholars and practitioners are feeling increasingly overwhelmed with the expanding set of legislation and case law available these days, which is assuming more and more of an international character. Consider, for example, European legislation, which is estimated to be 170,000 pages long, of which over 100,000 pages have been produced in the last ten years.

The European Union (EU) chose as one of its primary objectives to establish an integrated and standardized system of laws that applies in all member states. Furthermore, legislation is available in unstructured formats, which makes it difficult for users to cut through the information overload. As the law gets more complex, conflicting, and ever changing, more advanced methodologies are required for analyzing, representing and reasoning on legal knowledge.

Two main sequential steps may be identified in Legal AI systems: legal mining and legal reasoning.

Legal mining is the application of Natural Language Processing (NLP) methods to the source texts, in order to extract data, classify the documents fit to facilitate navigation and search.

Legal reasoning is concerned with the development of AI applications able to infer new knowledge and actions from the data mined via legal mining. In other words, legal reasoning aims at speed up the work that is currently mostly done by humans by automatizing repetitive operations, which are rather common in the legal domain, and performs automatic checks to support and monitor compliance assessment, risk analyses, etc.

The Horizon 2020 Marie Skłodowska-Curie Research and Innovation Staff Exchanges (RISE) project "MIREL - MIning and Reasoning with Legal texts" aims at bridging legal mining and legal reasoning, whose communities have previously w orked mostly in isolation.

In the past, research in Legal AI had primarily focused on legal mining, specifically on the design and implementations of legal document management systems for helping legal professionals to retrieve the information they are interested in.

Legal document management systems classify, index, and discover inter-links between legal documents by exploiting Natural Language Processing (NLP) tools such as parsers and statistical algorithms as well as semantic knowledge bases, such as legal ontologies in Web Ontology Language (OWL). This is often done by transforming the source legal documents into XML standards, where relevant information is tagged. The XML files are then archived and queried in subsequent phases. Although these techniques provide valid solutions to help navigate legislation and retrieve information, the overall usefulness and effectiveness of the systems are limited due to their focus on terminological issues and information retrieval while disregarding the specific semantic aspects of law, in particular its logical structure in terms of constitutive and regulative rules, which allows legal reasoning.

The major known problem to bridge mining and reasoning is the handling of multiple legal interpretations of the provisions. Being natural language ambiguous, legal interpretations are the different context-specific (pragmatic) interpretations of the terms and sentences occurring in the source texts. What makes legal texts so much dependent on human interpretation is that they are used in disputes that represent different interests, so that interpretation of terms tends to be stretched to the maximum. Since it is impossible to predict every possible context where the provisions will be deployed, legislators tend to use vague terms, which are used to account for the multitude of situations that should be covered by the abstract legislation, which often depend on the legal cases as they occur (precedent cases), and on the reflections of legal doctrine. It is eventually up to judges and other appointed authorities to decide the "final" interpretation of provisions in contexts. However, some contexts are clearly borderline, so that it is quite common that different judges adopt different legal interpretations, incompatible among themselves (sometimes even concerning identical cases). For legal reasoning, handling multiple interpretations, to be possibly computed via exclusive disjunction, of course introduces another layer of complexity.

Legal ontologies are the main instrument to bridge mining and reasoning from legal texts. Legal ontologies provide a comprehensive framework to model main legal concepts at stake, as they explicitly describe reality. In Artificial Intelligence, the objective is to provide people, or more typically artificial agents, with structured and navigable knowledge about classes, individuals, and their inter-relations, thus allowing them to share and reference knowledge about concepts in general or specific areas, abstracting from the languages in which these concepts are expressed. Ontologies are ultimately used in information technology applications for semantic searches, interoperability between systems, or to facilitate reasoning and problem solving. In Legal Informatics, ontologies are useful in a range of different scenarios. They could help legal practitioners and scholars keep up to date with continuous changes in the law and understand legal sub-languages outside their own areas of expertise or jurisdiction. They could help legislators draft legislation with clarity and consistency. Moreover, they could help identify the inter-relationship between general jurisdictions and specific related ones, e.g., between the jurisdiction of the European Union and the ones of the Member States, in order to foster harmonization.

The first two contributions of this special issue on Reasoning for Legal AI are concerned with computational approaches to deontic logic. Christoph Benzmüller, Ali Farjami, Paul Meder and Xavier Parent show how to provide a shallow semantical embedding in higher-order logic of the input/output logics called out2 and out4, as well as Åqvist's dyadic deontic logic called E. These embeddings are also encoded in Isabelle/HOL, which turns the system into a proof assistant for deontic logic reasoning. The experiments with this environment illustrate interactive and automated reasoning at the meta-level and the object-level. Moreover, the first embedding is applied to an example of moral luck.

The following two contributions describe concrete legal AI systems. Cleyton Rodrigues, Eunice Palmeira, Fred Freitas, Italo Oliveira and Ivan Varzinczak present the prototype LEGIS and discuss a proposal to handle legal normative exceptions and leverage inference Proofs Readability. The Sequent Calculus is used as a formal logic argumentation style to achieve a higher level of legibility. Guillaume Aucher, Jean Berbinau, and Marie-Laure Morin describe a Judgement Editor based on theoretical principles for Binary Decision Diagrams (BDDs).

Focusing on the compliance and reification problems, Guido Governatori and Antonino Rotolo introduce time and compensation mechanisms for checking legal compliance. Silvano Colombo Tosatto, Guido Governatori, Nic Van Beest and Francesco Olivieri provide efficient full compliance checking of concurrent components for business process models, and Régis Riveret, Antonino Rotolo and Giovanni Sartor provide a deontic argumentation framework towards doctrine reification.

Finally, three contributions focus on challenges in legal reasoning. Rafal Urbaniak argues that probabilistic legal decision standards still fail, and Réka Markovich argues that criminal law is not a limitation of the general applicability of the Hohfeldian theory of rights and duties and their correlativity. She presents an analysis of sanction in terms of rights and duties in order to resolve the seemingly paradoxical situation of legal systems in which one has the right to escape from prison. Moreover, in a second paper she discusses two limitations in legal knowledge-base constructing and formalizing law.

 Received August 2019

I/O Logic in HOL

Christoph Benzmüller
Freie Universität Berlin, Germany, and University of Luxembourg, Luxembourg
`c.benzmueller@gmail.com`

Ali Farjami
University of Luxembourg, Luxembourg
`ali.farjami@uni.lu`

Paul Meder
University of Luxembourg, Luxembourg
`paul.meder@uni.lu`

Xavier Parent
University of Luxembourg, Luxembourg
`xavier.parent@uni.lu`

Abstract

A shallow semantical embedding of Input/output logic in classical higher-order logic is presented, and shown to be faithful (sound and complete). This embedding has been implemented in Isabelle/HOL, a proof assistant tool. It is applied to a well-known example in moral philosophy, the example of moral luck.

Keywords: Input/output logic; Classical higher-order logic; Isabelle/HOL; Moral luck.

This work has been supported by the European Union's Horizon 2020 research and innovation programme under the Marie Skłodowska-Curie grant agreement No 690974 – MIREL – MIning and REasoning with Legal texts. Benzmüller has been funded by the Volkswagen Foundation under project CRAP – Consistent Rational Argumentation in Politics.

1 Introduction

Deontic logic is concerned with normative concepts like obligation, permission, and prohibition. On the one hand, we have the family of traditional deontic logics which includes Standard Deontic Logic (SDL), a modal logic of type **KD**, and Dyadic Deontic Logic (DDL) [1, 15, 16]. On the other hand, we have so-called "norm-based" deontic logics. The deontic operators are evaluated not with reference to a set of possible worlds but with reference to a set of norms. A particular framework that falls within this category is called Input/Output (I/O) logic [18]. It has gained recognition from the AI community, and has a dedicated chapter in the handbook of deontic logic [15]. The framework is expressive enough to support reasoning about constitutive, prescriptive and defeasible rules; these notions play an important role in the legal and ethical domains [13].

Our focus is on two I/O logics called *Basic Output* and *Basic Reusable Output*. We present an embedding of them into classical Higher-Order Logic (HOL), also known as simple type theory [14, 5], and study their automation. The syntax and semantics of HOL are well understood [3] and there exist automated proof tools for it; examples include Isabelle/HOL [21], LEO-II [10] and Leo-III [23]. Our approach is an indirect one. We take advantage of the known possibility of embedding I/O logic into modal logic, and we reuse the *shallow semantical embedding* of modal systems **K** and **KT** in HOL [8]. In related work, Benzmüller et al. [4, 6] developed analogous shallow semantical embeddings for some well-known dyadic deontic logics.

The embeddings presented in this article are faithful (sound and complete). They are also encoded in Isabelle/HOL to enable experiments. As an illustration of the kind of experiments the framework enables, we use a well-known example in moral philosophy, the example of moral luck [20]. This term refers to situations where an agent receives moral praise or blame for an action and its consequences even though he did not have full control over them. In particular, a classical scenario of moral luck, known as the *Drink and Drive* example, is used as an illustration.

The article is structured as follows: Section 2 gives a quick review of modal logic and higher-order logic, and Section 3 introduces I/O logic. The semantical embeddings of *Basic Output* and *Basic Reusable Output* in HOL are then described in Section 4. This section also shows the faithfulness of the embeddings. In Section 5 we apply the framework to the *Drink and Drive* example.

2 Preliminaries

In this section, we recap some important notions from modal logic and HOL.

2.1 Modal logic K

The language of **K** is obtained by supplementing the language of Propositional Logic (PL) with a modal operator $\Box$. It is generated as follows:

$$\varphi ::= p|\neg\varphi|\varphi \vee \varphi|\Box\varphi$$

where p denotes an atomic formula. Other logical connectives such as $\wedge$, $\rightarrow$ and $\Diamond$, are defined in the usual way. The axioms of system **K** consist of those of PL plus $\Box(\varphi \rightarrow \psi) \rightarrow (\Box\varphi \rightarrow \Box\psi)$, called axiom K. The rules of **K** are *Modus ponens* (from φ and $\varphi \rightarrow \psi$ infer ψ) and *Necessitation* (from φ infer $\Box\varphi$).

A Kripke model for **K** is a triple $M = \langle W, R, V \rangle$, where W is a non-empty set of possible worlds, R is a binary relation on W, called accessibility relation, and V is a function assigning a set of worlds to each atomic formula, that is, $V(p) \subseteq W$.

Truth of a formula φ in a model $M = \langle W, R, V \rangle$ and a world $s \in W$ is written as $M, s \models \varphi$. We define $V(\varphi) = \{s \in W | M, s \models \varphi\}$. The relation $\models$ is defined as follows:

$$
\begin{array}{lll}
M, s \models p & \text{if and only if} & s \in V(p) \\
M, s \models \neg\varphi & \text{if and only if} & M, s \not\models \varphi \text{ (that is, not } M, s \models \varphi) \\
M, s \models \varphi \vee \psi & \text{if and only if} & M, s \models \varphi \text{ or } M, s \models \psi \\
M, s \models \Box\varphi & \text{if and only if} & \text{for every } t \in W \text{ such that } sRt, M, t \models \varphi
\end{array}
$$

As usual, a modal formula φ is *true in a Kripke model* $M = \langle W, R, V \rangle$, i.e., $M \models \varphi$, if and only if for all worlds $s \in W$, we have $M, s \models \varphi$. A formula φ is *valid in a class $\mathcal{C}$ of Kripke models*, denoted as $\models_\mathcal{C} \varphi$, if and only if it is true in every model in class $\mathcal{C}$.

System **K** is determined by (i.e., is sound and complete with respect to) the class of all Kripke models. System **KT** is obtained from system **K** by adding the schema T : $\Box\varphi \rightarrow \varphi$ as an axiom. System **KT** is determined by the class of all Kripke models in which R is reflexive. We denote the class of all Kripke models and the class of Kripke models where R is reflexive as $\mathcal{C}_K$ and $\mathcal{C}_{KT}$, respectively.

Two other axiom schemas that can be added to **K** are 4 : $\Box\varphi \rightarrow \Box\Box\varphi$ and 5 : $\Diamond\varphi \rightarrow \Box\Diamond\varphi$. For instance, **K45** is an extension of **K** obtained by adding 4 and 5 as axioms. The schemas 4 and 5 are valid if R is *transitive* and *euclidean*, respectively.

2.2 Classical higher-order logic

HOL is based on simple typed λ-calculus. We assume that the set $\mathcal{T}$ of simple types is freely generated from a set of basic types $\{o, i\}$ using the function type constructor

$\rightarrow$. Type o denotes the set of Booleans whereas type i refers to a non-empty set of individuals.

For $\alpha, \beta, o \in \mathcal{T}$, the *language of HOL* is generated as follows:

$$s, t ::= p_\alpha \,|\, X_\alpha \,|\, (\lambda X_\alpha s_\beta)_{\alpha \to \beta} \,|\, (s_{\alpha \to \beta}\, t_\alpha)_\beta$$

where p_α represents a typed constant symbol (from a possibly infinite set $\mathcal{P}_\alpha$ of such constant symbols) and X_α represents a typed variable symbol (from a possibly infinite set $\mathcal{V}_\alpha$ of such symbols). $(\lambda X_\alpha s_\beta)_{\alpha \to \beta}$ and $(s_{\alpha \to \beta}\, t_\alpha)_\beta$ are called *abstraction* and *application*, respectively. HOL is a logic of terms in the sense that the *formulas of HOL* are given as terms of type o. Moreover, we require a sufficient number of primitive logical connectives in the signature of HOL, i.e., these logical connectives must be contained in the sets $\mathcal{P}_\alpha$ of constant symbols. The primitive logical connectives of choice in this paper are $\neg_{o \to o}$, $\vee_{o \to o \to o}$, $\Pi_{(\alpha \to o) \to o}$ and $=_{\alpha \to \alpha \to o}$. The symbols $\Pi_{(\alpha \to o) \to o}$ and $=_{\alpha \to \alpha \to o}$ generally assumed for each type $\alpha \in \mathcal{T}$. From the selected set of primitive connectives, other logical connectives can be introduced as abbreviations. Type information as well as brackets may be omitted if obvious from the context, and we may also use infix notation to improve readability. For example, we may write $(s \vee t)$ instead of $((\vee_{o \to o \to o}\, s_o)\, t_o)_o$. We often write $\forall X_\alpha s_o$ as syntactic sugar for $\Pi_{(\alpha \to o) \to o}(\lambda X_\alpha s_o)$.

The notions of *free variables*, α-*conversion*, $\beta\eta$-*equality* and *substitution* of a term s_α for a variable X_α in a term t_β, denoted as $[s/X]t$, are defined as usual.

The semantics of HOL are well understood and thoroughly documented [3]. In the remainder, the semantics of choice is Henkin's general models [17].

A *frame* D is a collection $\{D_\alpha\}_{\alpha \in \mathcal{T}}$ of nonempty sets D_α, such that $D_o = \{T, F\}$, denoting truth and falsehood, respectively. $D_{\alpha \to \beta}$ represents a collection of functions mapping D_α into D_β.

A *model* for HOL is a tuple $M = \langle D, I \rangle$, where D is a frame and I is a family of typed interpretation functions mapping constant symbols p_α to appropriate elements of D_α, called the *denotation* of p_α. The logical connectives $\neg$, $\vee$, Π and $=$ are always given in their expected standard denotations. A *variable assignment* g maps variables X_α to elements in D_α. $g[d/W]$ denotes the assignment that is identical to g, except for the variable W, which is now mapped to d. The *denotation* $\|s_\alpha\|^{M,g}$ of a HOL term s_α on a model $M = \langle D, I \rangle$ under assignment g is an element $d \in D_\alpha$ defined in the following way:

$$\|p_\alpha\|^{M,g} = I(p_\alpha)$$

$$\|X_\alpha\|^{M,g} = g(X_\alpha)$$

$$\|(s_{\alpha\to\beta}\,t_\alpha)_\beta\|^{M,g} = \|s_{\alpha\to\beta}\|^{M,g}(\|t_\alpha\|^{M,g})$$

$$\|(\lambda X_\alpha s_\beta)_{\alpha\to\beta}\|^{M,g} = \text{the function } f \text{ from } D_\alpha \text{ to } D_\beta \text{ such that}$$
$$f(d) = \|s_\beta\|^{M,g[d/X_\alpha]} \text{ for all } d \in D_\alpha$$

Since $I(\neg_{o\to o})$, $I(\vee_{o\to o\to o})$, $I(\Pi_{(\alpha\to o)\to o})$ and $I(=_{\alpha\to\alpha\to o})$ always denote the standard truth functions, we have:

1. $\|(\neg_{o\to o}\,s_o)_o\|^{M,g} = T$ iff $\|s_o\|^{M,g} = F$.

2. $\|((\vee_{o\to o\to o}\,s_o)\,t_o)_o\|^{M,g} = T$ iff $\|s_o\|^{M,g} = T$ or $\|t_o\|^{M,g} = T$.

3. $\|(\forall X_\alpha s_o)_o\|^{M,g} = \|(\Pi_{(\alpha\to o)\to o}(\lambda X_\alpha s_o))_o\|^{M,g} = T$ iff for all $d \in D_\alpha$ we have $\|s_o\|^{M,g[d/X_\alpha]} = T$.

4. $\|((=_{\alpha\to\alpha\to o}\,s_\alpha)\,t_\alpha)_o\|^{M,g} = T$ iff $\|s_\alpha\|^{M,g} = \|t_\alpha\|^{M,g}$.

A HOL formula s_o is *true* in a Henkin model M under the assignment g if and only if $\|s_o\|^{M,g} = T$. This is also expressed by the notation $M, g \models^{HOL} s_o$. A HOL formula s_o is called *valid* in M, denoted as $M \models^{HOL} s_o$, if and only if $M, g \models^{HOL} s_o$ for all assignments g. Moreover, a formula s_o is called *valid*, denoted as $\models^{HOL} s_o$, if and only if s_o is valid in all Henkin models M. Finally, we define $\Sigma \models^{HOL} s_o$ for a set of HOL formulas Σ if and only if $M \models^{HOL} s_o$ for all Henkin models M with $M \models^{HOL} t_o$ for all $t_o \in \Sigma$.

3 Input/Output Logic

Input/output logic was initially introduced by Makinson and van der Torre [18]. There are various I/O operations. In this paper we focus on two of them, called *"Basic Output"* and *"Basic Reusable Output"*.

3.1 Syntax

$G \subseteq \mathcal{L} \times \mathcal{L}$ is called a normative system, with $\mathcal{L}$ representing the set of all the formulas of propositional logic. A pair $(a, x) \in G$ is referred to as a conditional norm or obligation, where a and x are formulas of propositional logic. The pair (a, x) is read as "given a, it is obligatory that x". a is called the body and represents some situation or condition, whereas x is called the head and represents what is obligatory or desirable in that situation.

3.2 Semantics

For a set of formulas A, we define $G(A) = \{x \mid (a, x) \in G \text{ for some } a \in A\}$ and $Cn(A) = \{x \mid A \vdash x\}$ with $\vdash$ denoting the classical consequence relation. A set of formulas V is *maximal consistent* if it is consistent, and no proper extension of V is consistent. A set of formulas V is said to be *complete* if it is either *maximal consistent* or equal to $\mathcal{L}$.

Definition 1 (Basic Output). *Given a set of conditional norms G and an input set A of propositional formulas,*

$$out_2(G, A) = \bigcap \{Cn(G(V)) \mid A \subseteq V, V \text{ complete}\}$$

Definition 2 (Basic Reusable Output). *Given a set of conditional norms G and an input set A of propositional formulas,*

$$out_4(G, A) = \bigcap \{Cn(G(V)) \mid A \subseteq V \supseteq G(V), V \text{ complete}\}$$

Besides those traditional formulations of the operations, the paper [18] documents modal formulations for out_2 and out_4.

Theorem 1. *$x \in out_2(G, A)$ if and only if $x \in Cn(G(\mathcal{L}))$ and $G^\square \cup A \vdash_{\mathbf{S}} \square x$ for any modal logic $\mathbf{S}$ with $\mathbf{K_0} \subseteq \mathbf{S} \subseteq \mathbf{K45}$.*

Theorem 2. *$x \in out_4(G, A)$ if and only if $x \in Cn(G(\mathcal{L}))$ and $G^\square \cup A \vdash_{\mathbf{S}} \square x$ for any modal logic $\mathbf{S}$ with $\mathbf{K_0 T} \subseteq \mathbf{S} \subseteq \mathbf{KT45}$.*

$\mathbf{K_0}$ is a subsystem of system $\mathbf{K}$ with axiom K, *modus ponens* and the inference rule "from ψ, infer $\square\psi$, for all tautologies in propositional logic". $G^\square$ denotes the set containing all modal formulas of the form $b \to \square y$, such that $(b, y) \in G$. We have that $G^\square \cup A \vdash_{\mathbf{S}} \square x$ if for a finite subset Y of $G^\square \cup A$, it holds that $(\bigwedge Y \to \square x) \in \mathbf{S}$. The notation $\bigwedge Y$ stands for the conjunction of all the elements $y_1, y_2, \ldots, y_n$ in Y, i.e., $y_1 \wedge y_2 \wedge \cdots \wedge y_n$.

3.3 Proof theory

The proof theory of an I/O logic is specified via a number of derivation rules acting on pairs (a, x) of formulas. Given a set G of pairs, we write $(a, x) \in deriv_i(G)$ to say that (a, x) can be derived from G using those rules.

- (SI) Strengthening of the input: from (a, x) and $\vdash b \to a$, infer (b, x)

- (WO) Weakening of the output: from (a, x) and $\vdash x \to y$, infer (a, y)

- (AND) Conjunction of the output: from (a, x) and (a, y), infer $(a, x \wedge y)$

- (OR) Disjunction of the input: from (a, x) and (b, x), infer $(a \vee b, x)$

- (CT) Cumulative transitivity: from (a, x) and $(a \wedge x, y)$, infer (a, y)

The *Basic Output* is syntactically characterized by $deriv_2(G)$ that is closed under rules SI, WO, AND and OR. The *Basic Reusable Output* is determined by $deriv_4(G)$ that is closed under all of the five rules.

4 Shallow Semantical Embedding

The *shallow semantical embedding* approach proposed by Benzmüller [2] uses HOL as a meta-logic in order to represent and model the syntactic and semantical elements of a specific target logic. This methodology is documented and studied for Kripke semantics in [8] and for neighborhood semantics in [4]. This section presents shallow semantical embeddings of the I/O operations out_2 and out_4 in HOL and provides proofs for the soundness and completeness of both operations. To realize these embeddings, we below use the provided modal formulations of the operations; an alternative approach is studied in [7].

4.1 Semantical embedding of K and KT in HOL

We start by describing the semantical embeddings of **K** and **KT** in HOL. This material is taken from [8, 9].

By introducing a new type i to denote possible worlds, the formulas of **K** are identified with certain HOL terms (predicates) of type $i \rightarrow o$. The type $i \rightarrow o$ is abbreviated as τ in the remainder. This allows us to represent the formulas of **K** as functions from possible worlds to truth values in HOL and therefore the truth of a formula can explicitly be evaluated in a particular world. The HOL signature is assumed to further contain the constant symbol $r_{i \rightarrow i \rightarrow o}$. Moreover, for each atomic propositional symbol p^j of **K**, the HOL signature must contain the corresponding constant symbol p^j_τ. Without loss of generality, we assume that besides those symbols and the primitive logical connectives of HOL, no other constant symbols are given in the signature of HOL.

The mapping $\lfloor \cdot \rfloor$ translates a formula φ of **K** into a term $\lfloor \varphi \rfloor$ of HOL of type τ.

The mapping is defined recursively:

$$\begin{aligned}
\lfloor p^j \rfloor &= p^j_\tau \\
\lfloor \neg\varphi \rfloor &= \neg_{\tau\to\tau} \lfloor \varphi \rfloor \\
\lfloor \varphi \vee \psi \rfloor &= \vee_{\tau\to\tau\to\tau} \lfloor \varphi \rfloor \lfloor \psi \rfloor \\
\lfloor \Box\varphi \rfloor &= \Box_{\tau\to\tau} \lfloor \varphi \rfloor
\end{aligned}$$

$\neg_{\tau\to\tau}$, $\vee_{\tau\to\tau\to\tau}$ and $\Box_{\tau\to\tau}$ abbreviate the following terms of HOL:

$$\begin{aligned}
\neg_{\tau\to\tau} &= \lambda A_\tau \lambda X_i \neg(A\,X) \\
\vee_{\tau\to\tau\to\tau} &= \lambda A_\tau \lambda B_\tau \lambda X_i (A\,X \vee B\,X) \\
\Box_{\tau\to\tau} &= \lambda A_\tau \lambda X_i \forall Y_i (\neg(r_{i\to i\to o} X\,Y) \vee A\,Y)
\end{aligned}$$

Analyzing the truth of formula φ, represented by the HOL term $\lfloor \varphi \rfloor$, in a particular world w, represented by the term W_i, corresponds to evaluating the application $(\lfloor \varphi \rfloor W_i)$. In line with the previous work [9], we define $vld_{\tau\to o} = \lambda A_\tau \forall S_i(A\,S)$. With this definition, validity of a formula φ in $\mathbf{K}$ corresponds to the validity of the formula $(vld \lfloor \varphi \rfloor)$ in HOL, and vice versa.

To prove the soundness and completeness, that is, faithfulness, of the above embedding, a mapping from Kripke models into Henkin models is employed.

Lemma 1 (Kripke models $\Rightarrow$ Henkin models). *For every Kripke model $M = \langle W, R, V \rangle$ there exists a corresponding Henkin model H^M, such that for all formulas δ of $\mathbf{K}$, all assignments g and worlds s it holds:*

$$M, s \models \delta \text{ if and only if } \|\lfloor \delta \rfloor S_i\|^{H^M, g[s/S_i]} = T$$

Proof. See [8, 9]. $\qquad\square$

Lemma 2 (Henkin models $\Rightarrow$ Kripke models). *For every Henkin model $H = \langle \{D_\alpha\}_{\alpha\in\mathcal{T}}, I \rangle$ there exists a corresponding Kripke model M_H, such that for all formulas δ of $\mathbf{K}$, all assignments g and worlds s it holds:*

$$\|\lfloor \delta \rfloor S_i\|^{H, g[s/S_i]} = T \text{ if and only if } M_H, s \models \delta$$

Proof. See [8, 9]. $\qquad\square$

The following table summarizes the alignment of Kripke models and Henkin models. For the class of Kripke models $\langle W, R, V \rangle$ that validates some property, such as *reflexivity*, the corresponding class of Henkin models needs to validate a corresponding formula. In system $\mathbf{KT}$, for example, the class of Kripke models satisfies the property of *reflexivity*, which corresponds to axiom T. The counterpart of this property is represented as REF in HOL: $\forall X_i(r_{i \to i \to o} X_i X_i)$, where constant symbol $r_{i \to i \to o}$ denotes the accessibility relation.

Kripke model $\langle W, R, V \rangle$	Henkin model $\langle D, I \rangle$
Possible worlds $s \in W$	Set of individuals $s_i \in D_i$
Accessibility relation R	Binary predicates $r_{i \to i \to o}$
sRu	$Ir_{i \to i \to o}(s_i, u_i) = T$
Propositional letters p^j	Unary predicates $p^j_{i \to o}$
Valuation function $s \in V(p^j)$	Interpretation function $Ip^j_{i \to o}(s_i) = T$

These correspondences between Kripke and Henkin models include the assumptions that have been formulated at the beginning of this section.

Theorem 3 (Faithfulness of the embedding of system **K** in HOL for G and A finite).

$$\models_{\mathcal{C}_K} \varphi \text{ if and only if } \models^{HOL} vld\lfloor\varphi\rfloor$$

Proof. See [8, 9]. □

Theorem 4 (Faithfulness of the embedding of system **KT** in HOL for G and A finite).

$$\models_{\mathcal{C}_{KT}} \varphi \text{ if and only if } \{REF\} \models^{HOL} vld\lfloor\varphi\rfloor$$

Proof. See [8, 9]. □

4.2 Semantical embedding of I/O logic in HOL

In order to embed the operations out_2 and out_4 in HOL, we just use the corresponding modal formulations. We apply Theorem 3 and Theorem 4, respectively, to prove the faithfulness of the embeddings.

Theorem 5 (Faithfulness of the embedding of out_2 in HOL for G and A finite).

$$\varphi \in out_2(G, A)$$

if and only if

$$\models^{HOL} vld\lfloor\bigwedge(G^\square \cup A) \to \square\varphi\rfloor \text{ and } \models^{HOL} vld\lfloor\bigwedge G(\mathcal{L}) \to \varphi\rfloor$$

Proof. We choose **S** = **K** in Theorem 1 and then apply Theorem 3.

$$\varphi \in out_2(G, A)$$

if and only if

$$G^\Box \cup A \vdash_{\mathbf{K}} \Box\varphi \text{ and } \varphi \in Cn(G(\mathcal{L}))$$

if and only if

$$\models_{\mathcal{C}_K} \bigwedge (G^\Box \cup A) \to \Box\varphi \text{ and } \bigwedge G(\mathcal{L}) \vdash \varphi$$

if and only if

$$\models_{\mathcal{C}_K} \bigwedge (G^\Box \cup A) \to \Box\varphi \text{ and } \models_{\mathcal{C}_K} \bigwedge G(\mathcal{L}) \to \varphi$$

if and only if

$$\models^{HOL} vld \lfloor \bigwedge (G^\Box \cup A) \to \Box\varphi \rfloor \text{ and } \models^{HOL} vld \lfloor \bigwedge G(\mathcal{L}) \to \varphi \rfloor$$

$\Box$

Theorem 6 (Faithfulness of the embedding of out_4 in HOL for G and A finite).

$$\varphi \in out_4(G, A)$$

if and only if

$$\{REF\} \models^{HOL} vld \lfloor \bigwedge (G^\Box \cup A) \to \Box\varphi \rfloor \text{ and } \{REF\} \models^{HOL} vld \lfloor \bigwedge G(\mathcal{L}) \to \varphi \rfloor$$

Proof. We choose $\mathbf{S} = \mathbf{KT}$ in Theorem 2 and then apply Theorem 4.

$$\varphi \in out_4(G, A)$$

if and only if

$$G^\Box \cup A \vdash_{\mathbf{KT}} \Box\varphi \text{ and } \varphi \in Cn(G(\mathcal{L}))$$

if and only if

$$\models_{\mathcal{C}_{KT}} \bigwedge (G^\Box \cup A) \to \Box\varphi \text{ and } \bigwedge G(\mathcal{L}) \vdash \varphi$$

if and only if

$$\models_{\mathcal{C}_{KT}} \bigwedge (G^\Box \cup A) \to \Box\varphi \text{ and } \models_{\mathcal{C}_{KT}} \bigwedge G(\mathcal{L}) \to \varphi$$

if and only if

$$\{REF\} \models^{HOL} vld \lfloor \bigwedge (G^\Box \cup A) \to \Box\varphi \rfloor \text{ and } \{REF\} \models^{HOL} vld \lfloor \bigwedge G(\mathcal{L}) \to \varphi \rfloor$$

$\Box$

```
 1 theory IOL_out2 imports Main
 2 begin
 3 typedecl i (* type for possible worlds *)
 4 type_synonym τ = "(i⇒bool)"
 5 consts r :: "i⇒i⇒bool" (infixr "r"70) (* relation for a modal logic K *)
 6 definition knot   :: "τ⇒τ" ("¬_"[52]53)       where "¬φ ≡ λw. ¬φ(w)"
 7 definition kor    :: "τ⇒τ⇒τ" (infixr "∨"50)    where "φ∨ψ ≡ λw. φ(w) ∨ ψ(w)"
 8 definition kand   :: "τ⇒τ⇒τ" (infixr "∧"51)    where "φ∧ψ ≡ λw. φ(w) ∧ ψ(w)"
 9 definition kimp   :: "τ⇒τ⇒τ" (infixr "⟶"49)    where "φ⟶ψ ≡ λw. φ(w) ⟶ ψ(w)"
10 definition kbox   :: "τ⇒τ" ("□_k")             where "□_kφ ≡ λw. ∀v. w r v ⟶ φ(v)"
11 definition ktrue  :: "τ" ("⊤")                 where "⊤ ≡ λw. True"
12 definition kfalse :: "τ" ("⊥")                 where "⊥ ≡ λw. False"
13 definition kvalid :: "τ⇒bool" ("⌊_⌋"[8]109)    where "⌊p⌋ ≡ ∀w. p(w)" (* global validity *)
14
15 (* x ∈ out2(G,A) iff G^□∪A ⊢_K □x ∧ x ∈ Cn(G(L))  *)
16
17 consts a::τ b::τ e::τ
18
19 (* OR example: G = {(a,e),(b,e)}, e ∈ out2(G,{a∨b}) *)
20
21 (* G^□∪{a∨b} ⊢_K □e *)
22 lemma "⌊((a⟶□_ke)∧(b⟶□_ke)∧(a∨b)) ⟶ □_ke⌋" sledgehammer
23   using kand_def kimp_def kor_def kvalid_def by auto
24
25 (* e ∈ Cn(G(L)) *)
26 lemma "⌊(e∧e) ⟶ e⌋" sledgehammer by (simp add: kand_def kimp_def kvalid_def)
```

Figure 1: Semantical embedding of out_2 in Isabelle/HOL

4.3   Implementation of I/O logic in Isabelle/HOL

The semantical embeddings of the operations out_2 and out_4 in HOL as devised in the previous section have been implemented in the higher-order proof assistant tool Isabelle/HOL [21], see Fig. 1. We declare the type i to denote possible worlds and introduce the relevant connectives in lines 6–12.

Let the set of conditional norms G be composed of the elements (a, e) and (b, e), where a, b and e are propositional symbols, and let the input set A correspond to the singleton set containing $a \vee b$. By the rule of disjunction (OR), we should have that $e \in out_2(G, A)$. According to the provided translation, $e \in out_2(G, A)$ if and only if $G^\square \cup A \vdash_{\mathbf{K}} \square e$ and $e \in Cn(G(\mathcal{L}))$. Theorem 5 provides us now with higher-order formulations for both of these statements, i.e., $\models^{HOL} vld \lfloor \bigwedge (G^\square \cup A) \to \square e \rfloor$ and $\models^{HOL} vld \lfloor \bigwedge G(\mathcal{L}) \to e \rfloor$, respectively. Regarding the implementation, the propositional symbols a, b and e have to be declared as constants of type τ. The framework's integrated automatic theorem provers (ATPs), called via the Sledgehammer tool [12], are able to prove both statements. This is shown in Fig. 1, lines 22–23 and 26.

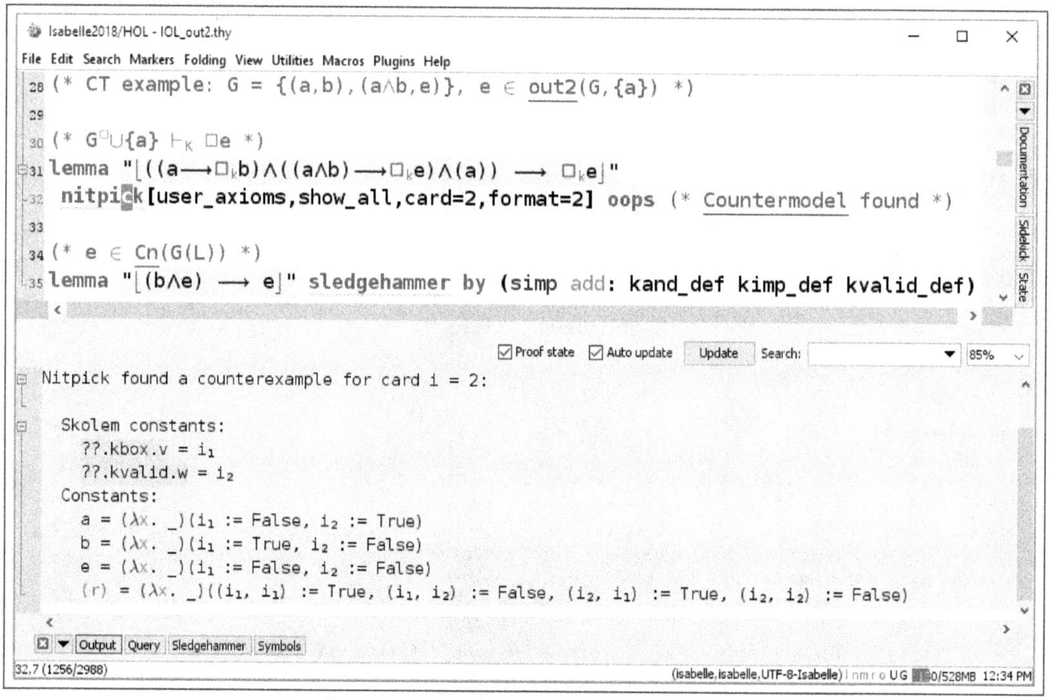

Figure 2: Failure of CT for out_2

Consider the set of conditional norms $G = \{(a, b), (a \wedge b, e)\}$ with the input set $A = \{a\}$. The rule of cumulative transitivity (CT) is not satisfied by the operation out_2. This can also be verified with our implementation. The model finder Nitpick [11] is able to generate a countermodel for the statement $G^\square \cup A \vdash_K \square e$ and therefore we were able to show that $e \notin out_2(G, A)$. In particular, Nitpick came up with a model M consisting of two possible worlds i_1 and i_2. We have that $V(a) = \{i_2\}$, $V(b) = \{i_1\}$ and $V(e) = \emptyset$. And $R = \{(i_1, i_1), (i_2, i_1)\}$. The formula $((a \rightarrow \square b) \wedge ((a \wedge b) \rightarrow \square e) \wedge a) \rightarrow \square e$ is not valid in this model. The formulation of the example and the generation of the countermodel is illustrated in Fig. 2.

The embedding of the operation out_4 refers to system **KT** which means that the corresponding class of Kripke models satisfies the property of reflexivity. In our implementation, the accessibility relation for this system is denoted by the constant r_t which we declare as reflexive. Due to this property, the Sledgehammer tool is able to prove the statement $G^\square \cup A \vdash_{KT} \square e$ and thus we can verify that $e \in out_4(G, A)$. Fig. 3 shows the encoding of the operation out_4 in Isabelle/HOL and the verification of the CT example.

```
1  theory IOL_out4 imports Main
2  begin
3  typedecl i (* type for possible worlds *)
4  type_synonym τ = "(i⇒bool)"
5  consts r_t :: "i⇒i⇒bool" (infixr "rt"70) (* relation for a modal logic KT *)
6  abbreviation reflexive where "reflexive r ≡ (∀x. r x x)"
7  axiomatization where ax_reflex_rt : "reflexive r_t"
8  definition knot   :: "τ⇒τ" ("¬_"[52]53)        where "¬φ ≡ λw. ¬φ(w)"
9  definition kor    :: "τ⇒τ⇒τ" (infixr "∨"50)    where "φ∨ψ ≡ λw. φ(w) ∨ ψ(w)"
10 definition kand   :: "τ⇒τ⇒τ" (infixr "∧"51)    where "φ∧ψ ≡ λw. φ(w) ∧ ψ(w)"
11 definition kimp   :: "τ⇒τ⇒τ" (infixr "⟶"49)    where "φ⟶ψ ≡ λw. φ(w) ⟶ ψ(w)"
12 definition kbox   :: "τ⇒τ" ("□_kt")            where "□_ktφ ≡ λw. ∀v. w rt v ⟶ φ(v)"
13 definition ktrue  :: "τ" ("⊤")                 where "⊤ ≡ λw. True"
14 definition kfalse :: "τ" ("⊥")                 where "⊥ ≡ λw. False"
15 definition kvalid :: "τ⇒bool" ("⌊_⌋"[8]109)    where "⌊p⌋ ≡ ∀w. p(w)" (* global validity *)
16
17 (* x ∈ out4(G,A) iff G^□∪A ⊢_KT □x ∧ x ∈ Cn(G(L)) *)
18
19 consts a::τ b::τ e::τ
20
21 (* CT example: G = {(a,b),(a∧b,e)}, e ∈ out4(G,{a}) *)
22
23 (* G^□∪{a} ⊢_KT □e *)
24 lemma "⌊((a⟶□_ktb)∧((a∧b)⟶□_kte)∧(a)) ⟶ □_kte⌋"
25   sledgehammer by (simp add: ax_reflex_rt kand_def kbox_def kimp_def kvalid_def)
26
27 (* e ∈ Cn(G(L)) *)
28 lemma "⌊(b∧e) ⟶ e⌋" sledgehammer by (simp add: kand_def kimp_def kvalid_def)
29 end
```

Figure 3: Semantical embedding of out_4 in Isabelle/HOL

5 Application: Moral Luck

The literature on moral luck [20] is addressing the question whether luck can ever make a moral difference or not. Examples involving moral luck are typical scenarios in which an agent is held accountable for his actions and its consequences even though it is clear that the agent was neither in full control of his actions nor its consequences. These examples are thus in conflict with the ethical principle that agents are not morally responsible for actions that they are unable to control.

The *Drink and Drive* [19] example highlights a classical scenario of moral luck. There exist many different variations of this example and a possible variant can be formulated as follows:

> Assume a situation where two persons, Ali and Paul, go out for a drink in the evening. Both of them go to the same bar, consume the same amount of alcoholic drinks and end up pretty drunk. At one point during the night, they both decide to leave the place. So they go to their own individual vehicles and hit the road in order to drive home. The roads are pretty deserted at that

time and Ali manages to drive home safely even with the high percentage of alcohol in his blood. Paul, in contrast, is facing something unexpected. Out of nowhere, a child appears in front of his car. Since he had a few drinks too much, his reaction time is impaired by the alcohol and it makes it impossible for him to stop and swerve to avoid hitting and killing the child.

Both *Ali* and *Paul*, made the blameworthy decision of driving while being drunk. But neither one of them had the intention to hit and kill anyone. Nevertheless, most people would tend to judge *Paul* more guilty than *Ali* simply because in his case a child got killed. However, both of them violated the same obligation, namely that one should not drive while being intoxicated and it was only a matter of luck that nobody got harmed or killed in the case of *Ali*. Therefore we say that *Ali* got morally lucky.

To formulate the *Drink and Drive* example in Isabelle/HOL, we first import the Isabelle/HOL file containing the implementation of the operation out_2. This can be done using the Isabelle/HOL command *imports* (cf. Fig. 4, line 1). Next, we have to declare three individuals, namely *Ali* and *Paul*, representing the two drivers, and *Child*, representing the *child* in the scenario (cf. Fig. 4, line 3). In lines 4–5, we define the constant symbols for the relevant propositions (state of affairs or action).

One associates with each driver a set of *Norms* and an *Input*. For the individual *Paul*, the set of *Norms G* is defined as follows: (cf. Fig. 4, lines 9–11)

- $(\top, \neg Kill\, Child \wedge \neg Hurt\, Child)$
  This norm states that it is forbidden to kill or even hurt the *child*.

- $(\top, Drive\_carefully\, Paul)$
  This norm states that *Paul* is obligated to drive carefully in any situation.

- $(\neg Drive\_carefully\, Paul, Stay\, Paul)$
  This norm states that if *Paul* does not drive carefully, he should stay (at his current location).

To complete the formalization for the individual *Paul*, we need to add the following facts to the *Input* set *A*: (cf. Fig. 4, lines 13–17)

- *Drunk Paul*; *Drive Paul*; *Jump Child*
  Paul is actually drunk; *Paul* drives home; The *child* jumps.

- *Drunk Paul* $\rightarrow \neg Drive\_carefully\, Paul$
  If *Paul* is drunk then he drives not carefully.

- $(\neg Drive\_carefully Paul \wedge Drive Paul \wedge Jump Child)$
  $\rightarrow (Kill Child \vee Hurt Child)$
  If *Paul* drives, but does not do it carefully, and the *child* jumps in front of his car then *Paul* will kill or hurt the *child*.

```
1  theory Drink_and_Drive  imports IOL_out2
2  begin
3  datatype indiv = Ali | Paul | Child (* Represent individuals Ali, Paul and Child *)
4  consts  Kill::"indiv⇒τ" Hurt::"indiv⇒τ" Drive_carefully::"indiv⇒τ" Stay::"indiv⇒τ"
5  Drunk::"indiv⇒τ" Jump::"indiv⇒τ" Drive::"indiv⇒τ"
6
7  axiomatization where
8  (* Norms *)
9  A0: "⌊T ⟶ □ₖ(¬ Kill Child ∧ ¬ Hurt Child)⌋" and
10 A1: "⌊T ⟶ □ₖ(Drive_carefully Paul)⌋" and
11 A2: "⌊¬ Drive_carefully Paul ⟶ □ₖ(Stay Paul)⌋"and
12 (* Input set *)
13 A3: "⌊Drunk Paul⌋" and
14 A4: "⌊Drive Paul⌋" and
15 A5: "⌊Jump Child⌋" and
16 A6: "⌊Drunk Paul ⟶ ¬ Drive_carefully Paul⌋" and
17 A7: "⌊(¬ Drive_carefully Paul ∧ Drive Paul ∧ Jump Child) ⟶ (Kill Child ∨ Hurt Child)⌋"
18
19 (* Consistency is confirmed by nitpick *)
20 lemma True nitpick [satisfy,user_axioms,show_all,expect=genuine] oops
21
22 lemma "⌊□ₖ(Stay Paul)⌋" using A2 A3 A6 sledgehammer by (simp add: kimp_def kvalid_def)
23
24 lemma "⌊□ₖ(Drive_carefully Paul) ∧ ¬ Drive_carefully Paul⌋" using A1 A3 A6
25   sledgehammer by (simp add: kand_def kimp_def ktrue_def kvalid_def)
26
27 lemma "⌊□ₖ(¬ Kill Child ∧ ¬ Hurt Child) ∧ (Kill Child ∨ Hurt Child)⌋" using A0 A3 A4 A5 A6 A7
28   sledgehammer by (simp add: kand_def kimp_def ktrue_def kvalid_def)
29 end
```

Figure 4: *Drink and Drive* scenario for *Paul* in Isabelle/HOL

Since Nitpick finds a model satisfying our statements, the formalization of the *Drink and Drive* is consistent; cf. Fig. 4, line 20.

Actually, we are able to derive the obligation that *Paul* should stay (at his current position) by using the norm A2 and the facts A3 and A6, meaning that we can derive $G^\square \cup A \vdash_{\mathbf{K}} \square Stay\_Paul$ and $Stay\_Paul \in Cn(G(\mathcal{L}))$. The first statement is proven by Sledgehammer tool; cf. Fig. 4, line 22. In this example, we skip checking the following (trivial) statements $X \in Cn(G(\mathcal{L}))$ for $X \in \{Stay\_Paul, Drive\_carefully\_Paul, \neg Kill Child \wedge \neg Hurt Child\}$ in Isabelle/HOL.

Furthermore, our implementation is capable of recognizing violations to norms, formally written as $\alpha \in out_2(G, A)$ and $\neg\alpha \in Cn(A)$. In particular, *Paul* violated the norms A0 and A1. For instance, the violation to A1 is proven by Sledgehammer;

cf. Fig. 4, lines 24–25. *Paul* did not drive carefully even though there is an obligation to do so, meaning that we have $G^\square \cup A \vdash_{\mathbf{K}} \square Drive\_carefully\_Paul$ (using A1), $Drive\_carefully\_Paul \in Cn(G(\mathcal{L}))$ and $\neg Drive\_carefully\_Paul \in Cn(A)$ (using A3 and A6).

```
1  theory Drink_and_Drive   imports IOL_out2
2  begin
3  datatype indiv = Ali | Paul | Child (* Represent individuals Ali, Paul and Child *)
4  consts   Kill::"indiv⇒τ" Hurt::"indiv⇒τ" Drive_carefully::"indiv⇒τ" Stay::"indiv⇒τ"
5  Drunk::"indiv⇒τ" Jump::"indiv⇒τ" Drive::"indiv⇒τ"
6
7  axiomatization where
8  (* Norms *)
9  A0: "⌊T ⟶ □ₖ(¬ Kill Child ∧ ¬ Hurt Child)⌋" and
10 A1: "⌊T ⟶ □ₖ(Drive_carefully Ali)⌋" and
11 A2: "⌊¬ Drive_carefully Ali ⟶ □ₖ(Stay Ali)⌋"and
12 (* Input set *)
13 A3: "⌊Drunk Ali⌋" and
14 A4: "⌊Drive Ali⌋" and
15 A6: "⌊Drunk Ali ⟶ ¬ Drive_carefully Ali⌋" and
16 A7: "⌊(¬ Drive_carefully Ali ∧ Drive Ali ∧ Jump Child) ⟶ (Kill Child ∨ Hurt Child)⌋"
17
18 (* Consistency is confirmed by nitpick *)
19 lemma True nitpick [satisfy,user_axioms,show_all,expect=genuine] oops
20
21 lemma "⌊□ₖ(Stay Ali)⌋"
22   using A2 A3 A6 sledgehammer by (simp add: kimp_def kvalid_def)
23
24 lemma "⌊□ₖ(Drive_carefully Ali) ∧ ¬ Drive_carefully Ali⌋"  using A1 A3 A6
25   sledgehammer by (simp add: kand_def kimp_def ktrue_def kvalid_def)
26
27 lemma"⌊□ₖ(¬ Kill Child ∧ ¬ Hurt Child) ∧ (Kill Child ∨ Hurt Child)⌋"
28   nitpick [user_axioms,show_all,expect=genuine] oops
29 end
```

Figure 5: *Drink and Drive* scenario for *Ali* in Isabelle/HOL

For the individual *Ali*, the set of *Norms* remains the same except that we adapted the name of the individual accordingly. However, in *Ali's* case, the *child* was not involved. Therefore, the *Input* set only consists of our facts: A3, A4, A6 and A7 (cf. Fig. 5, lines 13–16).

The formalization of *Ali's* scenario is consistent, again proven by Nitpick (cf. Figure 5, line 19). In contrast to *Paul*, *Ali* did not violated the norm A0 as Nitpick find a counter model for the corresponding statement (cf. Fig. 5, lines 27–28).

6   Conclusion

We have presented an embedding of two I/O operations in HOL and we have shown that each embedding is faithful, i.e., sound and complete. The work presented here continues a project started in Benzmüller et al. [4], and aiming at providing the theoretical foundation for the implementation and automation of deontic logic within existing theorem provers and proof assistants for HOL. Future research should investigate whether the provided implementation already supports non-trivial applications in practical normative reasoning such as legal reasoning or multi-agent systems, or whether further improvements are required. We could also employ our implementation to systematically study some meta-logical properties of I/O logic within Isabelle/HOL. Moreover, we could analogously implement intuitionistic I/O logic [22].

Acknowledgements

We thank an anonymous reviewer for valuable comments.

References

[1] Åqvist, L.: Deontic logic. In: Gabbay, D., and Guenthner, F., (eds.) *Handbook of Philosophical Logic*, pp. 147–264. Springer, Dordrecht (2002)

[2] Benzmüller, C.: Universal (meta-)logical reasoning: Recent successes. *Science of Computer Programming*, **172**, 48–62 (2019)

[3] Benzmüller, C., Brown, C., Kohlhase, M.: Higher-order semantics and extensionality. Journal of Symbolic Logic, **69**(4), 1027–1088 (2004)

[4] Benzmüller, C., Farjami, A., Parent, X.: A dyadic deontic logic in HOL. In: Broersen, J., Condoravdi, C., Nair, S., Pigozzi, G. (eds.) *Deontic Logic and Normative Systems — 14th International Conference, DEON 2018*, Utrecht, The Netherlands, 3-6 July, 2018, pp. 33–50, College Publications, UK (2018)

[5] Benzmüller, C., Andrews, P.B.: Church's type theory. In: Zalta, E.N. (ed.) *The Stanford Encyclopedia of Philosophy*. Metaphysics Research Lab, Stanford University, Summer 2019 Edition (2019)

[6] Benzmüller, C., Farjami, A., Parent, X.: Åqvist's dyadic deontic logic E in HOL. *Journal of Applied Logics – IfCoLoG Journal of Logics and their Applications*, this issue (2019)

[7] Benzmüller, C., Parent, X.: I/O logic in HOL – First steps. arXiv preprint, arXiv:1803.09681 [cs.AI] (2018)

[8] Benzmüller, C., Paulson, L.: Multimodal and intuitionistic logics in simple type theory. *The Logic Journal of the IGPL*, **18**(6), 881–892 (2010)

[9] Benzmüller, C., Paulson, L. C.: Quantified multimodal logics in simple type theory. *Logica Universalis* (Special Issue on Multimodal Logics), **7**(1), 7–20 (2013)

[10] Benzmüller, C., Sultana, N., Paulson, L. C., Theiß, F.: The higher-order prover LEO-II. *Journal of Automated Reasoning*, **55**(4), 389–404 (2015)

[11] Blanchette, J. C., Nipkow, T.: Nitpick: A counterexample generator for higher-order logic based on a relational model finder. In: Kaufmann, M., Paulson, L. C. (eds.) *International Conference on Interactive Theorem Proving 2010*, LNCS, vol. 6172 pp. 131–146, Springer (2010)

[12] Blanchette, J. C., Paulson, L. C.: Hammering away – A user's guide to Sledgehammer for Isabelle/HOL (2017)

[13] Boella, G., van der Torre, L.: Regulative and constitutive norms in normative multiagent systems. In: Dubois, D., Welty, C., Williams, M. (eds.) *Principles of Knowledge Representation and Reasoning: Proceedings of the Ninth International Conference (KR2004)*, Whistler, Canada, June 2-5, 2004, pp. 255–266, AAAI Press, USA (2004)

[14] Church, A.: A formulation of the simple theory of types. *Journal of Symbolic Logic*, **5**(2), 56–68 (1940)

[15] Gabbay, D., Horty, J., Parent, X., van der Meyden, R., van der Torre, L.: *Handbook of Deontic Logic and Normative Systems. Volume 1.* College Publications, UK (2013)

[16] Gabbay, D., Horty, J., Parent, X., van der Meyden, R., van der Torre, L.: *Handbook of Deontic Logic and Normative Systems. Volume 2.* College Publications, UK. (To appear)

[17] Henkin, L.: Completeness in the theory of types. *Journal of Symbolic Logic*, **5**(2), 81–91 (1950)

[18] Makinson, D., van der Torre, L.: Input/output logics. *Journal of Philosophical Logic*, **29**(4), 383–408 (2000)

[19] Nagel, T.: Moral luck. *Chapter in Mortal Questions,* Cambridge University Press, New York (1979)

[20] Nelkin, K.: Moral luck. In: Zalta, E.N. (ed.) *The Stanford Encyclopedia of Philosophy.* Metaphysics Research Lab, Stanford University (2013)

[21] Nipkow, T., Paulson, L. C., Wenzel., M.: *Isabelle/HOL — A proof assistant for higher-order logic*, volume 2283 of Lecture Notes in Computer Science. Springer (2002)

[22] Parent, X.: A modal translation of an intuitionistic I/O operation. Presented at the 7th Workshop on Intuitionistic Modal Logic and Applications (IMLA 2017), organized by V. de Paiva and S. Artemov at the University of Toulouse (France), 17-28 July, (2017)

[23] Steen, A., Benzmüller, C.: The higher-order prover Leo-III. In: Galmiche, D., Schulz, S., Sebastiani, R. (eds.) *Automated Reasoning. IJCAR 2018*, LNCS, vol. 10900, pp. 108-116, Springer (2018)

Received 29 November 2018

Åqvist's Dyadic Deontic Logic **E** in HOL

Christoph Benzmüller

Freie Universität Berlin, Germany, and University of Luxembourg, Luxembourg
`c.benzmueller@gmail.com`

Ali Farjami

University of Luxembourg, Luxembourg
`ali.farjami@uni.lu`

Xavier Parent

University of Luxembourg, Luxembourg
`xavier.parent@uni.lu`

Abstract

We devise a shallow semantical embedding of Åqvist's dyadic deontic logic **E** in classical higher-order logic. This embedding is shown to be faithful, viz. sound and complete. This embedding is also encoded in Isabelle/HOL, which turns this system into a proof assistant for deontic logic reasoning. The experiments with this environment provide evidence that this logic *implementation* fruitfully enables interactive and automated reasoning at the meta-level and the object-level.

Keywords: Dyadic deontic logic; Preference semantics; Classical higher-order logic; Semantical embedding; Automated reasoning.

1 Introduction

Normative notions such as obligation and permission are the subject of deontic logic [23] and conditional obligations are addressed in so-called *dyadic deontic logic.*

This work has been supported by the European Union's Horizon 2020 research and innovation programme under the Marie Skłodowska-Curie grant agreement No 690974 - MIREL - MIning and REasoning with Legal texts. Benzmüller has been funded by the Volkswagen Foundation under project CRAP — Consistent Rational Argumentation in Politics.

A landmark and historically important family of dyadic deontic logics has been proposed by B. Hansson [25]. These logics have been recast in the framework of possible world semantics by Åqvist [2]. They come with a preference semantics, in which a binary preference relation ranks the possible words in terms of betterness. The framework was motivated by the well-known paradoxes of *contrary-to-duty* (CTD) reasoning like Chisholm [19]'s paradox. In this paper, we focus on the class of all preference models, in which no specific properties (like reflexivity or transitivity) are required of the betterness relation. This class of models has a known axiomatic characterization, given by Åqvist's system **E** [30].

When applied as a meta-logical tool, *simple type theory* [20], aka classical Higher-Order Logic (HOL), can help to better understand semantical issues of embedded object logics. The syntax and semantics of HOL are well understood [10] and there exist automated proof tools for it; examples include Isabelle/HOL [28], LEO-II [15] and Leo-III [31].

In this paper we devise an *embedding* of **E** in HOL. This embedding utilizes the *shallow semantical embedding* approach that has been put forward by Benzmüller [7, 9] as a pragmatical solution towards universal logic reasoning. This approach uses HOL as (universal) meta-logic to specify, in a shallow way, the syntax and semantics of various object logics, in our case system **E**. The embedding has been encoded in Isabelle/HOL to enable syntactical and semantical experiments in deontic reasoning.

Benzmüller et al. [12] developed an analogous shallow semantical embedding for the dyadic deontic logic proposed by Carmo and Jones [18]. A core difference concerns the notion of semantics employed in both papers, which leads to different semantical embeddings. Instead of the semantics based on preference models as employed by Hansson [25] and Åqvist [2], a neighborhood semantics is employed by Carmo and Jones [18]. Moreover, this method has been applied to some more recent deontic frameworks like so-called I/O logic [13].

Deep semantical embeddings of non-classical logics have been studied in the related literature [22, 21]. The emphasis in these works typically is on interactive proofs of meta-logical properties. While meta-logical studies [11, 24] are also in reach for the methods presented here, our interest is in proof automation at object level, i.e., proof automation of Åqvist's system **E**. In other words, we are interested in practical normative reasoning applications of system **E** in which a high degree of automation at object level is required. Moreover, we are interested not only in the "propositional" system **E**, but also in quantified extensions of it. For this, we plan to accordingly adapt the achievements of previous works [14, 5]. Making deep semantical embeddings scale for quantified non-classical logics, on the contrary, seems more challenging and less promising regarding proof automation.

The article is structured as follows. Section 2 describes system **E** and Section 3 introduces HOL. The semantical embedding of **E** in HOL is then devised and studied in Section 4. This section also shows the faithfulness (viz. soundness and completeness) of the embedding. Section 5 discusses the implementation in Isabelle/HOL [28]. Section 6 concludes the paper.

2   Dyadic Deontic Logic **E**

The language of **E** is obtained by adding the following operators to the language of propositional logic: $\Box$ (for necessity); $\Diamond$ (for possibility); and $\bigcirc(-/-)$ (for conditional obligation) ; $P(-/-)$ (for conditional permission). $\bigcirc(\psi/\varphi)$ is read "If φ, then ψ is obligatory", and $P(\psi/\varphi)$ is read "If φ, then ψ is permitted". The set of well-formed formulas (wffs) is defined in the straightforward way. Iteration of the modal and deontic operators is permitted, and so are "mixed" formulas, e.g., $\bigcirc(q/p) \wedge p$. We put $\top =_{df} \neg q \vee q$, for some atomic wff q, and $\bot =_{df} \neg\top$. $\Diamond$ is the dual of $\Box$, viz. $\Diamond\varphi =_{df} \neg\Box\neg\varphi$. P is also the dual of $\bigcirc$, viz. $P(\psi/\varphi) =_{df} \neg\bigcirc(\neg\psi/\varphi)$.

We recall the main difference between the Kripke relational semantics for so-called Standard Deontic Logic (SDL) [23] and the semantics for **E**. The first one uses a binary classification of worlds into good/bad (or green/red). The second one allows for gradations between these two extremes. The closer a world is to ideality, the better it is.

A preference model is a structure $M = \langle W, \succeq, V \rangle$ where:

- W is a non-empty set of possible worlds (W is called "universe");

- $\succeq \subseteq W \times W$ (intuitively, $\succeq$ is a betterness or comparative goodness relation; "$s \succeq t$" can be read as "world s is at least as good as world t");

- V is a function assigning to each atomic wff a set of worlds, i.e., $V(p) \subseteq W$ (intuitively, $V(p)$ is the set of worlds at which p is true).

No specific properties (like reflexivity or transitivity) are required of the betterness relation.

Given a preference model $M = \langle W, \succeq, V \rangle$ and a world $s \in W$, we define the satisfaction relation $M, s \vDash \varphi$ (read as "world s satisfies φ in model M") by induction on the structure of φ as described below. Intuitively, the evaluation rule for the dyadic obligation operator puts $\bigcirc(\psi/\varphi)$ true whenever all the best φ-worlds are ψ-worlds. Here best is defined in terms of optimality rather than maximality [30]. A φ-world is optimal if it is at least as good as any other φ-world. We define $V^M(\varphi) = \{s \in W \mid M, s \vDash \varphi\}$ and $\text{opt}_{\succeq}(V^M(\varphi)) = \{s \in V^M(\varphi) \mid \forall t(t \vDash \varphi \rightarrow s \succeq t)\}$.

Whenever the model M is obvious from context, we write $V(\varphi)$ instead of $V^M(\varphi)$.

$M, s \models p$ if and only if $s \in V(p)$

$M, s \models \neg\varphi$ if and only if $M, s \not\models \varphi$ (that is, not $M, s \models \varphi$)

$M, s \models \varphi \vee \psi$ if and only if $M, s \models \varphi$ or $M, s \models \psi$

$M, s \models \Box\varphi$ if and only if $V(\varphi) = W$

$M, s \models \bigcirc(\psi/\varphi)$ if and only if $\mathrm{opt}_{\succeq}(V(\varphi)) \subseteq V(\psi)$

As usual, a formula φ is valid in a preference model $M = \langle W, \succeq, V \rangle$ (notation: $M \models \varphi$) if and only if, for all worlds $s \in W$, $M, s \models \varphi$. A formula φ is valid (notation: $\models \varphi$) if and only if it is valid in every preference model. The notions of semantic consequence and satisfiability in a model are defined as usual.

System **E** is defined by the following axioms and rules:

Axiom schemata for propositional logic	(PL)
S5-schemata for $\Box$ and $\Diamond$	(S5)
$\bigcirc(\psi_1 \to \psi_2/\varphi) \to (\bigcirc(\psi_1/\varphi) \to \bigcirc(\psi_2/\varphi))$	(COK)
$\bigcirc(\psi/\varphi) \to \Box\bigcirc(\psi/\varphi)$	(Abs)
$\Box\psi \to \bigcirc(\psi/\varphi)$	(Nec)
$\Box(\varphi_1 \leftrightarrow \varphi_2) \to (\bigcirc(\psi/\varphi_1) \leftrightarrow \bigcirc(\psi/\varphi_2))$	(Ext)
$\bigcirc(\varphi/\varphi)$	(Id)
$\bigcirc(\psi/\varphi_1 \wedge \varphi_2) \to \bigcirc(\varphi_2 \to \psi/\varphi_1)$	(Sh)
If $\vdash \varphi$ and $\vdash \varphi \to \psi$ then $\vdash \psi$	(MP)
If $\vdash \varphi$ then $\vdash \Box\varphi$	(N)

The notions of theoremhood, deducibility and consistency are defined as usual.

The following theorem tells us that system **E** is the weakest system that characterizes preference models. It also tells us that the assumptions of reflexivity and totalness of $\succeq$ do not modify the logic, in the sense that they do not add new validities (or theorems).

Theorem 1. *System* **E** *is sound and complete with respect to the class of all preference models. System* **E** *is also sound and complete with respect to the class of those in which $\succeq$ is reflexive, and with respect to the class of those in which $\succeq$ is total (for all $s, t \in W$, $s \succeq t$ or $t \succeq s$).*

Proof. See Parent [30]. $\qquad\square$

E is first in a family of three systems. Consider the condition of limitedness. Its role is to rule out infinite chains of strictly better worlds. Formally: if $V(\varphi) \neq \emptyset$, then $\mathrm{opt}_{\succeq}(V(\varphi)) \neq \emptyset$. Such a condition boosts the logic to system **F**, obtained by supplementing **E** with D*:

$$\Diamond\varphi \to (\bigcirc(\psi/\varphi) \to P(\psi/\varphi)) \tag{D*}$$

Similarly, the additional assumption of transitivity of $\succeq$ boosts the logic to system **G**, obtained by supplementing **F** with Sp:

$$(P(\psi/\varphi) \wedge \bigcirc((\psi \to \chi)/\varphi) \to \bigcirc(\chi/(\varphi \wedge \psi)) \tag{Sp}$$

None of **F** and **G** will concern us in this paper.

3 Classical Higher-Order Logic

In this section we introduce classical higher-order logic (HOL). The presentation, which has been adapted from [6], is rather detailed in order to keep the article sufficiently self-contained.

3.1 Syntax of HOL

To define the syntax of HOL, we first introduce the set T of *simple types*. We assume that T is freely generated from a set of *basic types* $BT \supseteq \{o, i\}$ using the function type constructor $\to$. Type o denotes the (bivalent) set of Booleans, and i a non-empty set of individuals.

For the definition of HOL, we start out with a family of denumerable sets of typed constant symbols $(C_\alpha)_{\alpha \in T}$, called the HOL *signature*, and a family of denumerable sets of typed variable symbols $(V_\alpha)_{\alpha \in T}$.[1] We employ Church-style typing, where each term t_α explicitly encodes its type information in subscript α.

The *language of HOL* is given as the smallest set of terms obeying the following conditions.

- Every typed constant symbol $c_\alpha \in C_\alpha$ is a HOL term of type α.

- Every typed variable symbol $X_\alpha \in V_\alpha$ is a HOL term of type α.

- If $s_{\alpha \to \beta}$ and t_α are HOL terms of types $\alpha \to \beta$ and α, respectively, then $(s_{\alpha \to \beta} t_\alpha)_\beta$, called *application*, is an HOL term of type β.

---

[1]For example in Section 4 we assume constant symbol r, with type $i \to i \to o$ as part of the signature.

- If $X_\alpha \in V_\alpha$ is a typed variable symbol and s_β is an HOL term of type β, then $(\lambda X_\alpha s_\beta)_{\alpha \to \beta}$, called *abstraction*, is an HOL term of type $\alpha \to \beta$.

The above definition encompasses the simply typed λ-calculus. In order to extend this base framework into logic HOL we simply ensure that the signature $(C_\alpha)_{\alpha \in T}$ provides a sufficient selection of primitive logical connectives. Without loss of generality, we here assume the following *primitive logical connectives* to be part of the signature: $\neg_{o \to o} \in C_{o \to o}$, $\vee_{o \to o \to o} \in C_{o \to o \to o}$, $\Pi_{(\alpha \to o) \to o} \in C_{(\alpha \to o) \to o}$ and $=_{\alpha \to \alpha \to \alpha} \in C_{\alpha \to \alpha \to \alpha}$, abbreviated as $=^\alpha$. The symbols $\Pi_{(\alpha \to o) \to o}$ and $=_{\alpha \to \alpha \to \alpha}$ are generally assumed for each type $\alpha \in T$. The denotation of the primitive logical connectives is fixed below according to their intended meaning. *Binder notation* $\forall X_\alpha \, s_o$ is used as an abbreviation for $(\Pi_{(\alpha \to o) \to o}(\lambda X_\alpha s_o))$. Universal quantification in HOL is thus modeled with the help of the logical constants $\Pi_{(\alpha \to o) \to o}$ to be used in combination with lambda-abstraction. That is, the only binding mechanism provided in HOL is lambda-abstraction.

HOL is a logic of terms in the sense that the *formulas of HOL* are given as the terms of type o. In addition to the primitive logical connectives selected above, we could assume *choice operators* $\epsilon_{(\alpha \to o) \to \alpha} \in C_{(\alpha \to o) \to \alpha}$ (for each type α) in the signature. We are not pursuing this here.

Type information, as well as brackets, may be omitted if obvious from the context, and we may also use infix notation to improve readability. For example, we may write $(s \vee t)$ instead of $((\vee_{o \to o \to o} s_o) t_o)$.

From the selected set of primitive connectives, other logical connectives can be introduced as abbreviations.[2] For example, we may define $s \wedge t := \neg(\neg s \vee \neg t)$, $s \to t := \neg s \vee t$, $s \longleftrightarrow t := (s \to t) \wedge (t \to s)$, $\top := (\lambda X_i X) = (\lambda X_i X)$, $\bot := \neg \top$ and $\exists X_\alpha s := \neg \forall X_\alpha \neg s$.

Each occurrence of a variable in a term is either bound by a λ or free. We use $free(s)$ to denote the set of variables with a free occurrence in s. We consider two terms to be *equal* if the terms are the same up to the names of bound variables, that is, we consider α-conversion implicitly.

Substitution of a term s_α for a variable X_α in a term t_β is denoted by $[s/X]t$. Since we consider α-conversion implicitly, we assume the bound variables of t to avoid variable capture.

Well-known operations and relations on HOL terms include $\beta\eta$-*normalization* and $\beta\eta$-*equality*, denoted by $s =_{\beta\eta} t$, β-*reduction* and η-*reduction*. A β-*redex* $(\lambda X s)t$ β-reduces to $[t/X]s$. An η-*redex* $\lambda X(sX)$, where $X \notin free(s)$, η-reduces to s. We

---

[2] As demonstrated by Andrews [8], we could, in fact, start out with only primitive equality in the signature (for all types α) and introduce all other logical connectives as abbreviations based on it.

write $s =_\beta t$ to mean s can be converted to t by a series of β-reductions and expansions. Similarly, $s =_{\beta\eta} t$ means s can be converted to t using both β and η.

3.2 Semantics of HOL

The semantics of HOL is well understood and thoroughly documented. The introduction provided next focuses on the aspects as needed for this article. For more details we refer to the literature [10].

The semantics of choice for the remainder is Henkin semantics, i.e., we work with Henkin's general models [26]. Henkin models and standard models are introduced next. We start out with introducing frame structures.

A *frame* D is a collection $\{D_\alpha\}_{\alpha \in T}$ of nonempty sets D_α, such that $D_o = \{T, F\}$ (for truth and falsehood). The $D_{\alpha \to \beta}$ are collections of functions mapping D_α into D_β.

A *model* for HOL is a tuple $M = \langle D, I \rangle$, where D is a frame, and I is a family of typed interpretation functions mapping constant symbols $p_\alpha \in C_\alpha$ to appropriate elements of D_α, called the *denotation of* p_α. The logical connectives $\neg$, $\vee$, Π and $=$ are always given their expected, standard denotations:[3]

- $I(\neg_{o \to o}) = not \in D_{o \to o}$ such that $not(T) = F$ and $not(F) = T$.

- $I(\vee_{o \to o \to o}) = or \in D_{o \to o \to o}$ such that $or(a, b) = T$ iff $(a = T$ or $b = T)$.

- $I(=_{\alpha \to \alpha \to o}) = id \in D_{\alpha \to \alpha \to o}$ such that for all $a, b \in D_\alpha$, $id(a, b) = T$ iff a is identical to b.

- $I(\Pi_{(\alpha \to o) \to o}) = all \in D_{(\alpha \to o) \to o}$ such that for all $s \in D_{\alpha \to o}$, $all(s) = T$ iff $s(a) = T$ for all $a \in D_\alpha$; i.e., s is the set of all objects of type α.

Variable assignments are a technical aid for the subsequent definition of an interpretation function $\|.\|^{M,g}$ for HOL terms. This interpretation function is parametric over a model M and a variable assignment g.

A *variable assignment* g maps variables X_α to elements in D_α. $g[d/W]$ denotes the assignment that is identical to g, except for variable W, which is now mapped to d.

The *denotation* $\|s_\alpha\|^{M,g}$ of an HOL term s_α on a model $M = \langle D, I \rangle$ under assignment g is an element $d \in D_\alpha$ defined in the following way:

---

[3]Since $=_{\alpha \to \alpha \to o}$ (for all types α) is in the signature, it is ensured that the domains $D_{\alpha \to \alpha \to o}$ contain the respective identity relations. This addresses an issue discovered by Andrews [1]: if such identity relations did not exist in the $D_{\alpha \to \alpha \to o}$, then Leibniz equality in Henkin semantics might not denote as intended.

$$\|p_\alpha\|^{M,g} = I(p_\alpha)$$

$$\|X_\alpha\|^{M,g} = g(X_\alpha)$$

$$\|(s_{\alpha\to\beta}\, t_\alpha)_\beta\|^{M,g} = \|s_{\alpha\to\beta}\|^{M,g}(\|t_\alpha\|^{M,g})$$

$$\|(\lambda X_\alpha s_\beta)_{\alpha\to\beta}\|^{M,g} = \text{the function } f \text{ from } D_\alpha \text{ to } D_\beta \text{ such that}$$
$$f(d) = \|s_\beta\|^{M,g[d/X_\alpha]} \text{ for all } d \in D_\alpha$$

A model $M = \langle D, I\rangle$ is called a *standard model* if and only if for all $\alpha, \beta \in T$ we have $D_{\alpha\to\beta} = \{f \mid f : D_\alpha \longrightarrow D_\beta\}$. In a *Henkin model (general model)* function spaces are not necessarily full. Instead it is only required that for all $\alpha, \beta \in T$, $D_{\alpha\to\beta} \subseteq \{f \mid f : D_\alpha \longrightarrow D_\beta\}$. However, it is required that the valuation function $\|\cdot\|^{M,g}$ from above is total, so that every term denotes. Note that this requirement, which is called *Denotatpflicht*, ensures that the function domains $D_{\alpha\to\beta}$ never become too sparse, that is, the denotations of the lambda-abstractions as devised above are always contained in them.

Corollary 1. *For any Henkin model $M = \langle D, I\rangle$ and variable assignment g:*

1. $\|(\neg_{o\to o}\, s_o)_o\|^{M,g} = T$ *iff* $\|s_o\|^{M,g} = F$.

2. $\|((\vee_{o\to o\to o}\, s_o)\, t_o)_o\|^{M,g} = T$ *iff* $\|s_o\|^{M,g} = T$ *or* $\|t_o\|^{M,g} = T$.

3. $\|((\wedge_{o\to o\to o}\, s_o)\, t_o)_o\|^{M,g} = T$ *iff* $\|s_o\|^{M,g} = T$ *and* $\|t_o\|^{M,g} = T$.

4. $\|((\to_{o\to o\to o}\, s_o)\, t_o)_o\|^{M,g} = T$ *iff*   *(if $\|s_o\|^{M,g} = T$ then $\|t_o\|^{M,g} = T$).*

5. $\|((\longleftrightarrow_{o\to o\to o}\, s_o)\, t_o)_o\|^{M,g} = T$ *iff*   *($\|s_o\|^{M,g} = T$ iff $\|t_o\|^{M,g} = T$).*

6. $\|\top\|^{M,g} = T$.

7. $\|\bot\|^{M,g} = F$.

8. $\|(\forall X_\alpha s_o)_o\|^{M,g} = T$ *iff*   *for all $d \in D_\alpha$ we have $\|s_o\|^{M,g[d/X_\alpha]} = T$.*

9. $\|(\exists X_\alpha s_o)_o\|^{M,g} = T$ *iff* *there exists $d \in D_\alpha$ such that $\|s_o\|^{M,g[d/X_\alpha]} = T$.*

Proof. The proof is straightforward, for instance we prove the first one. $\|(\neg_{o\to o}\, s_o)_o\|^{M,g} = T$ iff $\|\neg_{o\to o}\|^{M,g}(\|s_o\|^{M,g}) = T$ iff $not(\|s_o\|^{M,g}) = T$ iff $\|s_o\|^{M,g} = F$. $\qquad\square$

An HOL formula s_o is *true* in a Henkin model M under assignment g if and only if $\|s_o\|^{M,g} = T$; this is also expressed by writing that $M, g \models^{HOL} s_o$. An HOL formula s_o is called *valid* in M, which is expressed by writing that $M \models^{HOL} s_o$, if and only if $M, g \models^{HOL} s_o$ for all assignments g. Moreover, a formula s_o is called *valid*, expressed by writing that $\models^{HOL} s_o$, if and only if s_o is valid in all Henkin models M.

4 Embedding E into HOL

4.1 Semantical embedding

The formulas of **E** are identified in our semantical embedding with certain HOL terms (predicates) of type $i \to o$. They can be applied to terms of type i, which are assumed to denote possible worlds. That is, the HOL type i is now identified with a (non-empty) set of worlds. Type $i \to o$ is abbreviated as τ in the remainder. The HOL signature is assumed to contain the constant symbol $r_{i \to \tau}$. Moreover, for each atomic propositional symbol p^j of **E**, the HOL signature must contain the corresponding constant symbol p^j_τ. Without loss of generality, we assume that besides those symbols and the primitive logical connectives of HOL, no other constant symbols are given in the signature of HOL.

The mapping $\lfloor \cdot \rfloor$ translates a formula φ of **E** into a term $\lfloor \varphi \rfloor$ of HOL of type τ. The mapping is defined recursively:

$$
\begin{aligned}
\lfloor p^j \rfloor &= p^j_\tau \\
\lfloor \neg \varphi \rfloor &= \neg_{\tau \to \tau} \lfloor \varphi \rfloor \\
\lfloor \varphi \vee \psi \rfloor &= \vee_{\tau \to \tau \to \tau} \lfloor \varphi \rfloor \lfloor \psi \rfloor \\
\lfloor \Box \varphi \rfloor &= \Box_{\tau \to \tau} \lfloor \varphi \rfloor \\
\lfloor \bigcirc(\psi / \varphi) \rfloor &= \bigcirc_{\tau \to \tau \to \tau} \lfloor \varphi \rfloor \lfloor \psi \rfloor
\end{aligned}
$$

$\neg_{\tau \to \tau}$, $\vee_{\tau \to \tau \to \tau}$, $\Box_{\tau \to \tau}$ and $\bigcirc_{\tau \to \tau \to \tau}$ abbreviate the following terms of HOL:

$$
\begin{aligned}
\neg_{\tau \to \tau} &= \lambda A_\tau \lambda X_i \neg (A\,X) \\
\vee_{\tau \to \tau \to \tau} &= \lambda A_\tau \lambda B_\tau \lambda X_i (A\,X \vee B\,X) \\
\Box_{\tau \to \tau} &= \lambda A_\tau \lambda X_i \forall Y_i (A\,Y) \\
\bigcirc_{\tau \to \tau \to \tau} &= \lambda A_\tau \lambda B_\tau \lambda X_i \forall W_i ((\lambda V_i (A\,V \wedge (\forall Y_i (A\,Y \to r_{i \to \tau} V\,Y)))) \, W \to B\,W)^4
\end{aligned}
$$

Analyzing the truth of formula φ, represented by the HOL term $\lfloor \varphi \rfloor$, in a particular world w, represented by the term W_i, corresponds to evaluating the application $(\lfloor \varphi \rfloor W_i)$. In line with previous work [14], we define $vld_{\tau \to o} = \lambda A_\tau \forall S_i (A\,S)$. With this definition, validity of a formula φ in **E** corresponds to the validity of the formula $(vld \lfloor \varphi \rfloor)$ in HOL, and vice versa.

---

[4]If $\mathrm{opt}_\succeq(A)$ is taken as a abbreviation for $\lambda V_i (AV \wedge (\forall Y_i (AY \to r_{i \to \tau} V\,Y)))$, then this can be simplified to $\bigcirc_{\tau \to \tau \to \tau} = \lambda A_\tau \lambda B_\tau \lambda X_i (\mathrm{opt}_\succeq(A) \subseteq B)$.

4.2   Soundness and completeness

To prove the soundness and completeness, that is, faithfulness, of the above embedding, a mapping from preference models into Henkin models is employed.

Definition 1 (Preference model $\Rightarrow$ Henkin model). *Let $M = \langle W, \succeq, V \rangle$ be a preference model. Let $p^1, ..., p^m$ for $m \geq 1$ be atomic propositional symbols and $\lfloor p^j \rfloor = p_\tau^j$ for $j = 1, ..., m$. A Henkin model $H^M = \langle \{D_\alpha\}_{\alpha \in T}, I \rangle$ for M is defined as follows: D_i is chosen as the set of possible worlds W and all other sets $D_{\alpha \to \beta}$ are chosen as (not necessarily full) sets of functions from D_α to D_β. For all $D_{\alpha \to \beta}$ the rule that every term $t_{\alpha \to \beta}$ must have a denotation in $D_{\alpha \to \beta}$ must be obeyed, in particular, it is required that D_τ and $D_{i \to \tau}$ contain the elements Ip_τ^j and $Ir_{i \to \tau}$. Interpretation I is constructed as follows:*

1. *For $1 \leq i \leq m$, $Ip_\tau^j \in D_\tau$ is chosen such that $Ip_\tau^j(s) = T$ iff $s \in V(p^j)$ in M.*

2. *$Ir_{i \to \tau} \in D_{i \to \tau}$ is chosen such that $Ir_{i \to \tau}(s, u) = T$ iff $s \succeq u$ in M.*

Since we assume that there are no other symbols (besides the r, the p^j and the primitive logical connectives) in the signature of HOL, I is a total function. Moreover, the above construction guarantees that H^M is a Henkin model: $\langle D, I \rangle$ is a frame, and the choice of I in combination with the Denotatpflicht ensures that for arbitrary assignments g, $\|.\|^{H^M, g}$ is a total evaluation function.

Lemma 1. *Let H^M be a Henkin model for a preference model M. For all formulas δ of $\mathbf{E}$, all assignments g and worlds s it holds:*

$$M, s \models \delta \text{ if and only if } \| \lfloor \delta \rfloor \, S_i \|^{H^M, g[s/S_i]} = T$$

Proof. See appendix. $\square$

Lemma 2 (Henkin model $\Rightarrow$ Preference model). *For every Henkin model $H = \langle \{D_\alpha\}_{\alpha \in T}, I \rangle$ there exists a corresponding preference model M. Corresponding here means that for all formulas δ of $\mathbf{E}$ and for all assignments g and worlds s,*

$$\| \lfloor \delta \rfloor S_i \|^{H, g[s/S_i]} = T \text{ if and only if } M, s \models \delta$$

Proof. Suppose that $H = \langle \{D_\alpha\}_{\alpha \in T}, I \rangle$ is a Henkin model. Without loss of generality, we can assume that the domains of H are denumerable [26]. We construct the corresponding preference model M as follows:

- $W = D_i$.

- $s \succeq u$ for $s, u \in W$ iff $Ir_{i \to \tau}(s, u) = T$.

- $s \in V(p_\tau^j)$ iff $Ip_\tau^j(s) = T$ for all p^j.

Moreover, the above construction ensures that H is a Henkin model for M. Hence, Lemma 1 applies. This ensures that for all formulas δ of **E**, for all assignments g and all worlds s we have $\|\lfloor \delta \rfloor S_i\|^{H,g[s/S_i]} = T$ if and only if $M, s \vDash \delta$. $\qquad\square$

Theorem 2 (Soundness and completeness of the embedding).

$$\vDash \varphi \text{ if and only if } \vDash^{HOL} vld \lfloor \varphi \rfloor$$

Proof. (Soundness, $\leftarrow$) The proof is by contraposition. Assume $\nvDash \varphi$, i.e, there is a preference model $M = \langle W, \succeq, V \rangle$, and a world $s \in W$, such that $M, s \nvDash \varphi$. By Lemma 1 for an arbitrary assignment g it holds that $\|\lfloor \varphi \rfloor S_i\|^{H^M, g[s/S_i]} = F$ in Henkin model $H^M = \langle \{D_\alpha\}_{\alpha \in T}, I \rangle$. Thus, by definition of $\|.\|$, it holds that $\|\forall S_i(\lfloor \varphi \rfloor S_i)\|^{H^M, g} = \|vld \lfloor \varphi \rfloor\|^{H^M, g} = F$. Hence, $H^M \nvDash^{HOL} vld \lfloor \varphi \rfloor$. By definition $\nvDash^{HOL} vld \lfloor \varphi \rfloor$.

(Completeness, $\rightarrow$) The proof is again by contraposition. Assume $\nvDash^{HOL} vld \lfloor \varphi \rfloor$, i.e., there is a Henkin model $H = \langle \{D_\alpha\}_{\alpha \in T}, I \rangle$ and an assignment g such that $\|vld \lfloor \varphi \rfloor\|^{H,g} = F$. By Lemma 2, there is a preference model M such that $M \nvDash \varphi$. Hence, $\nvDash \varphi$. $\qquad\square$

Remark: In contrast to a deep logical embedding, in which the syntactical structure and the semantics of logic L would be formalized in full detail (using e.g., structural induction and recursion), only the core differences in the semantics of both system **E** and meta-logic HOL have been explicitly encoded in our shallow semantical embedding. In a certain sense we have thus shown, that system **E** can, in fact, be identified and handled as a natural fragment of HOL.

5 Implementation in Isabelle/HOL

5.1 Implementation

The semantical embedding as devised in Section 4 has been implemented in the higher-order proof assistant Isabelle/HOL [28]. Figure 1 displays the respective encoding. Some explanations are in order:

- On line 3, the type i for possible words is introduced
- On line 4, the type τ for formulas is introduced
- On line 5, a designated constant for the actual world (aw) is introduced
- On line 6, the constant r is introduced. r encodes the preference relation $\succeq$
- Lines 8–14 define the Boolean connectives in the usual way

```
1  theory DDLE imports Main
2  begin
3  typedecl i (* type for possible worlds *)
4  type_synonym τ = "(i⇒bool)" (* type for formulas *)
5  consts aw::i (* actual world *)
6  consts r :: "i⇒τ"  (infixr "r" 70) (* comparative goodness relation *)
7
8  definition ddetop         :: "τ" ("⊤")                 where "⊤ ≡ λw. True"
9  definition ddebot         :: "τ" ("⊥")                 where "⊥ ≡ λw. False"
10 definition ddeneg         :: "τ⇒τ" ("¬_"[52]53)        where "¬φ ≡ λw. ¬φ(w)"
11 definition ddeand         :: "τ⇒τ⇒τ" (infixr"∧"51) where "φ∧ψ ≡ λw. φ(w)∧ψ(w)"
12 definition ddeor          :: "τ⇒τ⇒τ" (infixr"∨"50) where "φ∨ψ ≡ λw. φ(w)∨ψ(w)"
13 definition ddeimp         :: "τ⇒τ⇒τ" (infixr"→"49) where "φ→ψ ≡ λw. φ(w)⟶ψ(w)"
14 definition ddeequivt      :: "τ⇒τ⇒τ" (infixr"↔"48) where "φ↔ψ ≡ λw. φ(w)⟷ψ(w)"
15
16 definition ddebox         :: "τ⇒τ" ("□")               where "□φ ≡ λw. ∀v. φ(v)"
17 definition ddediomond     :: "τ⇒τ" ("◇")               where "◇φ ≡ λw. ∃v. φ(v)"
18
19 definition ddeopt :: "τ⇒τ" ("opt<_>") (* obligation/permission operators *)
20   where "opt<φ> ≡ (λv. ( (φ)(v) ∧ (∀x. ((φ)(x)  ⟶  v r x)) ) )"
21 abbreviation(input) msubset :: "τ⇒τ⇒bool" (infix "⊆" 53)
22   where "φ ⊆ ψ ≡ ∀x. φ x ⟶ ψ x"
23 definition ddecond :: "τ⇒τ⇒τ" ("○<_|_>")
24   where "○<ψ|φ> ≡ λw. opt<φ> ⊆ ψ"
25 definition ddeperm :: "τ⇒τ⇒τ" ("P<_|_>")
26   where "P<ψ|φ> ≡ ¬○<¬ψ|φ>"
27
28 definition ddevalid :: "τ⇒bool" ("⌊_⌋"[8]109) (* global validity *)
29   where "⌊p⌋ ≡ ∀w. p w"
30 definition ddeactual :: "τ⇒bool" ("⌊_⌋ₗ"[7]105) (* local validity *)
31   where "⌊p⌋ₗ ≡ p(aw)"
32
33 lemma True nitpick [satisfy,user_axioms,show_all,expect=genuine] oops (* consistency check *)
34 end
```

Figure 1: Shallow semantical embedding of system **E** in Isabelle/HOL

- Lines 16 and 17 introduce the alethic operators $\square$ and $\lozenge$
- The dyadic deontic operators are handled in lines 19–26. Lines 19–20 define the notion of optimal φ-world, and lines 23–26 define the dyadic operators using this notion.
- Lines 28–31 introduce the notion of global validity (i.e, truth in all worlds) and local validity (truth at the actual world).

A sample query is run on line 33. The model finder Nitpick [16] confirms the consistency of the definitions.

In the remainder of this section, we illustrate how the implementation in Isabelle/HOL can be used.

5.2 CTD scenarios

In this section we apply the framework to one of the benchmark problems of deontic logic, the problem of CTD reasoning. We give two examples of CTD scenarios discussed in the deontic logic literature: Chisholm's scenario [19]; Reykjavic's scenario [4].

Chisholm's scenario. The scenario involves the following four sentences:
1. It ought to be that a certain man goes (to the assistance of his neighbours);
2. It ought to be that if he goes he tells them he is coming;
3. If he does not go, he ought not to tell them he is coming;
4. He does not go.

We briefly recall the problem raised by CTDs in SDL. (For more on CTDs the reader may wish to consult [27].) When representing a conditional obligation sentence $\bigcirc(\psi/\varphi)$ in SDL, one separates the contribution of *if ... then* and that of *ought*. The *ought* operator can then take either wide scope ("It ought to be the case that, if φ, then ψ") or narrow scope ("If φ, then it ought to be the case that ψ"). There are thus different possible formalisations of the scenario depending on the choice being made. It turns out that none rendering is satisfactory. The formalisation of these sentences is either inconsistent or the sentences are logically dependent. The Chisholm set is therefore called a paradox.

System **E** is known to provide a solution to Chisholm's paradox: the formalisation of 1–4 is consistent, and each sentence remains logically independent one from the others. These two facts are confirmed by our implementation. This is documented further by Figure 2. On line 1, the theory embedding **E** in Isabelle/HOL (as described in Figure 1) is loaded. On lines 11–14, the Chisholm scenario is encoded. On line 17, a consistency check query is run. Nitpick confirms consistency of 1–4, and outputs the Henkin model described in Figure 3. One can easily read off the preference model this Henkin model encodes. We have

- $W = \{i_1, i_2, i_3, i_4\}$
- $\succeq = \{(i_3, i_1), (i_3, i_2)\}$
- $V(go) = \emptyset$ and $V(tell) = \{i_4\}$.

On the one hand, $i_4 \models \neg go$. On the other hand, each obligation in Chisholm's set (ax1, ax2 and ax3) is vacuously true, because the set of best antecedent-worlds is empty:

$$\mathrm{opt}_{\succeq}(V(\top)) = \mathrm{opt}_{\succeq}(V(\neg go)) = \mathrm{opt}_{\succeq}(V(go)) = \emptyset$$

The above model does the job, but it is not a very interesting one. One can enforce some aspects of the outputted model, by putting a number of suitable constraints on this one, like those shown on lines 19 and 20. First, all the possible

```
1  theory Chisholm_Scenario imports DDLE
2  begin
3  (* Defining some parameters for Nitpick *)
4  nitpick_params [user_axioms,show_all,expect=genuine,format=2]
5
6  (* Constants for the Chisholm scenario *)
7  consts go :: "τ"  tell :: "τ"
8
9  context (*Chisholm scenario is formalized *)
10 assumes
11    ax1: "⌊○<go|T>⌋" and (* It ought to be that a certain man goes. *)
12    ax2: "⌊○<tell|go>⌋" and (* It ought to be that if he goes he tells them he is coming. *)
13    ax3: "⌊○<¬tell|¬go>⌋" and (* If he does not go, he ought not to tell them he is coming. *)
14    ax4: "⌊¬go⌋ι" (* He does not go. *)
15
16 begin (* Consistency is confirmed by Nitpick *)
17 lemma True nitpick [satisfy,card=4] oops
18 lemma assumes
19      "⌊◇(go ∧ tell)⌋" and "⌊◇(go ∧ ¬tell)⌋" and "⌊◇(¬go ∧ tell)⌋" and "⌊◇(¬go ∧ ¬tell)⌋"
20      and limitedness: "(∀φ. (∃x. (φ)x) ⟶ (∃x. opt<φ>x))"
21      shows True nitpick [satisfy,card=4]  oops
22 end
23
24 (* Independence confirmed by Nitpick; countermodels are produced *)
25 lemma assumes "⌊○<go|T> ∧ ○<tell|go> ∧ ○<¬tell|¬go>⌋ι" shows "⌊¬go⌋ι" nitpick oops
26 lemma assumes "⌊○<tell|go> ∧ ○<¬tell|¬go> ∧ ¬go⌋ι" shows "⌊○<go|T>⌋ι" nitpick oops
27 lemma assumes "⌊○<go|T> ∧ ○<¬tell|¬go> ∧ ¬go⌋ι" shows "⌊○<tell|go>⌋ι" nitpick oops
28 lemma assumes "⌊○<go|T> ∧ ○<tell|go>∧ ¬go⌋ι" shows "⌊○<¬tell|¬go>⌋ι" nitpick oops
29 end
```

Figure 2: Chisholm's paradox in Isabelle/HOL

```
Nitpicking formula...
Nitpick found a model for card i = 4:

  Constants:
    go = (λx. _)(i₁ := False, i₂ := False, i₃ := False, i₄ := False)
    tell = (λx. _)(i₁ := False, i₂ := False, i₃ := False, i₄ := True)
    aw = i₄
    (r) =
      (λx. _)
      ((i₁, i₁) := False, (i₁, i₂) := False, (i₁, i₃) := False, (i₁, i₄) := False, (i₂, i₁) := False,
       (i₂, i₂) := False, (i₂, i₃) := False, (i₂, i₄) := False, (i₃, i₁) := True, (i₃, i₂) := True,
       (i₃, i₃) := False, (i₃, i₄) := False, (i₄, i₁) := False, (i₄, i₂) := False, (i₄, i₃) := False,
       (i₄, i₄) := False)
```

Figure 3: Henkin model for the Chisholm scenario

truth assignments for the relevant propositional letters must be considered (line 19).
Second, the condition of limitedness (which boots us to system **F**) must be verified
(line 20). The combination of these two constraints has the effect of preventing the
obligations ax1, ax2 and ax3 from being vacuously true.

Lines 25 to 28 in Figure 2 confirm that the representation of the scenario in **E** meets the requirement of independence. This is confirmed by showing that no sentence follows logically for the other three.

Reykjavic's scenario. It consists of the following five sentences:
1. You should not tell the secret to Reagan;
2. You should not tell the secret to Gorbachev;
3. You should tell Reagan if you tell Gorbachev;
4. You should tell Gorbachev if you tell Reagan;
5. You tell the secret to Gorbachev.

```
 1 theory Reykjavic_Scenario imports DDLE
 2 begin
 3 consts (* We introduce special constants *)
 4 tell_Reagan:: "τ"  tell_Gorbachev:: "τ"
 5 context (*Reykjavic scenario*)
 6 assumes
 7   ax1:"⌊○<¬ tell_Reagan|T>⌋" and (* You should not tell the secret to Reagan. *)
 8   ax2:"⌊○<¬tell_Gorbachev|T>⌋" and (* You should not tell the secret to Gorbachev. *)
 9   ax3:"⌊○<tell_Reagan|tell_Gorbachev>⌋" and (* You should tell Reagan if you tell Gorbachev. *)
10   ax4:"⌊○<tell_Gorbachev|tell_Reagan>⌋" and (* You should tell Gorbachev if you tell Reagan. *)
11   ax5:"⌊tell_Gorbachev⌋τ" (* You tell the secret to Gorbachev. *)
12
13 begin
14 lemma True  nitpick [satisfy,user_axioms,show_all,expect=genuine,card=2,format=2] oops
15 end  (* Consistency is confirmed by Nitpick *)
16 end
```

Figure 4: The Reykjavic scenario in system **E**

On line 14 in Figure 4 Nitpick confirms that the set of formulas ax1–ax5 (=the representation of 1–5 in **E**) is consistent.

5.3 Automatic verification of validities

Automatic verification of valid formulas is also possible. In Figure 5 Isabelle/HOL confirms the validity of each and every axiom and primitive rule of **E** by using the Sledgehammer tool [17] that gives access to automatic theorem provers (ATPs). Figure 6 gives the example of four "reduction" laws identified by Belanyek et al. [3]. They use these reduction laws to establish a more general result concerning iterated modalities in **G**, to the effect that any formula containing nested modal operators is equivalent to some formula with no nesting. The reduction laws are:

$$\bigcirc(\varphi|(\pi \vee (\chi \wedge \bigcirc(\gamma|\eta)))) \leftrightarrow ((\bigcirc(\gamma|\eta) \wedge \bigcirc(\varphi|(\pi \vee \chi))) \vee (\neg \bigcirc(\gamma|\eta) \wedge \bigcirc(\varphi|\pi)))$$
$$\bigcirc(\varphi|(\pi \vee (\chi \wedge \neg \bigcirc(\gamma|\eta)))) \leftrightarrow ((\neg \bigcirc(\gamma|\eta) \wedge \bigcirc(\varphi|(\pi \vee \chi))) \vee (\bigcirc(\gamma|\eta) \wedge \bigcirc(\varphi|\pi)))$$
$$\bigcirc(\pi \vee (\chi \wedge \bigcirc(\gamma|\eta))|\psi) \leftrightarrow ((\bigcirc(\gamma|\eta) \wedge \bigcirc(\pi \vee \chi|\psi)) \vee (\neg \bigcirc(\gamma|\eta) \wedge \bigcirc(\pi|\psi)))$$
$$\bigcirc(\pi \vee (\chi \wedge \neg \bigcirc(\gamma|\eta))|\psi) \leftrightarrow ((\neg \bigcirc(\gamma|\eta) \wedge \bigcirc(\pi \vee \chi|\psi)) \vee (\bigcirc(\gamma|\eta) \wedge \bigcirc(\pi|\psi)))$$

```
1  theory Axioms imports DDLE
2  begin
3  lemma COK:"⌊○<(ψ₁→ψ₂)|φ> → (○<ψ₁|φ> → ○<ψ₂|φ>)⌋" sledgehammer
4    by (simp add: ddecond_def ddeimp_def ddevalid_def)
5
6  lemma Abs:"⌊○<ψ|φ> → □○<ψ|φ>⌋" sledgehammer
7    by (simp add: ddebox_def ddecond_def ddeimp_def ddevalid_def)
8
9  lemma Nec:"⌊□ψ → ○<ψ|φ>⌋" sledgehammer
10   by (simp add: ddebox_def ddecond_def ddeimp_def ddevalid_def)
11
12 lemma Ext:"⌊□(φ₁↔φ₂) → (○<ψ|φ₁> ↔ ○<ψ|φ₂>)⌋" unfolding Defs sledgehammer
13   by (simp add: ddecond_def ddeopt_def)
14
15 lemma Id:"⌊○<φ|φ>⌋" sledgehammer
16   by (simp add: ddecond_def ddeopt_def ddevalid_def)
17
18 lemma Sh:"⌊○<ψ|φ₁∧φ₂> → ○<(φ₂→ψ)|φ₁>⌋" sledgehammer
19   by (simp add: ddeand_def ddecond_def ddeimp_def ddeopt_def ddevalid_def)
20
21 lemma MP:"(⌊φ⌋∧⌊φ→ψ⌋)⟹⌊ψ⌋" unfolding Defs sledgehammer by simp
22
23 lemma N:"⌊φ⌋⟹⌊□φ⌋" unfolding Defs sledgehammer by simp
24 end
```

Figure 5: Verifying the validity of the axioms and rules of system **E**

On lines 3-13 in Figure 6 , Isabelle/HOL confirms that the proofs of these equivalences carry over from **G** to **E**. However, the more general result concerning iterated

```
1  theory Reduction_laws  imports DDLE
2  begin
3  lemma "⌊○<φ|(π∨(χ∧○<γ|η>))> ↔ ((○<γ|η>∧○<φ|(π∨χ)>)∨(¬○<γ|η>∧○<φ|π>))⌋"
4    unfolding Defs sledgehammer by (smt ddecond_def ddeopt_def)
5
6  lemma "⌊○<φ|(π∨(χ∧¬○<γ|η>))> ↔ ((¬○<γ|η>∧○<φ|(π∨χ)>)∨(○<γ|η>∧○<φ|π>))⌋"
7    unfolding Defs sledgehammer by (smt ddecond_def ddeopt_def)
8
9  lemma "⌊○<(π∨(χ∧○<γ|η>))|ψ> ↔ ((○<γ|η>∧○<(π∨χ)|ψ>)∨(¬○<γ|η>∧○<π|ψ>))⌋"
10   unfolding Defs sledgehammer using ddecond_def by auto
11
12 lemma "⌊○<(π∨(χ∧¬○<γ|η>))|ψ> ↔ ((¬○<γ|η>∧○<(π∨χ)|ψ>)∨(○<γ|η>∧○<π|ψ>))⌋"
13   unfolding Defs sledgehammer using ddecond_def by auto
14
15 lemma "⌊□φ ↔ ○<⊥|¬φ>⌋" nitpick [satisfy,user_axioms,show_all,expect=genuine] oops
16 end
```

Figure 6: Reduction laws in system **E**

modalities does not. To establish that one, the authors appeal to the fact that in **G**, $\Box$ is definable in terms of $\bigcirc(-/-)$: $\Box\varphi \leftrightarrow \bigcirc(\bot/\neg\varphi)$. Nitpick confirms that this equivalence is falsifiable in the class of all preference models (line 15).

5.4 Correspondence theory

The aim of correspondence theory is to establish connections between properties of Kripke frames and the formulas in modal logic that are true in all Kripke frames with these properties. Figure 7 shows some first experimentations in correspondence theory. Lines 8–9 tell us that limitedness is equivalent with (and thus corresponds to) D*. Lines 11–13 tell us that limitedness and transitivity are conjointly enough to get both D* and Sp. However, on lines 15–16, Isabelle/HOL fails to show that they are necessary conditions too. The problem is with the proof of the property of transitivity (lines 23–24). The good news is: we do not get a counter-model to the implication (calls for countermodel search with nitpick are not displayed here).

```
 1 theory  Correspondence_theory  imports DDLE
 2 begin
 3 abbreviation limitedness  where "limitedness ≡ (∀φ. (∃x. (φ)x) ⟶ (∃x. opt<φ>x))"
 4 abbreviation Dstar_valid  where "Dstar_valid ≡ (∀φ ψ. ⌊◇φ → (○<ψ|φ> → ¬○<¬ψ|φ>)⌋)"
 5 abbreviation transitivity where "transitivity ≡ (∀x y z. (x r y ∧ y r z) ⟶ x r z)"
 6 abbreviation Sp_valid    where "Sp_valid ≡ (∀φ ψ χ. ⌊(¬○<¬ψ|φ> ∧ ○<ψ→χ|φ>) → ○<χ|φ∧ψ>⌋)"
 7
 8 lemma "limitedness ⟷ Dstar_valid"
 9  unfolding ddecond_def ddediomond_def ddeimp_def ddeneg_def ddevalid_def by auto
10
11 lemma "(limitedness ∧ transitivity) ⟶ (Sp_valid ∧ Dstar_valid)"
12  unfolding ddecond_def ddediomond_def ddeimp_def ddeneg_def ddeand_def ddevalid_def ddeopt_def
13  sledgehammer by smt (*This direction is provable*)
14
15 lemma "(Sp_valid ∧ Dstar_valid) ⟶ (limitedness ∧ transitivity)"
16  unfolding ddecond_def ddediomond_def ddeimp_def ddeneg_def ddeand_def ddevalid_def ddeopt_def oops
17  (*This direction unfortunately not yet, but we also do not get a countermodel*)
18
19 lemma "(Sp_valid ∧ Dstar_valid) ⟶ limitedness"
20  unfolding ddecond_def ddediomond_def ddeimp_def ddeneg_def ddeand_def ddevalid_def ddeopt_def
21  sledgehammer by auto (*Splitting the conjunction, limitednedd is easy for the ATPs*)
22
23 lemma "(Sp_valid ∧ Dstar_valid) ⟶ transitivity"
24  unfolding ddecond_def ddediomond_def ddeimp_def ddeneg_def ddeand_def ddevalid_def oops
25  (*Splitting the conjunction, transitivity is too hard for the ATPs*)
26  (*This direction unfortunately not yet, but we also do not get a countermodel*)
27 end
```

Figure 7: Experiments in correspondence theory

6 Conclusion

A shallow semantical embedding of Åqvist's dyadic deontic logic **E** in classical higher-order logic has been presented and shown to be faithful (sound and complete). The work presented here and in Benzmüller et al. [12] provides the theoretical foundation for the implementation and automation of dyadic deontic logic within existing theorem provers and proof assistants for HOL. We do not define new logics. Instead, we provide an empirical infrastructure for assessing practical aspects of ambitious, state-of-the-art deontic logics; this has not been done before.

We end this paper by listing a number of topics for future research. First, it would be worthwhile to study the shallow semantical embedding of the stronger systems **F** and **G** in HOL. Second, it would be interesting to look at the three systems from the point of view of a semantics defining best in terms of maximality rather than optimality [29, 30]. Third, we could employ our implementation to systematically inspect and verify some meta-logical properties of these systems within Isabelle/HOL. Fourth, it would be interesting to study the quantified extensions of these systems. Previous work has focused on monadic modal logic and conditional logic [5, 6, 14]. Last, but not least, experiments could investigate whether the provided implementation already supports non-trivial applications in practical normative reasoning, or whether further improvements are required.

Acknowledgements

We thank an anonymous reviewer for valuable comments.

References

[1] Andrews, P.B.: General models and extensionality. *Journal of Symbolic Logic* **37**(2), 395–397 (1972)

[2] Åqvist, L.: Deontic logic. In: Gabbay, D., Guenthner, F. (eds.) *Handbook of Philosophical Logic*, pp. 147–264, Springer, Dordrecht (2002)

[3] Belanyek, A., Grossi, D., van der Hoek, W.: A note on nesting in dyadic deontic logic. arXiv preprint, arXiv:1710.03481 (2017)

[4] Belzer., M.: A logic of deliberation. In: Kehler, T. (ed.) *Proceedings of the Fifth National Conference on Artificial Intelligence*, pp. 38–43 (1986)

[5] Benzmüller, C.: Automating quantified conditional logics in HOL. In: Rossi, F. (ed.) *23rd International Joint Conference on Artificial Intelligence*, IJCAI-13, Beijing, China, pp. 746–753, AAAI Press (2013)

[6] Benzmüller, C.: Cut-elimination for quantified conditional logic. *Journal of Philosophical Logic*, **46**(3), 333–353 (2017)

[7] Benzmüller, C.: Universal (meta-)logical reasoning: Recent successes. *Science of Computer Programming*, **172**, 48–62 (2019)

[8] Benzmüller, C., Andrews, P.B.: Church's type theory. In: Zalta, E.N. (ed.) *The Stanford Encyclopedia of Philosophy*. Metaphysics Research Lab, Stanford University, Summer 2019 Edition (2019)

[9] Benzmüller, C.: Universal (meta-)logical reasoning: The Wise Men Puzzle (Isabelle/HOL dataset). *Data in Brief*, **24**, no. 103774 (2019)

[10] Benzmüller, C., Brown, C., Kohlhase, M.: Higher-order semantics and extensionality. *Journal of Symbolic Logic*, **69**(4), 1027–1088 (2004)

[11] Benzmüller, C., Claus, M., Sultana, N.: Systematic verification of the modal logic cube in Isabelle/HOL. In: Kaliszyk, C., Paskevich, A. (eds.) *Workshop on Proof Exchange for Theorem Proving, PxTP 2015*, Berlin, Germany, EPTCS, vol. 186, pp. 24–41 (2015)

[12] Benzmüller, C., Farjami, A., Parent., X.: A dyadic deontic logic in HOL. In: Broersen, J., Condoravdi, C., Nair, S., Pigozzi, G. (eds.) *Deontic Logic and Normative Systems — 14th International Conference, DEON 2018*, Utrecht, The Netherlands, 3-6 July, 2018, pp. 33–50, College Publications, UK (2018)

[13] Benzmüller, C., Farjami, A., Meder, P., Parent., X.: I/O logic in HOL. *Journal of Applied Logics – IfCoLoG Journal of Logics and their Applications*, this issue (2019).

[14] Benzmüller, C., Paulson, L. C.: Quantified multimodal logics in simple type theory. *Logica Universalis* (Special Issue on Multimodal Logics), **7**(1), 7–20 (2013)

[15] Benzmüller, C., Sultana, N., Paulson, L. C., Theiß, F.: The higher-order prover LEO-II. *Journal of Automated Reasoning*, **55**(4), 389–404 (2015)

[16] Blanchette, J.C., Nipkow, T.: Nitpick: A counterexample generator for higher-order logic based on a relational model finder. In: Kaufmann, M., Paulson, L. C. (eds.) *International Conference on Interactive Theorem Proving 2010*, LNCS, vol. 6172, pp. 131–146, Springer (2010)

[17] Blanchette, J. C., Paulson, L. C.: Hammering away - A user's guide to Sledgehammer for Isabelle/HOL (2017)

[18] Carmo, J., Jones, A.: Completeness and decidability results for a logic of contrary-to-duty conditionals. *Journal of Logic and Computation* **23**(3), 585–626 (2013)

[19] Chisholm, R. M.: Contrary-to-duty imperatives and deontic logic. *Analysis*, **24**(2), 33–36 (1963)

[20] Church, A.: A formulation of the simple theory of types. *Journal of Symbolic Logic*, **5**(2), 56–68 (1940)

[21] Doczkal, C., Bard, J.: Completeness and decidability of converse PDL in the constructive type theory of Coq. In: Andronick, J., Felty, A. P. (eds.) *International Conference on Certified Programs and Proofs, CPP 2018*, Los Angeles, USA, Proceedings of the 7th ACM SIGPLAN, pp. 42–52, ACM, New York, USA (2018)

[22] Doczkal, C., Smolka., G.: Completeness and decidability results for CTL in constructive type theory. *Journal of Automated Reasoning*, **56**(32), 343–365 (2016)

[23] Gabbay, D., Horty, J., Parent, X., van der Meyden, R., van der Torre, L.: *Handbook of*

Deontic Logic and Normative Systems. Volume 1. College Publications, UK (2013)

[24] Kirchner, D., Benzmüller, C., Zalta, E.: Mechanizing principia logico-metaphysica in functional type theory. CoRR https://arxiv.org/abs/1711.06542 (2017)

[25] Hansson, B.: An analysis of some deontic logics. *Noûs*, 373–398 (1969)

[26] Henkin, L.: Completeness in the theory of types. *Journal of Symbolic Logic*, **5**(2), 81–91 (1950)

[27] Hilpinen, R., McNamara, P.: Deontic logic. In [23, pp. 3–136]

[28] Nipkow, T., Paulson, L. C., Wenzel., M.: *Isabelle/HOL — A proof assistant for higher-order logic.* volume 2283 of Lecture Notes in Computer Science, Springer (2002)

[29] Parent, X.: Maximality vs optimality in dyadic deontic logic - Completeness results for systems in Hansson's tradition. *Journal of Philosophical Logic*, **43**(6), 1101–1128 (2014)

[30] Parent, X.: Completeness of Åqvist's systems E and F. *The Review of Symbolic Logic*, **8**(1), 164–177 (2015)

[31] Steen, A., Benzmüller, C.: The higher-order prover Leo-III. In: Galmiche, D., Schulz, S., Sebastiani, R. (eds.) *Automated Reasoning. IJCAR 2018*, LNCS, vol. 10900, pp. 108-116, Springer (2018)

Appendix

Proof of Lemma 1

In the proof we implicitly employ curring and uncurring, and we associate sets with their characteristic functions. Throughout the proof whenever possible we omit types in order to avoid making the notation too cumbersome. The proof of Lemma 1 is by induction on the structure of δ. We start with the case where δ is p^j. We have

$$\|\lfloor p^j \rfloor S\|^{H^M, g[s/S_i]} = T$$
$$\Leftrightarrow \quad \|p_\tau^j S\|^{H^M, g[s/S_i]} = T$$
$$\Leftrightarrow \quad I p_\tau^j(s) = T$$
$$\Leftrightarrow \quad s \in V(p^j) \quad \text{(by definition of } H^M)$$
$$\Leftrightarrow \quad M, s \vDash p^j$$

In the inductive cases we make use of the following **induction hypothesis**: *For sentences δ' structurally smaller than δ we have: For all assignments g and states s, $\|\lfloor\delta'\rfloor S\|^{H^M, g[s/S_i]} = T$ if and only if $M, s \vDash \delta'$.*

We consider each inductive case in turn:

(a) $\delta = \varphi \vee \psi$. In this case:

$$\|\lfloor \varphi \vee \psi \rfloor S\|^{H^M, g[s/S_i]} = T$$
$$\Leftrightarrow \quad \|(\lfloor\varphi\rfloor \vee_{\tau\to\tau\to\tau} \lfloor\psi\rfloor)S\|^{H^M, g[s/S_i]} = T$$
$$\Leftrightarrow \quad \|(\lfloor\varphi\rfloor S) \vee (\lfloor\psi\rfloor S)\|^{H^M, g[s/S_i]} = T \quad ((\lfloor\varphi\rfloor \vee_{\tau\to\tau\to\tau} \lfloor\psi\rfloor)S =_{\beta\eta} (\lfloor\varphi\rfloor S) \vee (\lfloor\psi\rfloor S))$$

$\Leftrightarrow \quad \|\lfloor\varphi\rfloor S\|^{H^M, g[s/S_i]} = T \text{ or } \|\lfloor\psi\rfloor S\|^{H^M, g[s/S_i]} = T$

$\Leftrightarrow \quad M, s \vDash \varphi \text{ or } M, s \vDash \psi \quad \text{(by induction hypothesis)}$

$\Leftrightarrow \quad M, s \vDash \varphi \vee \psi$

(b) $\delta = \neg\varphi$. In this case:

$\quad \|\lfloor\neg\varphi\rfloor S\|^{H^M, g[s/S_i]} = T$

$\Leftrightarrow \quad \|(\neg_{\tau\to\tau}\lfloor\varphi\rfloor)S\|^{H^M, g[s/S_i]} = T$

$\Leftrightarrow \quad \|\neg(\lfloor\varphi\rfloor)S)\|^{H^M, g[s/S_i]} = T \quad ((\neg_{\tau\to\tau}\lfloor\varphi\rfloor)S =_{\beta\eta} \neg(\lfloor\varphi\rfloor S))$

$\Leftrightarrow \quad \|\lfloor\varphi\rfloor S\|^{H^M, g[s/S_i]} = F$

$\Leftrightarrow \quad M, s \nvDash \varphi \quad \text{(by induction hypothesis)}$

$\Leftrightarrow \quad M, s \vDash \neg\varphi$

(c) $\delta = \Box\varphi$. We have the following chain of equivalences:

$\quad \|\lfloor\Box\varphi\rfloor S\|^{H^M, g[s/S_i]} = T$

$\Leftrightarrow \quad \|(\lambda X \forall Y(\lfloor\varphi\rfloor Y))S\|^{H^M, g[s/S_i]} = T$

$\Leftrightarrow \quad \|\forall Y(\lfloor\varphi\rfloor Y)\|^{H^M, g[s/S_i]} = T$

$\Leftrightarrow \quad \text{For all } a \in D_i \text{ we have } \|\lfloor\varphi\rfloor Y\|^{H^M, g[s/S_i][a/Y_i]} = T$

$\Leftrightarrow \quad \text{For all } a \in D_i \text{ we have } \|\lfloor\varphi\rfloor Y\|^{H^M, g[a/Y_i]} = T \quad (S \notin free(\lfloor\varphi\rfloor) = \emptyset)$

$\Leftrightarrow \quad \text{For all } a \in D_i \text{ we have } M, a \vDash \varphi \quad \text{(by induction hypothesis)}$

$\Leftrightarrow \quad M, s \vDash \Box\varphi$

(d) $\delta = \bigcirc(\psi/\varphi)$. We have the following chain of equivalences:

$\quad \|\lfloor\bigcirc(\psi/\varphi)\rfloor S\|^{H^M, g[s/S_i]} = T$

$\Leftrightarrow \quad \|(\lambda X \forall W((\lambda V(\lfloor\varphi\rfloor V \wedge (\forall Y(\lfloor\varphi\rfloor Y \to r\, V\, Y)))) W \to \lfloor\psi\rfloor W))S\|^{H^M, g[s/S_i]} = T$

$\Leftrightarrow \quad \|\forall W((\lambda V(\lfloor\varphi\rfloor V \wedge (\forall Y(\lfloor\varphi\rfloor Y \to r\, V\, Y)))) W \to \lfloor\psi\rfloor W)\|^{H^M, g[s/S_i]} = T$

$\Leftrightarrow \quad \text{For all } u \in D_i \text{ we have:}$

$\quad \|(\lambda V(\lfloor\varphi\rfloor V \wedge (\forall Y(\lfloor\varphi\rfloor Y \to r\, V\, Y)))) W \to \lfloor\psi\rfloor W\|^{H^M, g[s/S_i][u/W_i]} = T$

$\Leftrightarrow \quad \text{For all } u \in D_i \text{ we have:}$

$\quad \text{If } \|(\lambda V(\lfloor\varphi\rfloor V \wedge (\forall Y(\lfloor\varphi\rfloor Y \to r\, V\, Y)))) W\|^{H^M, g[s/S_i][u/W_i]} = T,$

$\quad \text{then } \|\lfloor\psi\rfloor W\|^{H^M, g[s/S_i][u/W_i]} = T$

$\Leftrightarrow \quad \text{For all } u \in D_i \text{ we have:}$

$\quad \text{If } \|\lfloor\varphi\rfloor W\|^{H^M, g[s/S_i][u/W_i]} = T \text{ and}$

$\quad \|\forall Y(\lfloor\varphi\rfloor Y \to r\, W\, Y)\|^{H^M, g[s/S_i][u/W_i]} = T,$

$\quad \text{then } \|\lfloor\psi\rfloor V\|^{H^M, g[s/S_i][u/W_i]} = T$

$\Leftrightarrow \quad \text{For all } u \in D_i \text{ we have:}$

$\quad \text{If } \|\lfloor\varphi\rfloor W\|^{H^M, g[s/S_i][u/W_i]} = T \text{ and}$

$\quad \text{for all } t \in D_i \text{ we have } \|\lfloor\varphi\rfloor Y \to r\, W\, Y\|^{H^M, g[s/S_i][u/W_i][t/Y_i]} = T,$

then $\| \lfloor \psi \rfloor W \|^{H^M, g[s/S_i][u/W_i]} = T$

$\Leftrightarrow$ For all $u \in D_i$ we have:

If $\| \lfloor \varphi \rfloor W \|^{H^M, g[s/S_i][u/W_i]} = T$ and

for all $t \in D_i$ we have $\| \lfloor \varphi \rfloor Y \|^{H^M, g[s/S_i][u/W_i][t/Y_i]} = T$ implies $Ir_{i \to \tau}(u, t) = T$,

then $\| \lfloor \psi \rfloor W \|^{H^M, g[s/S_i][u/W_i]} = T$

$\Leftrightarrow$ For all $u \in D_i$ we have:

If $u \in V(\varphi)$ and

for all $t \in D_i$ we have $t \in V(\varphi)$ implies $u \succeq t$,

then $u \in V(\psi)$ (**see the justification ***)

$\Leftrightarrow$ $\mathrm{opt}_{\succeq}(V(\varphi)) \subseteq V(\psi)$

$\Leftrightarrow$ $M, s \models \bigcirc(\psi/\varphi)$

Justification *: What we need to show is: $\| \lfloor \varphi \rfloor \|^{H^M, g[s/S_i]}$ is identified with $V(\varphi)$ (analogously ψ). By induction hypothesis, for all assignments g and states s, we have $\| \lfloor \varphi \rfloor S \|^{H^M, g[s/S_i]} = T$ if and only if $M, s \models \varphi$. Expanding the details of this equivalence we have: for all assignments g and states s

$$s \in \| \lfloor \varphi \rfloor \|^{H^M, g[s/S_i]} \quad \text{(functions to type } o \text{ are associated with sets)}$$

$$\Leftrightarrow \| \lfloor \varphi \rfloor \|^{H^M, g[s/S_i]}(s) = T$$

$$\Leftrightarrow \| \lfloor \varphi \rfloor \|^{H^M, g[s/S_i]} \| S \|^{H^M, g[s/S_i]} = T$$

$$\Leftrightarrow \| \lfloor \varphi \rfloor S \|^{H^M, g[s/S_i]} = T$$

$$\Leftrightarrow M, s \models \varphi$$

$$\Leftrightarrow s \in V(\varphi)$$

Hence, $s \in \| \lfloor \varphi \rfloor \|^{H^M, g[s/S_i]}$ if and only if $s \in V(\varphi)$.

By extensionality we thus know that $\| \lfloor \varphi \rfloor \|^{H^M, g[s/S_i]}$ is identified with $V(\varphi)$. Moreover, since H^M obeys the Denotatpflicht we know that $V(\varphi) \in D_\tau$.

Received 29 November 2018

LEGIS: A Proposal to Handle Legal Normative Exceptions and Leverage Inference Proofs Readability

Cleyton Mário de Oliveira Rodrigues
University of Pernambuco, Garanhuns–PE, Brazil
Center of Informatics, Federal University of Pernambuco, Recife–PE, Brazil
cleyton.rodrigues@upe.br

Eunice Palmeira
Federal Institute of Alagoas (IFAL), Maceió–AL, Brazil
eunicepalmeira@ifal.edu.br

Fred Freitas
Center of Informatics, Federal University of Pernambuco, Recife–PE, Brazil
fred@cin.ufpe.br

Italo Oliveira
Faculty of Law, Federal University of Pernambuco, Recife–PE, Brazil
italojsoliveira1@gmail.com

Ivan Varzinczak
Centre de Recherche en Informatique de Lens, Université d'Artois, Lens, France
varzinczak@cril.fr

Abstract

Although the representation of normative texts and simulation of legal acts are commonly interdisciplinary themes in the field of Artificial Intelligence and Law (AI & Law), some questions remain open or are yet explored. Among them, we can mention the formalization of the legal body in the face of explicit or implicit exceptions in the juridical reasoning, and the treatment of readability issues, in exposing or justifying decision-making. In this paper, we present the prototype LEGIS and discuss about a proposal to simulate legal action on two fronts. We adopt a non-monotonic semantics for knowledge representation that is appropriate to the singularities of the legal realm, the Preferential Semantics, and propose a transformation to a formal logic argumentation style, the Sequent Calculus, in order to raise the inference proofs to a level of legibility not yet conveniently attained by conventional reasoners.

Vol. 6 No. 5 2019
Journal of Applied Logics — IfCoLog Journal of Logics and their Applications

1 Introduction

The interdisciplinary field of AI & Law has witnessed the construction of conceptual on-tologies capable of mapping the complexity of the legal domain, and of simulating legal actions based on normative texts. Despite intense research in recent years, some subareas still require further investigation. Two not fully resolved issues in this universe are the in-ability to produce a coherent system w.r.t. the judicial reality (for example, able to handle exceptions between written rules), and the technical language used by the formalisms of knowledge representation, which undermines the understanding of those who would use the system in practice.

Figure 1 pictures a peculiar situation where a juridical normative knowledge-based sys-tem would be quite applicable. An agent A deliberately kills an agent B; without further information, the situation normally leads to a simple homicide classification with basic prison sentences. Additional circumstances, such as behavior driven either by frivolous[1] or moral reasons, would increase or decrease the calculus of the punishment, respectively. In addition, more exclusive circumstances, such as those related to sex-based hate, may lead to specific homicide extensions (in this case, a Femicide), overriding previous generic in-ferences. In this perspective, we argue that a system capable of reasoning over the legal corpus covering possible exceptions, as well as being able to respond (in a controlled nat-ural language) about crimes, penalties, and conflicts between norms, is unprecedented and necessary. Although in different proportions, such a practical system would be fruitful for different users, ranging from ordinary laymen, passing law students to lawyers and judges.

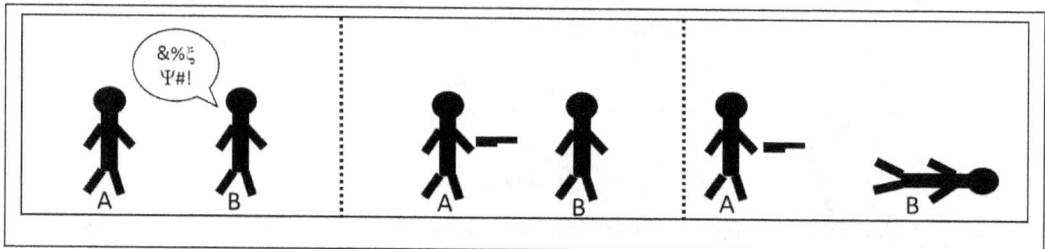

Figure 1: An arbitrary homicide situation

This paper, therefore, proposes the development of a prototype, known as LEGIS (the acronym for LEgal analysIS), addressing the aforementioned issues. We focus on norma-tive legal knowledge, that is, that derived from written legal rules and from legal principles; the ontological basis of LEGIS represents a portion of the Brazilian Penal code. In this context, the exceptions dealt with are those that occur between crimes surrounded by spe-cific circumstances (e.g., an infanticide) in relation to more "normal" crimes (e.g., a typical

---

[1]In Figure 1, the murder was motivated by a silly discussion among the agents, that is, a shallow reason.

homicide). Besides, the second problem addressed consists of clarifying the users about reasoning and decision-making. For that purpose, LEGIS is soon encompassing an approach that transforms connection-based proofs over the Semantic Web language OWL [1] into sequent calculus' proofs. Such proofs are already quite close to natural language, and an additional translation to text will certainly serve users, e.g., to justify their arguments better while relying on LEGIS.

The paper is organized as follows. Section 2 presents the architecture proposal for LEGIS. Section 3 highlights how exceptions can arise in legal texts. Section 4 shows the syntax, semantics, and reasoning tasks for the Classical and Preferential Description Logic. An axiomatization of the Penal Code based on these logics are discussed in Section 5. In Section 6, we briefly introduce the Connection Calculus proof search, and how to transform connection proofs into more intelligible Sequent Proofs. Finally, conclusions about LEGIS and the ongoing works are discussed in Section 7.

2 LEGIS Proposal

LEGIS is a collaborative effort aimed at reasoning on legal norms. In this perspective, the project unfolds in some dimensions, such as: classical vs. defeasible knowledge bases; monotonic vs. non-monotonic approaches to reasoning; practical implementations with parsimonious use of resources (time, memory, ...); and a justification module for inference proofs. A holistic view of LEGIS extensions is highlighted in Figure 2.

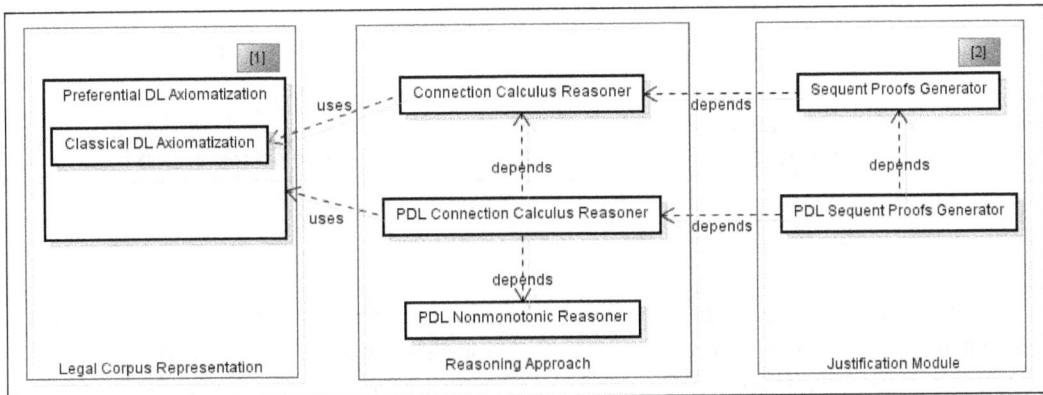

Figure 2: The Holistic View of LEGIS Prototype

Roughly, the idea of LEGIS can be broken down into three levels: one for the representation of knowledge of the legal realm, another for the reasoning strategies, and a third module to provide explanations of the inference proofs closer to natural language. In

Figure 2, the knowledge base should be used by the reasoning module, which provides a customized entry for the transformation process embedded within the justification module.

In particular, in order to attain the previous issues, this work has a two-fold purpose. The first concerns the use of a non-monotonic logical semantics capable of representing and reasoning with exceptions, common in legal texts (labeled in Figure 2 with the identifier [1]). The ontological basis of LEGIS is formed by so-called classical ontologies, i.e. those which reason according to the principles of classical logic, as well as by defeasible ontologies, which allow provisional inferences to be removed as possibly contradictory information is added. These ontologies conceptualize norms of the Brazilian legal system, and the exceptions dealt here involve those that happen between a more general norm and a more specific one. Norms, in turn, are specialized into written rules and general principles. Another goal is the use of a formal logic argumentation in order to make the inference proofs more intelligible to the end user (labeled in Figure 2 with the identifier [2]). This purpose lies in the efficiency vs. readability trade-off regarding the inference engines. So far, LEGIS' reasoning tasks include lawsuit simulation, classification of criminal behavior, and penalty calculus. LEGIS' architecture is illustrated in Figure 3.

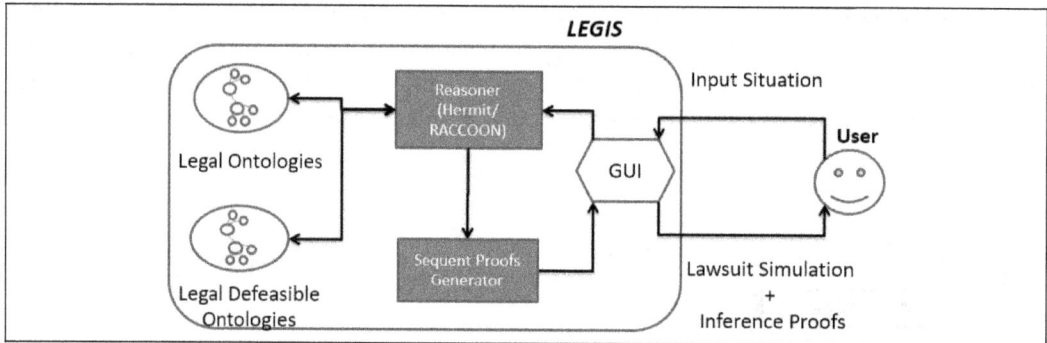

Figure 3: LEGIS Architecture

Through a Graphical User Interface (GUI), an user poses an arbitrary situation. By means of specific reasoners, consistent OWL ontologies will serve as a basis for classifying input instances, even using defeasible axioms. In case of using an inference engine based on the connection calculus (such as RACCOON, an OWL reasoner based on a Description Logic connection calculus, developed under our group [2]), it is possible to transform connection proofs into Sequent proofs (rooted in Sequent Calculus [3]), returning the simulation result and a more readable proof of the inferences made. In the following sections we detail how LEGIS addresses these specific issues.

3   Exceptions in Legal Regulations

Mapping legal normative knowledge into a mathematical formalism free of ambiguities demands time and effort. Some inherent peculiarities of the domain make it especially challenging. Potential sources of anomalies are, for example, the volume of data, the heterogeneity of legal sources, and the legal jargon itself, which uses syntactic inversions, referential ambiguities, and vague terms (open-textured concepts) [4]. In addition, legal systems often present singularities from their political, social and cultural contexts, which makes hard to find a general formalism suited to all of them.

On one hand, it is unfeasible to draw up a normative document capable of anticipating all possible and relevant circumstances. This is why in some cases there are general legal principles that may override the rules, in order to avoid injustices or unwanted conclusion from the literal and direct application of rules. On the other hand, exceptions can be explicitly added throughout the text to accommodate potential specificities of a more general case. In addition to the legal domain peculiarities aforementioned, which may lead to exceptions, Atienza and Manero (2012) [5] argue that the very interplay among laws with legal principles, and apparent conflicts between rules can lead to exceptions and even lack of consensus among lawyers themselves, creating defeasible scenarios of regulation.

In order to illustrate such situation, we exemplify a scenario where an agent's conduct matches the typification of crimes against property, as well as the Trifling principle which removes any criminal liability if the subtracted good is of irrelevant value. The crimes against property correspond to the protected legal interest in the crimes set out in Articles 155-180 of the Brazilian Penal Code[2]. The Trifling principle is entirely related to the globally accepted principle known as *De Minimis Non Curat Lex* [6], in which a behavior with extremely low transgression of the law is not classified as illegal. We transcribe the related legislation below, followed by two didactic examples.

- In Portuguese:

    - **Furto**: *Subtrair, para si ou para outrem, coisa alheia móvel. (Art. 155).*

- In English:

    - **Theft**: *To take a chattel[3], for himself or others.* (Art. 155).

Example 3.1. Will is in a restaurant, and momentarily leaves his wallet on the table to go to the bathroom. John, as he walks past Will's desk, grabs his wallet and leaves.

---

[2]http://www.planalto.gov.br/ccivil_03/decreto-lei/Del2848compilado.htm
[3]An item of personal property that is movable.

Example 3.2. John is a family man who is unemployed. John often asks for money from passersby near a bakery. Taking advantage of the distraction of an attendant in this bakery, and very hungry, John steals two loaves that were on a nearby counter.

The behavior in Example 3.1 matches a typical theft crime. Regarding the behavior in Example 3.2, at first sight, the conduct falls under article 155 of the Penal Code. Nevertheless when analyzing the patrimonial issue, emerge questions like: Does somebody suffer serious injury? Was the bakery impoverished? How much do two loaves cost? Was the act previously planned? From a material point of view, the action becomes atypical, as it does not apply a very serious legal injury, thus not involving criminal charges. The principle of insignificance overrides the theft imputability. Therefore, it is assumed that the legal texts are defeasible.

Although debates on the use of non-monotonic legal reasoning persist today [7], in the 1980s, Gardner (1987) [8] elicited the minimum requirements for legal reasoning accordingly what happens in legal practice: ability to reason with cases, and to handle open-textured predicates, exceptions, conflicts between rules, besides the ability to handle change and non-monotonicity.

The research developed under the umbrella of AI & Law has relied on full synergy with Description Logic formalism. However, as legal regulations are somehow defeasible, open to implicit exceptions; the inferences made in the legal field are not completely linear, they are usually overruled by new information acquired. Therefore, generalizations are only valid for more typical cases. In the following section, we briefly introduce the syntax and semantics of Description Logic, as well as a DL extension addressing a defeasible subsumption constructor to axiomatize exceptions for typical situations. In addition, we discuss a portion of the axiomatization of the Brazilian Criminal domain.

4 Description Logics

4.1 Classical Description Logic

Description Logics (DLs) [9] are a family of formalisms to knowledge representation and reasoning, able to balance the trade-off between expressiveness and decidability for classical monotonic logic. DLs can be seen as subsets of First-order Logic (FOL), in particular, a well-behaved fragment of L2 FOL (first order predicates with 2 variables). DLs accommodate a range of different flavours, each with its own requirements of decidability and expressiveness. For the sake of clarity, throughout the text, we focus on the sublanguage $\mathcal{ALC}$ (Attributive Language with Complements) which allows axiomatizing an arbitrary domain through conjunction, disjunction, negation, existential and universal restriction constructors.

4.1.1 DL Syntax

A DL language is structured in terms of elementary building blocks, i.e., atomic concepts (A), atomic roles (R), and Individuals (I). Complex concept expressions (C, D) may be constructed on these basic descriptions. In especial, $\mathcal{ALC}$ grammar allows the following concept expressions:

$$C, D ::= A \mid C \sqcap C \mid C \sqcup C \mid \neg C \mid \top \mid \bot \mid \exists R.C \mid \forall R.C$$

An $\mathcal{ALC}$ knowledge base ($\mathcal{KB} := \langle \mathcal{A}, \mathcal{T} \rangle$) is conveniently divided into two disjoint components, one comprising terminological axioms ($\mathcal{T}$), such as concept inclusion and equivalence ($C \sqsubseteq D$ and $C \equiv D$, respectively) and the second with assertional axioms ($\mathcal{A}$), such as concept and role assertion ($a : C$ and $(a, b) : R$, where a, b are members of the set of individuals). Hereinafter, we will refer to these components as TBox and ABox, respectively.

4.1.2 DL Semantics

As for its semantics, DL is based on the Open-World Assumption (OWA) [10], since in practice it is inevitably common to handle in the knowledge base with incomplete information. DL semantics is built on top of FOL interpretations, as described in [9]. In short, an Interpretation ($\mathcal{I}$) is a tuple $\langle \Delta^{\mathcal{I}}, \cdot^{\mathcal{I}} \rangle$, where $\Delta^{\mathcal{I}}$ represents the non-empty set known as the domain of $\mathcal{I}$; and $\cdot^{\mathcal{I}}$ is a function that maps concepts to subsets of $\Delta^{\mathcal{I}}$, relations to subsets of $\Delta^{\mathcal{I}} \times \Delta^{\mathcal{I}}$, and each individual name a to an element $a^{\mathcal{I}} \in \Delta^{\mathcal{I}}$, from which we can ascribe the following semantics for the $\mathcal{ALC}$ constructors:

- Individual Name (a): $a^{\mathcal{I}}$;

- Atomic Role (R): $R^{\mathcal{I}}$;

- Atomic Concept (A) : $A^{\mathcal{I}}$;

- Intersection ($C \sqcap D$): $C^{\mathcal{I}} \cap D^{\mathcal{I}}$;

- Union ($C \sqcup D$): $C^{\mathcal{I}} \cup D^{\mathcal{I}}$

- Complement ($\neg C$): $\Delta^{\mathcal{I}} \backslash C^{\mathcal{I}}$;

- Top Concept ($\top$): $\Delta^{\mathcal{I}}$;

- Bottom Concept ($\bot$): $\emptyset$;

- Existential Restriction ($\exists R.C$): $\{a \in \Delta^{\mathcal{I}} \mid \exists b, (a, b) \in R^{\mathcal{I}}, b \in C^{\mathcal{I}}\}$;

- Universal Restriction ($\forall R.C$): $\{a \in \Delta^{\mathcal{I}} \mid \forall b, (a, b) \in R^{\mathcal{I}} \Rightarrow b \in C^{\mathcal{I}}\}$;

- Subsumption ($C \sqsubseteq D$): $C^{\mathcal{I}} \subseteq D^{\mathcal{I}}$;

- Equivalence ($C \equiv D$): $C^{\mathcal{I}} = D^{\mathcal{I}}$;

- Concept Assertion ($C(a)$): $a^{\mathcal{I}} \in C^{\mathcal{I}}$;

- Role Assertion ($R(a,b)$): $\langle a^{\mathcal{I}}, b^{\mathcal{I}} \rangle \in R^{\mathcal{I}}$

4.1.3 DL Reasoning Tasks

From the first-order interpretations, some reasoning tasks [9] are available in DL, such as *Concept Satisfiability* and *Logical Implication*. Given an arbitrary concept C, C is satisfiable iff it admits a model. An interpretation $\mathcal{I}$ is a model of a concept C if $C^{\mathcal{I}} \neq \emptyset$. Likewise, an interpretation $\mathcal{I}$ is a model of a general concept subsumption ($C \sqsubseteq D$) if $C^{\mathcal{I}} \subseteq D^{\mathcal{I}}$.

Some reasoning tasks can be applied directly to the knowledge base as a whole, in both TBox and ABox sub components. Taking into account the terminological part, we emphasize the following inference tasks:

- *Knowledge Base Satisfiability*: Given a knowledge base $\mathcal{KB}$, and two concepts C and D, $\mathcal{KB}$ is satisfiable if it admits a model, that is, an Interpretation $\mathcal{I}$, which for every axiom $C \sqsubseteq D$ in $\mathcal{KB}$, $C^{\mathcal{I}} \subseteq D^{\mathcal{I}}$.

- *Concept Satisfiability w.r.t. Knowledge Base ($\mathcal{KB} \not\models C \equiv \bot$)*: Given a knowledge base $\mathcal{KB}$, and a concept C, C is satisfiable w.r.t. $\mathcal{KB}$ if there is an Interpretation $\mathcal{I}$, which is a model for $\mathcal{KB}$, and further a model for C, that is, $C^{\mathcal{I}} \neq \emptyset$.

- *Logical Implication ($\mathcal{KB} \models C \sqsubseteq D$)*: Given a knowledge base $\mathcal{KB}$, and two concepts C and D, D subsumes C, if for all models $\mathcal{I}$ of $\mathcal{KB}$, $C^{\mathcal{I}} \subseteq D^{\mathcal{I}}$.

For the assertional component, the following reasoning tasks stand out:

- *Concept Instantiation ($\mathcal{KB} \models x : C$)*: Given a knowledge base $\mathcal{KB}$, and an individual x, x is an instance of concept C w.r.t. $\mathcal{KB}$ if $x^{\mathcal{I}} \in C^{\mathcal{I}}$ holds for all models $\mathcal{I}$ of $\mathcal{KB}$;

- *Role Name Instantiation ($\mathcal{KB} \models (x,y) : R$)*: Given a knowledge base $\mathcal{KB}$, and some individuals x, y, the pair of individuals (x, y) is an instance of role name R w.r.t. $\mathcal{KB}$ if $\langle x^{\mathcal{I}}, y^{\mathcal{I}} \rangle \in R^{\mathcal{I}}$ holds for all models $\mathcal{I}$ of $\mathcal{KB}$;

From the considerations made so far, we emphasize that an interpretation $\mathcal{I}$ is a model of a $\mathcal{KB} := \langle \mathcal{A}, \mathcal{T} \rangle$ if $\mathcal{I}$ is a model of $\mathcal{T}$ and a model of $\mathcal{A}$.

Example 4.1. To exemplify the classical DL, let us say that something that has a criminal act is a crime. In addition, a theft has an action of subtraction[4], and that any role "has" is associated with a criminal act. $\mathcal{KB}_{crime}$ represents the DL axioms:

$$\mathcal{KB}_{crime} = \left\{ \begin{array}{l} \exists has.CriminalAct \sqsubseteq Crime \\ Theft \sqsubseteq \exists has.Subtraction \sqcap \forall has.CriminalAct \end{array} \right\}$$

From $\mathcal{KB}_{crime}$, by Logical Implication inference task, we have:

$$\mathcal{KB}_{crime} \models Theft \sqsubseteq Crime$$

4.2 Preferential Description Logic

DL entailment is non-ampliative and non-defeasible, features that are sought for when reasoning with incomplete information and (potential) exceptions. In order to cope with exceptionality, Britz et al. (2011) [11] introduced the Preferential Description Logic (PDL), a DL extension addressing a defeasible subsumption constructor ($\sqsubseteq$). The principal idea is to organize the elements of a domain in degrees of normality, from bottom (the most typical) to up. We note that this kind of knowledge base stratification is fully aligned with the way humans actually reason under incomplete information. In carrying out the reasoning, humans being do not explicitly think of all special cases that would prevent a conclusion from being drawn. Instead, we base our reasoning only on the information at our disposal and provisionally jump to the conclusion. It is only when we come across new information that we accommodate it with the previous knowledge we had and, usually, we do it in a non-disruptive way.

4.2.1 Preferential DL Syntax and Semantics

By extending the DL semantics with non-monotonic reasoning, Britz et al. (2011) [11] have proposed a partial order to set out the levels of typicality. Therefore, the semantics of Preferential DL is organized in terms of strictly partially-ordered structures, $\mathcal{P} := \langle \Delta^{\mathcal{P}}, \cdot^{\mathcal{P}}, <^{\mathcal{P}} \rangle$, where: $\langle \Delta^{\mathcal{P}}, \cdot^{\mathcal{P}} \rangle$ is an ordinary DL interpretation; and, $<^{\mathcal{P}}$ is a irreflexive, anti-symmetric and transitive partial order on $\Delta^{\mathcal{P}}$. Therefore, given a preferential DL interpretation $\mathcal{P}$ and a defeasible subsumption statement $C \sqsubseteq D$, the semantics of this defeasible axiom is given by:

$$\mathcal{P} \Vdash C \sqsubseteq D \text{ iff } \min_{<^{\mathcal{P}}}(C^{\mathcal{P}}) \subseteq D^{\mathcal{P}}$$

The intuition is that objects lower down in $<^{\mathcal{P}}$ are more **normal** than those higher up. Thus, $\min_{<^{\mathcal{P}}}(C^{\mathcal{P}})$ denotes the most typical elements in $C^{\mathcal{P}}$. In order to explain this preferential semantics, Figure 4 pictures a domain stratified in levels of typicality addressing

---

[4]In our context, "subtraction" is a convenient synonym for stealing.

the criminal domain. We have introduced the concept of EVENT, that is, a category of elements that happens in time, such as an action. EVENT1 maps approximately to Example 3.1, while EVENT2 focuses on the violation addressed in the Example 3.2. In this sense, instead of axiomatizing that events in which an item was subtracted from someone is definitely a theft, it is said that "typically" (that is, in the most normal case), these events are thefts. In such domains, these normal cases is organized in the lower part (EVENT1) of the preferential interpretation of EVENT OF SUBTRACTION domain. In the higher level, EVENT2 is a subtraction of an object (LOAF) whose value is so derisory that the Trifle Principle would be triggered to ward off any indication of crime. Thus, regarding this domain, we have the preferential domain $\mathcal{P}$ defined in terms of $\langle \Delta^{\mathcal{P}}, .^{\mathcal{P}}, <^{\mathcal{P}} \rangle$, where:

$$\mathcal{P}: \left\{ \begin{array}{l} \Delta^{\mathcal{P}} = \{event1, event2, loaf, wallet\} \\ \text{THEFT}^{\mathcal{P}} = \{event1\} \\ \text{NONCRIMINAL}^{\mathcal{P}} = \{event2\} \\ \text{OBJECT}^{\mathcal{P}} = \{wallet, loaf\} \\ \text{WORTHLESSOBJECT}^{\mathcal{P}} = \{loaf\} \\ <^{\mathcal{P}} = \{(event1, event2)\} \end{array} \right\}$$

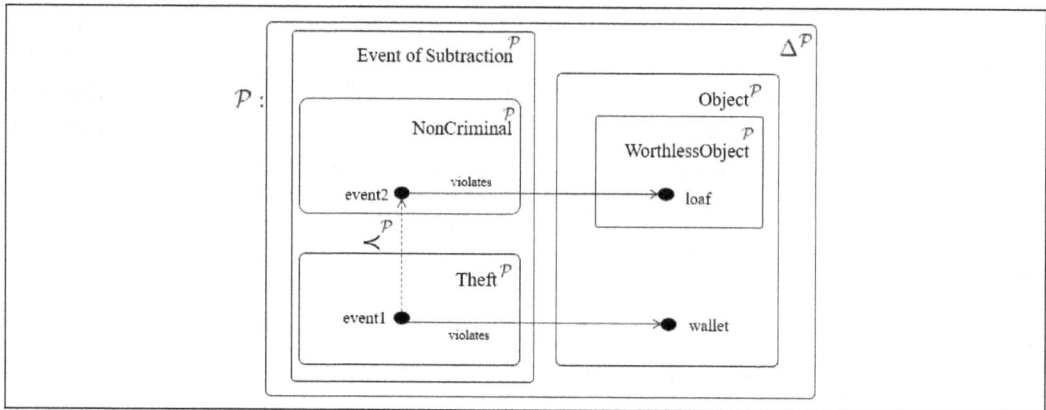

Figure 4: Hierarchy of Item Subtraction Events

Accordingly, we say that **normally**, an item subtraction event is a theft:

$$\text{EVENTOFSUBTRACTION} \sqsubseteq_{\sim} \text{THEFT}$$

4.2.2 Preferential and Rational Entailment

To provide reasoning capabilities within a defeasible knowledge base, the newly introduced subsumption constructor also allows for inference tasks, namely the Preferential and Rational entailment tasks. A subsumption relation $C \sqsubseteq_{\sim} D$ is preferentially entailed by a given defeasible knowledge base $\mathcal{KB}$ iff $C \mathrel{\underset{\sim}{\sqsubseteq}} D$ is a statement of the preferential closure of $\mathcal{KB}$ [11], i.e., it is a derivation from $\mathcal{KB}$ using the following rules of Preferential Subsumption (derived from the KLM theory [12]):

$$
\left\{
\begin{array}{ll}
\text{REFLEXIVITY}: C \mathrel{\underset{\sim}{\sqsubseteq}} C & \text{LEFT LOGICAL EQUIVALENCE}: \dfrac{C \equiv D, C \mathrel{\underset{\sim}{\sqsubseteq}} E}{D \mathrel{\underset{\sim}{\sqsubseteq}} E} \\[2ex]
\text{AND}: \dfrac{C \mathrel{\underset{\sim}{\sqsubseteq}} D, C \mathrel{\underset{\sim}{\sqsubseteq}} E}{C \mathrel{\underset{\sim}{\sqsubseteq}} D \sqcap E} & \text{OR}: \dfrac{C \mathrel{\underset{\sim}{\sqsubseteq}} E, D \mathrel{\underset{\sim}{\sqsubseteq}} E}{C \sqcup D \mathrel{\underset{\sim}{\sqsubseteq}} E} \\[2ex]
\text{RIGHT WEAKENING}: \dfrac{C \mathrel{\underset{\sim}{\sqsubseteq}} D, D \sqsubseteq E}{C \mathrel{\underset{\sim}{\sqsubseteq}} E} & \text{CAUTIOUS MONOTONICITY}: \dfrac{C \mathrel{\underset{\sim}{\sqsubseteq}} D, C \mathrel{\underset{\sim}{\sqsubseteq}} E}{C \sqcap D \mathrel{\underset{\sim}{\sqsubseteq}} E}
\end{array}
\right\}
$$

As usual, inferences in the legal domain should be ampliative beyond retractable. Nevertheless, the preferential entailment does not cover such requirement, since there is no way, from the provided properties, to have $\mathcal{P} \Vdash C \sqcap E \mathrel{\underset{\sim}{\sqsubseteq}} D$ from $\mathcal{P} \Vdash C \mathrel{\underset{\sim}{\sqsubseteq}} D$. In order to accomplish this, Britz et al. (2011) [11] define further the Rational entailment task. Therefore, an additional property, the Rational Monotonicity (RM), should also be ensured by the defeasible subsumption constructor:

$$
\text{RATIONAL MONOTONICITY}: \frac{C \mathrel{\underset{\sim}{\sqsubseteq}} D, C \mathrel{\underset{\sim}{\not\sqsubseteq}} \neg E}{C \sqcap E \mathrel{\underset{\sim}{\sqsubseteq}} D}
$$

It is worth to mention that we are referring to one of the fundamental principles of rationality in non-monotonic reasoning, namely the principle of *presumption of typicality*, formalized by Lehmann (1995) [13]. Briefly, the principle of presumption of typicality is at the heart of a form of ampliative reasoning and states that we shall always assume that we are dealing with the most typical possible situation compatible with the information at our disposal. RM is a necessary condition to model the presumption of typicality; therefore, in the absence of opposite information, RM property infers that individuals are as typical as possible (plausible, though provisional inferences). In this sense, a subsumption relation $C \mathrel{\underset{\sim}{\sqsubseteq}} D$ is rationally entailed by a defeasible knowledge base $\mathcal{KB}$ [11], if $C \mathrel{\underset{\sim}{\sqsubseteq}} D$ is an axiom inferred by the above-mentioned properties including Rational Monotonicity. Suppose, for example, the following TBox:

$$
\mathcal{T} = \left\{
\begin{array}{lcl}
\text{EventOfSubtraction} & \sqsubseteq & \text{Theft} \\
\text{Theft} & \sqsubseteq & \text{Crime} \\
\text{EventOfSubtraction} & \sqsubseteq & \neg\exists\text{violates.WorthlessObject}
\end{array}
\right\}
$$

By the Right Weakening property, we have:

$$
[1] \frac{\left\{ \text{EventOfSubtraction} \mathrel{\underset{\sim}{\sqsubseteq}} \text{Theft}, \ \text{Theft} \sqsubseteq \text{Crime} \right\}}{\models \text{EventOfSubtraction} \mathrel{\underset{\sim}{\sqsubseteq}} \text{Crime}}
$$

In the same way, considering the result of the inference in [1], for any concept expression D, since $\text{EventOfSubtraction} \mathrel{\underset{\sim}{\not\sqsubseteq}} \neg D$, we have by the Rational Monotonicity:

$$[2] \frac{\left\{ \text{ EventOfSubtraction} \sqsubseteq \text{Crime, } \text{EventOfSubtraction} \not\sqsubseteq \neg D \right\}}{\vDash \text{EventOfSubtraction} \sqcap D \sqsubseteq \text{Crime}}$$

Obviously, by the same conditions, we can not consider that we will be dealing with the most typical situations possible, considering that D is $\exists$violates.WorthlessObject. Therefore, an arbitrary reasoner **cannot** infer:

$$\mathcal{T} \vDash \text{EventOfSubtraction} \sqcap \exists \text{violates.WorthlessObject} \sqsubseteq \text{Crime}$$

5 A Proposal to Axiomatize the Legal Domain

In this section, we explore how it is possible to axiomatize the criminal domain, taking into account the exceptions between the norms through the (Preferential) Description Logic. We are not interested in presenting a complete axiomatization of the penal code, but rather how we can extend a legal corpus base with defeasible axioms. The full ontology can be found at https://github.com/cleytonrodrigues/Tese.

Throughout the development of the legal conceptual model, we seek to align our domain with some foundational (or upper) ontology, favoring the ontological adequacy, that is, the degree of closeness to reality [14]. As the upper ontology, we chose to stick to UFO (Unified Foundational Ontology [15]) for grounding our concepts with the UFO categories, thus avoiding typical mistakes while building our ontology hierarchy. UFO is a collection of domain-independent ontologies that makes explicit as much as possible the assumptions and rationales w.r.t. the commonsense, through a rich axiomatization of the vocabulary used. In particular, UFO is based on the ontologies of universals, besides providing a profile with constraints that govern how to construct ontologically valid models that are consistent with reality. In the definition of its categories, UFO incorporates, among others, the principle of identity (which provides for the possibility of judging two entities as being the same, i.e, sortals and non-sortals types), besides the principle of rigidity, which investigates whether a type can be instantiated imperiously in all contexts or not (derived from [16]). Table 1 shows a part of this profile.

In particular, UFO addresses the dichotomy between endurant (UFO-A subontology) and perdurant (UFO-B subontology) categories, as shown at the top of Figure 5. For the first component, we have those entities that persist in time (as an agent, an object), and for the second, those which occur in time (i.e., framed by a time interval), as an event. Endurants may be existentially independent (Substantial) or exist only when associated with another entity (Moment). A notoriously complex type of endurant is Situation, a portion of reality recognized as a whole, a state of affairs. In practice, situations are fulfilled by other

Stereotype	Type	Constraint
«kind»	Rigid Sortal	Supertype cannot be a member of «subkind», «phase», «role», «roleMixin».
«subkind»	Rigid Sortal	Supertype cannot be a member of «phase», «role», «roleMixin», and there must be exactly one «kind» as the supertype
«phase»	Anti-Rigid Sortal	Instantiated only in certain contexts, and defined as part of a partition. There must be exactly one «kind» as the supertype.
«role»	Anti-Rigid Sortal	Instantiated only in certain contexts, and dependent on an external relationship. Cardinality on the opposite side of the «role» type should be ≥ 1. There must be exactly one «kind» as the supertype.

Table 1: Modeling Profile of UFO [15]

endurants, including other minor situations. Dependent moment instances may be tied to either a single entity – Intrinsic Moment –, or to an assortment of these: a Relator.

Another reason for choosing this top-level ontology comes from the fact that it provides an ontology of social entities, known as UFO-C [17]. The legal domain is conceived as a description of social reality, where a group of individuals behaves according to a set of State-approved rules that either allow, forbid, or force them to act under some specific circumstances. UFO-C already considers some assumptions of the legal universe. Agents and objects are part of UFO-C subontology. However, unlike an inanimate object, an agent creates actions (Action Contribution).

Figure 6 illustrates a brief overview of the conceptualization of Crime, regarding the Brazilian Criminal Law. Actually, it is an update of the studies discussed in [18] and [19]. The engineering of this conceptual model was elaborated according to a middle-out approach [20], where intermediate categories of elements are identified first. These are then specialized to match the concepts extracted from legal texts, and generalized towards more generic concepts extracted from the UFO foundational ontology. A crime is a kind of event having as the central kernel an action (we do not consider the crimes of omission, that is, those where an agent had a legal obligation to act, but decides not to accomplish it). CRIMINALACT, therefore, represents these legal actions performed by an OFFENDER, who violates some object of the VICTIM. A LEGALOBJECT can be abstract (honor, life, public peace), or physical (patrimony). An event, in general, starts from an earlier situation towards a result. It is worth mentioning that other important criminal entities, such as space-temporal

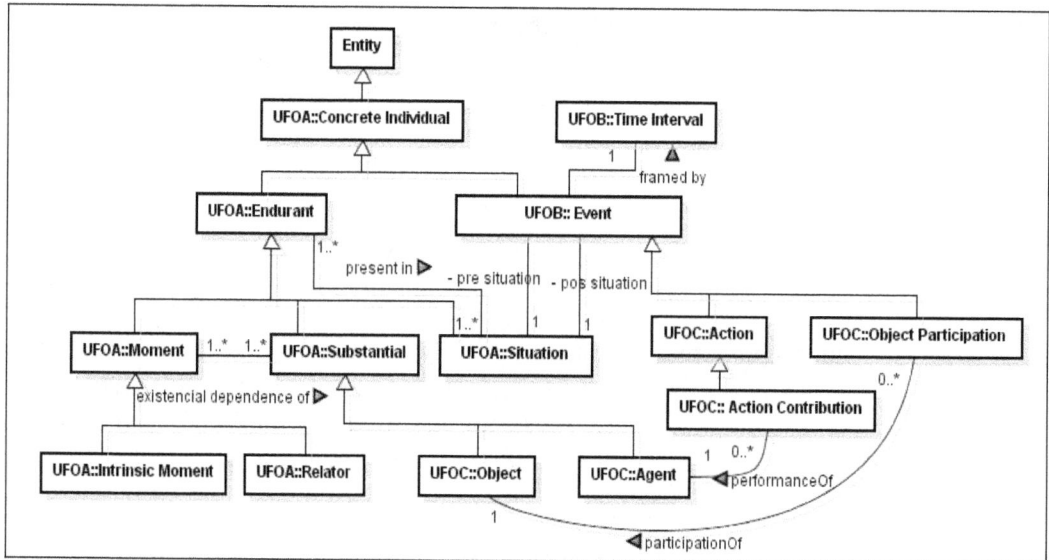

Figure 5: UFO Concepts

occurrence, deontic notions of prohibition/permission, norms and punishments are outside the scope of this work; therefore, they are not displayed in the model of Figure 6[5]. Next, we present a DL axiomatization with pure classical axioms. Then, we show an elaborated base enriched with defeasible axioms, highlighting the problems resolved.

5.1 A Pedagogical Example in DL

In order to make clear the problems arising from the exceptions in the legal texts, we axiomatized a knowledge base related to example 3.2. The base holds the terminological axioms ($\mathcal{T}$) and the assertional ones ($\mathcal{A}$). The terminological axioms map descriptions of a Theft (a subkind of criminal event). In addition, this TBox addresses further a slightly modified set of circumstances that rule out the classification of a crime. For the latter case, we consider the aforementioned Trifle principle.

---

[5]Additional information can be found at https://github.com/cleytonrodrigues/Tese

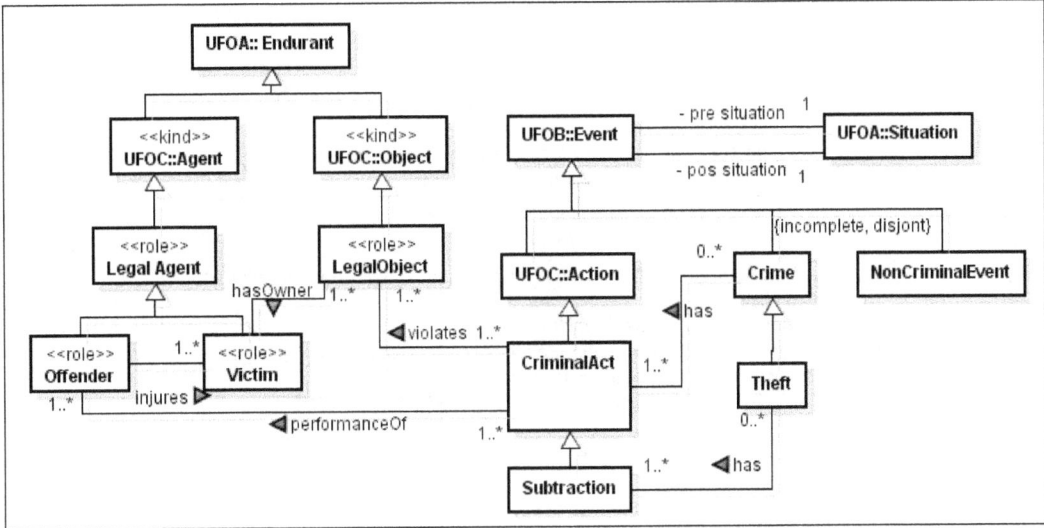

Figure 6: Conceptualization of a Criminal Event

$$\mathcal{T} = \left\{ \begin{array}{lll}
\text{Crime} & \equiv & \text{Event} \sqcap \exists\text{has.CriminalAct} \\
\text{CriminalAct} & \equiv & \text{Action} \sqcap \exists\text{performanceOf.Offender} \sqcap \exists\text{violates.LegalObject} \\
\text{Offender} & \sqsubseteq & \exists\text{injures.Victim} \\
\text{Subtraction} & \sqsubseteq & \text{CriminalAct} \\
\text{Event} & \sqcap & \exists\text{has.Subtraction} \sqcap \exists\text{violates.ChattelObject} \sqsubseteq \text{Theft,} \\
\text{Event} & \sqcap & \exists\text{has.Subtraction} \sqcap \exists\text{violates.ChattelObject} \\
& & \sqcap\exists\text{violates.WorthlessObject} \sqsubseteq \text{NonCriminalEvent,} \\
\text{Theft} & \sqsubseteq & \text{Crime,} \\
\text{NonCriminalEvent} & \sqsubseteq & \neg\text{Crime}
\end{array} \right.$$

$$\mathcal{A} = \left\{ \begin{array}{l}
\text{Event(johnBehavior), Subtraction(loafSubtraction), ChattelObject(loaf),} \\
\text{WorthlessObject(loaf), has(johnBehavior, loafSubtraction),} \\
\text{violates(johnBehavior, loaf).}
\end{array} \right\}$$

An EVENT carried out by means of a SUBTRACTION, violating a CHATTELOBJECT classifies the behavior as a theft. New information acquired as the despicable value of the object (WORTHLESSOBJECT) should refute the previous inference, causing a retraction of the knowledge base. Under the new condition, the event no longer meets the typical theft. It is not possible to keep both inferences, because they are disjoint. It is therefore suggested that the Trifle principle is an exception to the normal case.

Classical DL, therefore, does not address what happens in legal practice. Considering the ABox from the same example, John's behavior would be classified as a theft and a non-criminal event, making the knowledge base inconsistent, i.e., $\mathcal{KB} \models \top \sqsubseteq \bot$, since:

$$[3] \frac{\left\{\begin{array}{l} \text{Event} \sqcap \exists\text{has.Subtraction} \sqcap \exists\text{violates.ChattelObject} \sqsubseteq \text{Theft,} \\ \text{Event(johnBehavior),} \\ \text{Subtraction(loafSubtraction), has(johnBehavior, loafSubtraction),} \\ \text{violates(johnBehavior, loaf), ChattelObject(loaf)} \end{array}\right\}}{\models \text{Theft(johnBehavior)}}$$

$$[4] \frac{\left\{\begin{array}{l} \text{Event} \sqcap \exists\text{has.Subtraction} \sqcap \exists\text{violates.ChattelObject} \sqcap \exists\text{violates.WorthlessObject} \\ \quad \sqsubseteq \text{NonCriminalEvent,} \\ \text{Event(johnBehavior), WorthlessObject(loaf),} \\ \text{Subtraction(loafSubtraction), has(johnBehavior, loafSubtraction),} \\ \text{violates(johnBehavior, loaf), ChattelObject(loaf)} \end{array}\right\}}{\models \text{NonCriminalEvent(johnBehavior)}}$$

$$\mathcal{KB} \models \{\text{Theft(johnBehavior), NonCriminalEvent(johnBehavior), NonCriminalEvent} \sqsubseteq \neg\text{Theft}\}$$

We therefore need a non-monotonic extension of DL Logic capable of dealing satisfactorily with exceptions, as in the interplay between principles and legal laws. Therefore, based on the Preferential DL semantics, the terminological component w.r.t. the john's behavior needs to be slightly modified. In particular, it is necessary to axiomatize that: (1) an event with a subtraction of a chattel object is **typically** a theft, and (2) these events do not **typically** violate a worthless object, and (3) an event with a trifling value object subtraction is **typically** a non-criminal event. The new TBox is shown as follows (the other axioms in $\mathcal{T}$ remain unchanged):

$$\mathcal{T} = \left\{\begin{array}{l} \text{Event} \sqcap \exists\text{has.Subtraction} \sqcap \exists\text{violates.ChattelObject} \sqsubseteq\!\!\!\sim \text{Theft, (1)} \\ \text{Event} \sqcap \exists\text{has.Subtraction} \sqcap \exists\text{violates.ChattelObject} \\ \quad \sqsubseteq\!\!\!\sim \neg\exists\text{violates.WorthlessObject, (2)} \\ \text{Event} \sqcap \exists\text{has.Subtraction} \sqcap \exists\text{violates.ChattelObject} \\ \quad \sqcap \exists\text{violates.WorthlessObject} \sqsubseteq\!\!\!\sim \text{NonCriminalEvent, (3)} \\ \text{Theft} \sqsubseteq \text{Crime,} \\ \text{NonCriminalEvent} \sqsubseteq \neg\text{Crime} \end{array}\right\}$$

Back to Example 3.2, the Rational Monotonicity property rightly prevents John's behavior from being classified as a Theft, but we still have:

$$[5] \frac{\left\{\begin{array}{l} \text{Event} \sqcap \exists\text{has.Subtraction} \sqcap \exists\text{violates.ChattelObject} \sqcap \exists\text{violates.WorthlessObject} \\ \quad \sqsubseteq\!\!\!\sim \text{NonCriminalEvent,} \\ \text{Event(johnBehavior), WorthlessObject(loaf),} \\ \text{Subtraction(loafSubtraction), has(johnBehavior, loafSubtraction),} \\ \text{violates(johnBehavior, loaf), ChattelObject(loaf)} \end{array}\right\}}{\models \text{NonCriminalEvent(johnBehavior)}}$$

In the following section, we discuss the second objective of this study, as highlighted in Figure 2; specifically, the development of the LEGIS module that is capable of producing more readable inference proofs.

6   A Proposal to Sequent Proofs Generator

As previously discussed, it is not enough to develop systems of legal simulation, without guaranteeing an understandable proof verification. Therefore, we discuss an approach based on a formal logic argumentation, in order to provide legible inferences proofs. Nevertheless, the proposal showed here deals only with Classical DL. The extension addressing the Preferential counterpart is discussed in the final remarks.

Freitas and Otten (2016) [1] have proposed a Connection Calculus for the Description Logic $\mathcal{ALC}$ (DL connection method $\mathcal{ALC}$ θ-CM), in the search for a reasoning method that makes a parsimonious usage of memory. In addition, an efficient implementation of this reasoning, known as RACCOON, was developed by Melo Filho et al. (2017) [2]. RACCOON is also highlighted in Figure 3 as an inference engine capable of parsing and reasoning about OWL 2 $\mathcal{ALC}$ ontologies. However, Proof Calculus is far from easy to assimilate. Sequent Calculus [3], a calculus for expressing line-by-line logical arguments, is a more intuitive proof logic. Therefore, we propose in the next subsections a transformation of $\mathcal{ALC}$ Connection Proofs into $\mathcal{ALC}$ Sequent Proofs.

6.1   Non-clausal $\mathcal{ALC}$ θ-Connection Proofs

Our focus is on the non-clausal $\mathcal{ALC}$ θ-Connection Calculus, which is based on the Connection Calculus [21]. Connection calculus is a clear and effective inference method applied successfully over First-Order Logic (FOL). The main idea of connection calculus is checking paths through the FOL formula represented as a matrix, with the purpose of connecting a literal P with its complement $\neg P$. Each pair sets up a connection, which coincides with a tautology in the search branch being examined; therefore, one formula is valid if each path through its matrix representation has a connection. However, before attempting to find a proof, connection calculus converts a formula into a disjunctive normal form (or clausal form), while non-clausal $\mathcal{ALC}$ θ-connection calculus works directly on the structure of the original formula, hence avoiding any translation steps. The later uses $\mathcal{ALC}$ formula with polarity and non-clausal matrices.

An $\mathcal{ALC}$ formula can be expressed as a literal L, or by a disjunction $(C \sqcup D)$, or an universal restriction $(\forall R.C)$, or a conjunction $(C \sqcap D)$, or an existential restriction $(\exists R.C)$. C and D are arbitrary concept expressions and L is either an atomic concept or role, possibly negated or instantiated. The *polarity* is denoted by F^p, where F is an $\mathcal{ALC}$ formula and p is the polarity ($p \in \{0, 1\}$). It is used to represent negation in a matrix, i.e. if F and

$\neg F$ are $\mathcal{ALC}$ formulae, F has polarity 0 and $\neg F$ has polarity 1 (represented by F^0 and F^1, respectively). The non-clausal matrix is a set of clauses, and each clause is a set of literals and (sub)matrices. The matrix of F^p, denoted by $M(F^p)$, is defined inductively according to Table 2. Therefore, the F matrix is $M(F^0)$. Connection Calculus provides further a graphical representation, in which clauses are organized horizontally, while literals and (sub-)matrices of each clause are arranged vertically. A matrix M can be simplified by replacing matrices and clauses of the form $M = \{\ldots, \{X_1, \ldots, X_n\}, \ldots\}$ within M by $M' = \{\ldots, X_1, \ldots, X_n, \ldots\}$. Restrictions are represented by lines; restrictions with indexes (i.e., the notation $L_{i,j}$) are horizontal lines; restrictions without indexes are vertical lines.

Type	F^p	$M(F^p)$	Type	F^p	$M(F^p)$
Atomic	A^0	$\{\{A^0\}\}$	β	$(C \sqcap D)^0$	$\{\{M(C^0), M(D^0)\}\}$
	A^1	$\{\{A^1\}\}$		$(C \sqcup D)^1$	$\{\{M(C^1), M(D^1)\}\}$
α	$(\neg C)^0$	$M(C^1)$		$(C \sqsubseteq D)^1$	$\{\{M(C^0), M(D^1)\}\}$
	$(\neg C)^1$	$M(C^0)$	γ	$(\forall R.D)^1$	$\{\{M(\underline{R}^0), M(\underline{D}^1)\}\}$
	$(C \sqcap D)^1$	$\{\{M(C^1)\}, \{M(D^1)\}\}$		$(\exists R.D)^0$	$\{\{M(\underline{R}^0), M(\underline{D}^0)\}\}$
	$(C \sqcup D)^0$	$\{\{M(C^0)\}, \{M(D^0)\}\}$	δ	$(\forall R.D)^0$	$\{\{M(\underline{R}^1)\}, \{M(\underline{D}^0)\}\}$
	$(C \sqsubseteq D)^0$	$\{\{M(C^1)\}, \{M(D^0)\}\}$		$(\exists R.D)^1$	$\{\{M(\underline{R}^1)\}, \{M(\underline{D}^1)\}\}$
	$(C \models D)^0$	$\{\{M(C^1)\}, \{M(D^0)\}\}$			

Table 2: Matrix of a formula $\mathcal{ALC}$ F^p.

In order to define the non-clausal matrix of an arbitrary $\mathcal{ALC}$ formula, the process starts by the root position ($\models$ or $\sqsubseteq$), which has polarity 0. For example, suppose the example 6.1 drawn from the universe of crime and theft, and the query F_1:

Example 6.1.

$$\left.\begin{array}{l} (\exists \text{has.CriminalAct} \sqsubseteq \text{Crime}) \ \sqcap \\ (\text{Theft} \sqsubseteq \exists \text{has.Subtraction} \sqcap \forall \text{has.CriminalAct}) \end{array}\right\} \begin{array}{l} \models \text{Theft} \sqsubseteq \\ \text{Crime} \end{array}$$

The simplified non-clausal matrix M_1 of F_1 is:

$$\{ \{\text{has}^0, \text{CriminalAct}^0, \text{Crime}^1\},$$
$$\{ \text{Theft}^0, \{\{\text{has}_1^1\}, \{\text{Subtraction}_1^1\}, \{\text{has}^0, \underline{\text{CriminalAct}^1}\}\} \},$$
$$\{\text{Theft(johnBehavior)}^1\}, \{\text{Crime(johnBehavior)}^0\} \}$$

Its graphical representation (without polarity notation) is shown in Figure 7. The validation process consists in checking paths through DL formulae, represented as a matrix

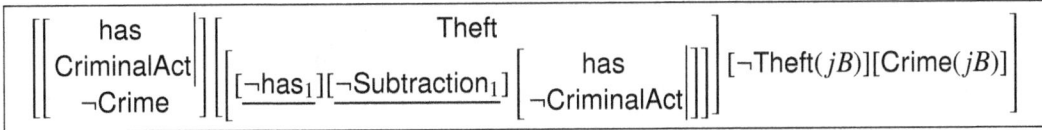

Figure 7: Graphical representation of non-clausal matrix for F_1.

with the purpose of connecting a literal P with its complement $\neg P$, which are in different clauses. Therefore, a *path* is a disjunction of literals of the form $P_1 \sqcup \ldots \sqcup P_n$.

Stemming from query F_1 and its graphical matrix representation, the non-clausal $\mathcal{ALC}$ θ-connection proof is depicted in Figure 8. This process is guided by an active path, a subset of a path being investigated through the matrix. It consists of a set of literals that have been connected to reach the current path of proof. In the first step, a clause of the consequent side is selected, $Crime(jB)$, and through an extension step, $Crime(jB)$ is connected to $\neg Crime(jB)$ applying the θ-substitution, which assigns each (possibly omitted) variable an individual or another variable. All remaining paths through the second matrix of the first clause have to be investigated. In order to accomplish this, the second proof extension step connects $CriminalAct$ to $\neg CriminalAct$. The third step connects has to $\neg has$. Likewise, $Theft$ is connected with $\neg Theft(jB)$. Finally, a reduction step connects has to $\neg has$ literal in the active path. This ends the proof showing that every path through the related matrix contains a θ-complementary connection. Therefore, the $\mathcal{ALC}$ query is valid.

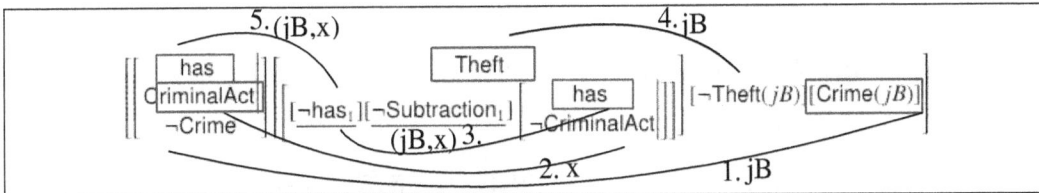

Figure 8: $\mathcal{ALC}$ θ-connection proof using the graphical representation.

6.2 Translating $\mathcal{ALC}$ Connection Proofs into $\mathcal{ALC}$ Sequent Proofs

Given an $\mathcal{ALC}$ formula and its $\mathcal{ALC}$ non-clausal matrix proof, the conversion procedure begins by representing the $\mathcal{ALC}$ formula in its corresponding syntactic tree, where each node can have up to two child nodes. Every node is structured in terms of: (i) a position that identifies each element (predicate or connective) in the formula and is denoted by $a_0, a_1, a_2, \ldots$; (ii) a label consisting of a connective or a logical quantifier (or the predicate itself, if it is atomic); (iii) a polarity (0 or 1), determined by the label and polarity of its parent nodes (root position has polarity 0); and (iv) a type labelled by one Greek letter (α, β, α', β', γ or δ), which is determined by its polarity and its label. Leaf node has no type.

Polarity and type of a node are presented in Table 3. The first entry, $(C \sqcap D)^1$ for example, means that a node labeled with $\sqcap$ and polarity 1 has type α and its successor nodes have polarity 1. The syntactic tree for F_1 is shown in Figure 9. The literals names are abbreviated due to space limitations.

Type α			Type β			Type δ		
$(C \sqcap D)^1$	C^1	D^1	$(C \sqcap D)^0$	C^0	D^0	$(\forall R.C)^0$	R^1	C^0
$(C \sqcup D)^0$	C^0	D^0	$(C \sqcup D)^1$	C^1	D^1	$(\exists R.C)^1$	R^1	C^1
$(\neg C)^1$	C^0							
$(\neg C)^0$	C^1							
Type α'			Type β'			Type γ		
$(C \sqsubseteq D)^0$	C^1	D^0	$(C \sqsubseteq D)^1$	C^0	D^1	$(\forall R.C)^1$	R^0	C^1
$(C \models D)^0$	C^1	D^0				$(\exists R.C)^0$	R^0	C^0

Table 3: Polarity and types of nodes for $\mathcal{ALC}$

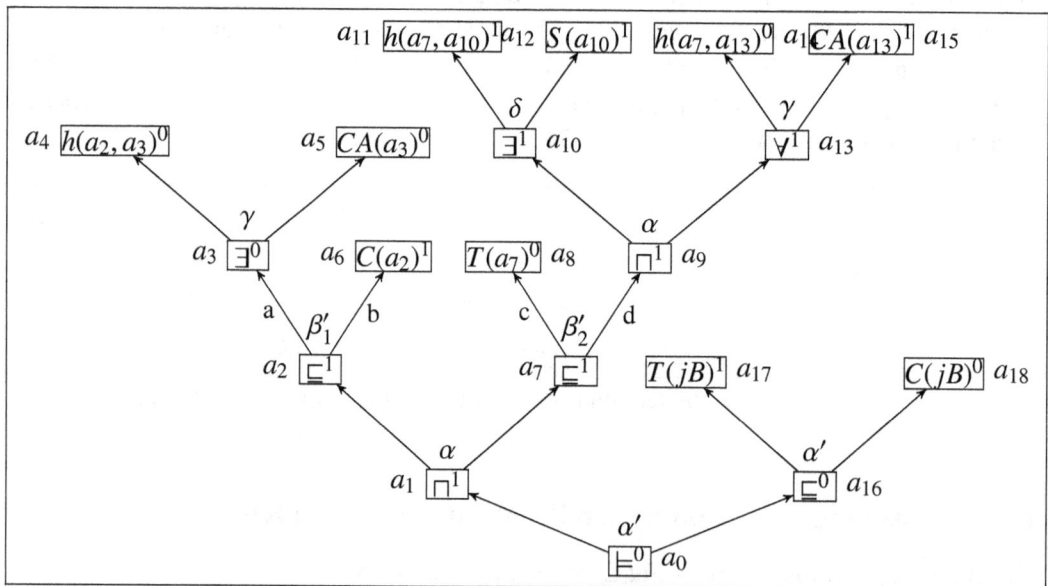

Figure 9: Formula Tree for F_1.

Back to translation, a position is assigned to each corresponding elements in the non-clausal matrix, as shown in Figure 10. After this, the $\mathcal{ALC}$ non-clausal matrix proof is read, and for each connection found, the tree is examined in order to find leaf nodes that correspond to the connection. The paths between the root node and these nodes in the tree are then analyzed to determine the order of the nodes to be worked on, and thus building a (partial) sequent proof structure. This structure provides information about the ordering in which a given formula F has to be transformed by the rules of the sequent calculus. In addition, it brings out information about branches in the sequent given by positions of type β and β', as shown in Figure 10.

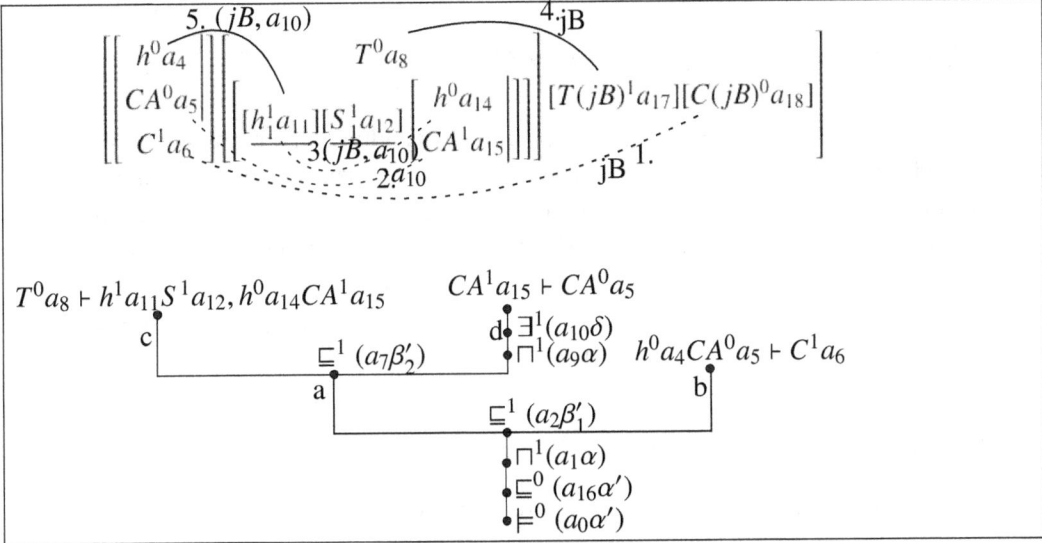

$$5.\ (jB, a_{10}) \qquad\qquad 4.jB$$

$$\begin{bmatrix} \begin{bmatrix} h^0 a_4 \\ CA^0 a_5 \\ C^1 a_6 \end{bmatrix} & [h^1_1 a_{11}][S^1_1 a_{12}] & T^0 a_8 & \begin{bmatrix} h^0 a_{14} \\ CA^1 a_{15} \end{bmatrix} & [T(jB)^1 a_{17}][C(jB)^0 a_{18}] \end{bmatrix}$$

$$3.(jB, a_{10}) \qquad 2.a_{10} \qquad\qquad jB\ 1.$$

$$T^0 a_8 \vdash h^1 a_{11} S^1 a_{12}, h^0 a_{14} CA^1 a_{15} \qquad CA^1 a_{15} \vdash CA^0 a_5$$

$$c \qquad\qquad d\ \exists^1(a_{10}\delta)$$
$$\sqsubseteq^1(a_7\beta'_2) \qquad \sqcap^1(a_9\alpha)\quad h^0 a_4 CA^0 a_5 \vdash C^1 a_6$$
$$a \qquad\qquad\qquad b$$
$$\sqsubseteq^1(a_2\beta'_1)$$
$$\sqcap^1(a_1\alpha)$$
$$\sqsubseteq^0(a_{16}\alpha')$$
$$\models^0(a_0\alpha')$$

Figure 10: Matrix and structure of the Sequent Proof for F_1.

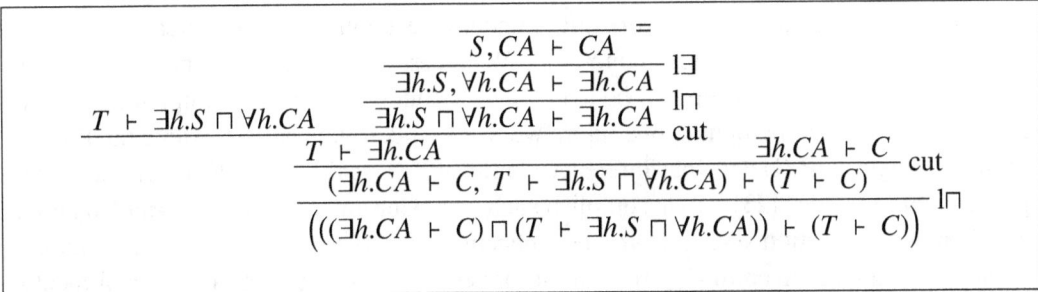

$$\cfrac{\cfrac{\cfrac{\overline{S, CA \vdash CA}}{\exists h.S, \forall h.CA \vdash \exists h.CA}{}^{\exists}}{\exists h.S \sqcap \forall h.CA \vdash \exists h.CA}{}^{\sqcap} \quad \cfrac{}{\exists h.CA \vdash C}}{\cfrac{\cfrac{T \vdash \exists h.S \sqcap \forall h.CA \qquad \exists h.S \sqcap \forall h.CA \vdash \exists h.CA}{T \vdash \exists h.CA}{}^{cut} \qquad \exists h.CA \vdash C}{(\exists h.CA \vdash C, T \vdash \exists h.S \sqcap \forall h.CA) \vdash (T \vdash C)}{}^{cut}}}{\big(((\exists h.CA \vdash C) \sqcap (T \vdash \exists h.S \sqcap \forall h.CA)) \vdash (T \vdash C)\big)}{}^{\sqcap}$$

Figure 11: Complete representation of the resulting $\mathcal{ALC}$ Sequent Proof for F_1.

Furthermore, a complete sequent proof (Figure 11) is constructed from the partial sequent proof obtained in the process and by the correspondence between the node and the rules of the sequent, described in Tables 4 and 5 (appendix A). Rules of Sequent Calculus for $\mathcal{ALC}$ [22] is described in appendix B.

The conversion method might be used in practical applications, in areas that employ DL reasoning and generate descriptions on natural language inferences for lay users. Proof conversion can help users understand why a particular situation is characterized as a criminal event, making its use viable in practice, if an additional translation from these sequents to natural language is accomplished. Our research group is already working in this second translation, which will be available soon.

7 Final Remarks and Ongoing Works

LEGIS is a legal action simulation proposal that should address an ontological basis of legal norms and principles, some inference mechanisms for efficient reasoning, and a justification module capable of generating intuitive proofs. In the present study, we proposed an axiomatization based on the Preferential Description Logic to address the possible levels of exception between the norms. In addition, the conversion process between connection and sequent proofs highlighted is complete. Since the prototype is the (partial) result of a joint effort, other activities have also been carried out in related studies:

- an ontology for Crimes against Life [19];

- an ontology for Crimes against Property [18]; and,

- an implementation for the Connection calculus for the Description Logic $\mathcal{ALC}$ [2];

With respect to the ongoing works, it is under investigation how to extend the current OWL 2 reasoners to enable non-monotonic inferences, according to Preferential DL perspective. Similarly, a next step will be to engineer a connection calculus implementation to Preferential $\mathcal{ALC}$. In particular, one future work is to investigate how to extend the Protégé reasoner plugin DIP (Defeasible Inference Platform) – a scalable implementation for the preferential semantics [23] – to accomplish such task. Currently, we are also implementing an automatic translation system from the connections proofs to sequent proofs and another to natural language. According, a sequent proof generator for Preferential DL entailment is also expected.

Finally, we intend to make LEGIS available as a web-based front-end system through which it is possible to perform functional and accessible legal simulations by the mapped ontologies. We hope that the results obtained so far can improve the layperson's legal understanding and assist in the labor-intensive task of lawsuits performed by professional lawyers. A prototype is available at `https://github.com/cleytonrodrigues/Tese`. Currently, for arbitrary situations, the prototype is able to infer about the presence of some crime, the violated norms, and the penalties imposed.

Acknowledgement

This research is part of project APQ-0550-1.03/16 (*Reconciling Description logic and non-monotonic reasoning in the legal domain*), supported by Fundação de Amparo à Ciência e Tecnologia do Estado de Pernambuco, by the Institut National de Recherche en Informatique et Automatique and by the Centre National de la Recherche Scientifique.

References

[1] Freitas, F., Otten, J.: A connection calculus for the description logic $\mathcal{ALC}$. In Khoury, R., Drummond, C., eds.: Advances in Artificial Intelligence, Cham, Springer International Publishing (2016) 243–256

[2] Melo Filho, D., Freitas, F., Otten, J.: Raccoon: A connection reasoner for the description logic alc. In Eiter, T., Sands, D., eds.: LPAR-21. 21st International Conference on Logic for Programming, Artificial Intelligence and Reasoning. Volume 46 of EPiC Series in Computing., EasyChair (2017) 200–211

[3] Gentzen, G.: Untersuchungen über das logische Schließen II. Mathematische Zeitschrift **39** (1935)

[4] Rissland, E.L., Ashley, K.D., Loui, R.: Ai and law: A fruitful synergy. Artificial Intelligence **150**(1) (2003) 1 – 15

[5] Atienza, M., Manero, J.R.: Rules, principles and defeasibility. In Beltrán, J.F., Ratti, G.B., eds.: The Logic of Legal Requirements - Essays on Defeseability. Oxford University Press, Oxford, United Kingdom (2012) 238–253

[6] Samuels, A.: De minimis non curat lex. Statute Law Review **6**(1) (1985) 167–169

[7] Araszkiewicz, M. In: Legal Rules: Defeasible or Indefeasible? Springer International Publishing, Cham (2015) 415–431

[8] Gardner, A.: An Artificial Intelligence Approach to Legal Reasoning. MIT Press, Cambridge, MA, USA (1987)

[9] Baader, F., Calvanese, D., McGuinness, D.L., Nardi, D., Patel-Schneider, P.F.: The Description Logic Handbook: Theory, Implementation and Applications. 2nd edn. Cambridge University Press, New York, NY, USA (2010)

[10] Reiter, R. In: On Closed World Data Bases. Springer US, Boston, MA (1978) 55–76

[11] Britz, K., Meyer, T., Varzinczak, I.: Semantic foundation for preferential description logics. In Wang, D., Reynolds, M., eds.: AI 2011: Advances in Artificial Intelligence, Berlin, Heidelberg, Springer Berlin Heidelberg (2011) 491–500

[12] Kraus, S., Lehmann, D., Magidor, M.: Nonmonotonic reasoning, preferential models and cumulative logics. Artificial Intelligence **44**(1-2) (1990) 167–207

[13] Lehmann, D.: Another perspective on default reasoning. Annals of Mathematics and Artificial Intelligence **15**(1) (1995) 61–82

[14] Bringuente, A.C.O., de Almeida Falbo, R., Guizzardi, G.: Using a foundational ontology for reengineering a software process ontology. JIDM **2**(3) (2011) 511–526

[15] Guizzardi, G.: Ontological foundations for structural conceptual models. PhD thesis, University of Twente (2005)

[16] Guarino, N., Welty, C.: An overview of ontoclean. In Staab, S., Studer, R., eds.: Handbook on Ontologies. International Handbooks on Information Systems. Springer Berlin Heidelberg (2004) 151–171

[17] Guizzardi, G., Wagner, G.: A unified foundational ontology and some applications of it in business modeling. In Grundspenkis, J., Kirikova, M., eds.: CAiSE Workshops (3), Faculty of Computer Science and Information Technology, Riga Technical University, Riga, Latvia (2004) 129–143

[18] Rodrigues, C.M.d.O., Azevedo, R.R., Freitas, F.L.G., da Silva, E.P., da Silva Barros, P.V.: An ontological approach for simulating legal action in the brazilian penal code. In: Proceedings of the 30th Annual ACM Symposium on Applied Computing. SAC '15, New York, NY, USA, ACM (2015) 376–381

[19] Rodrigues, C.M.d.O., de Freitas, F.L.G., da Silva Oliveira, I.J.: An ontological approach to the three-phase method of imposing penalties in the brazilian criminal code. In: 2017 Brazilian Conference on Intelligent Systems, Uberlândia, Brazil, October 2-5. (2017) 414–419

[20] Uschold, M., Gruninger, M.: Ontologies: Principles, Methods, Applications. Volume 11. (1996)

[21] Bibel, W.: Automated theorem proving. 2 edn. Künstliche Intelligenz. Vieweg, Braunschweig, Germany (1987)

[22] Borgida, A., Franconi, E., Horrocks, I.: Explaining ALC Subsumption. ECAI - European Conference on Artificial Intelligence (2000)

[23] Casini, G., Meyer, T.A., Moodley, K., Sattler, U., Varzinczak, I.J.: Introducing defeasibility into OWL ontologies. In: The Semantic Web - ISWC 2015 - 14th International Semantic Web Conference, Bethlehem, PA, USA, October 11-15, 2015, Proceedings, Part II. (2015) 409–426

[24] Girard, J.Y., Taylor, P., Lafont, Y.: Proofs and Types. Cambridge University Press, New York, NY, USA (1989)

A Matrices for Translating $\mathcal{ALC}$ Connection Proofs into $\mathcal{ALC}$ Sequent Proofs

B A Sequent Calculus for $\mathcal{ALC}$

The calculus consists of three parts, where the first two describe sets of rules, while the latter describes a set of axioms, see figure 12, and the Cut Elimination Theorem is applied according to the proposition 1.

Proposition 1. Cut Elimination Theorem [24]. Let S be a set of sequents (axioms) and s

Type α	Rule	Type β	Rule
$\neg^1$	r¬¬	$\sqcap^0$	l¬⊓
$\neg^0$	l¬¬	$\sqcup^1$	r¬⊔
$\sqcap^1$	r¬⊓	**Type δ**	**Rule**
$\sqcup^0$	l¬⊔	$\forall^0$	l¬∀
		$\exists^1$	r¬∃

Table 4: Correspondence between label, polarity and type of a node, preceded by a node labeled by a negation, with the sequent rules

Type α	Rule	Type β	Rule	Type δ	Rule
$\sqcap^1$	l⊓	$\sqcap^0$	r⊓	$\forall^0$	r∀
$\sqcup^0$	r⊔	$\sqcup^1$	l⊔	$\exists^1$	l∃
$\neg^1$	∅				
$\neg^0$	∅				
Type α'	**Rule**	**Type β'**	**Rule**	**Type γ**	**Rule**
$\sqsubseteq^0$	∅	$\sqsubseteq^1$	$\dfrac{\Gamma \vdash \Delta, A \qquad A, \Sigma \vdash \Pi}{\Gamma, \Sigma \vdash \Delta, \Pi}$	$\forall^1$	∅
$\vDash^0$	∅			$\exists^0$	∅

Table 5: Correspondence between label, polarity and type of a node, not preceded by negation, with the sequent rules

an individual sequent. $S \vdash_{SC} s$, if and only if, there is a proof in SC of s whose leaves are either logical or sequent axioms obtained by the substitution of S-belonging sequents , where the cut rule, $\frac{\Gamma \vdash \Delta, A \quad A, \Sigma \vdash \Pi}{\Gamma, \Sigma \vdash \Delta, \Pi}$, is only applied with a premise being an axiom.

Received 1 December 2018

Rules for the propositional formulae

$$\frac{X,a,b \vdash Y}{X,\, a \sqcap b \vdash Y} \quad (l\sqcap)$$

$$\frac{X \vdash a,Y \quad X \vdash b,Y}{X,\, \vdash a \sqcap b, Y} \quad (r\sqcap)$$

$$\frac{X,\neg a \vdash Y \quad X,\neg b \vdash Y}{X,\neg(a \sqcap b) \vdash Y} \quad (l\neg\sqcap)$$

$$\frac{X \vdash \neg a, \neg b, Y}{X \vdash \neg(a \sqcap b), Y} \quad (r\neg\sqcap)$$

$$\frac{X,a \vdash Y \quad X,b \vdash Y}{X,\, a \sqcup b \vdash Y} \quad (l\sqcup)$$

$$\frac{X \vdash a,b,Y}{X \vdash a \sqcup b, Y} \quad (r\sqcup)$$

$$\frac{X,\neg a,\neg b \vdash Y}{X,\neg(a \sqcup b) \vdash Y} \quad (l\neg\sqcup)$$

$$\frac{X \vdash \neg a, Y \quad X \vdash \neg b, Y}{X \vdash \neg(a \sqcup b), Y} \quad (r\neg\sqcup)$$

$$\frac{X,a \vdash Y}{X,\neg\neg a \vdash Y} \quad (l\neg\neg)$$

$$\frac{X \vdash a, Y}{X \vdash \neg\neg a, Y} \quad (r\neg\neg)$$

Rules for quantified formulae

$$\frac{X' \vdash b, Y'}{X \vdash \forall r.b, Y} \quad (r\forall)$$

$$\frac{X', b \vdash Y'}{X,\, \exists r.b \vdash Y} \quad (l\exists)$$

$$\frac{X', \neg b \vdash Y'}{X, \neg\forall r.b \vdash Y} \quad (l\neg\forall)$$

$$\frac{X' \vdash \neg b, Y'}{X \vdash \neg\exists r.b, Y} \quad (r\neg\exists)$$

where $X' = \{a \mid \forall r.a \in X\} \cup \{\neg a \mid \neg\exists r.a \in X\}$, and
$Y' = \{a \mid \exists r.a \in Y\} \cup \{\neg a \mid \neg\forall r.a \in Y\}$

Termination axioms

$X, a \vdash a, Y$	$(=)$		$X, \neg a \vdash \neg a, Y$	$(=)$
$X, a, \neg a \vdash Y$	$(l\uparrow)$		$X \vdash a, \neg a, Y$	$(r\uparrow)$
$X, \bot \vdash Y$	$(l\bot)$		$X \vdash \top, Y$	$(l\top)$

Figure 12: Rules of Sequent Calculus for $\mathcal{ALC}$ [22].

Received 1 December 2018

Principles for a Judgement Editor Based on Binary Decision Diagrams

Guillaume Aucher
Univ Rennes, CNRS, IRISA
guillaume.aucher@irisa.fr

Jean Berbinau
French Association of IT Judicial Experts
(ex- & honorary member)
jean.berbinau@m4x.org

Marie-Laure Morin
CNRS, Cour de cassation (Former trial judge)
mlmorin@orange.fr

Vol. 6 No. 5 2019
Journal of Applied Logics — IfCoLog Journal of Logics and their Applications

Abstract

We introduce the theoretical principles that underlie the design of a software tool which could be used by judges for making decisions about litigations and for writing judgements. The tool is based on Binary Decision Diagrams (BDD), which are graphical representations of truth–valued functions associated to propositional formulas. Given a type of litigation, the tool asks questions to the judge; each question is represented by a propositional atom. Their answers, true or false, allow to evaluate the truth value of the formula which encodes the overall recommendation of the software about the litigation. Our approach combines some sort of 'theoretical' or 'legal' reasoning dealing with the core of the litigation itself together with some sort of 'procedural' reasoning dealing with the protocol that has to be followed by the judge during the trial: some questions must necessarily be examined and sometimes in a specific order. That is why we consider extensions of BDD called Multi-BDD. They are BDD with multiple roots corresponding to the different specific issues that must necessarily be addressed by the judge during the trial. We illustrate our ideas on a case study dealing with French trade union elections which has been used throughout our project with the Cour de cassation. We also introduce the prototype developed during our project and a link with restricted access to try it out.

1   Introduction

The systematic form of many legal systems derives from Roman law. Romans of antiquity were the first to integrate and apply the methods and rigor of Greek philosophy and logic to the law, especially Gallius, Mucius Scaevola in the 2nd century BC and subsequent jurisconsults of the imperial period. The rediscovery of Roman law via Justinien's compilations in the 12th–13th century in northern Italy and France followed by its interaction and confrontation with the humanists of the 15th–16th century rang the knell of custom and unwritten law and greatly contributed to shape the landscape of law in continental Europe [39]. It is therefore not surprising that the structures of Roman law and of many subsequent and current legal systems are closely related to logic and mathematical theories. Over the centuries, many jurists have brought to the fore the systematic and deductive power of legal systems, such as for instance the Roman Cicero (106–43 BC, *"jus in artem redigere"*), the French Domat (1625–1696), the German Leibniz (1646–1716) or the Italian Beccaria (1738–1794).

Logic, sometimes viewed as the "calculus of computer science" [28, 42], then became a natural theoretical background to address new kinds of legal matters. In the new societal and virtual context raised by the recent technological developments of internet and computers, law must still be enforced, transparent, accountable and

understandable by anybody more than ever. As it turns out, an important amount
of works at the intersection of logic, law and artificial intelligence has emerged over
the last decades. We are now going to briefly summarize it and we refer the reader
to the handbooks [22, 18] or survey articles [15, 35, 27] for more details and pointers.

1.1   A status quo in logic and law

Originally, logic was intended to be used for the representation of law in a clear and
unambiguous manner. On top of this representation, some kind of reasoning could
then take place to infer some information. Sergot and Kowalski [37] were pioneers in
that field, with the use of logic programming that they applied to the formalization of
the British Nationality Act. However, they encountered difficulties with the Prolog
treatment of negation as failure and with the lack of deontic operators [29]. In
Prolog, something is said to be false if it is not known or cannot be inferred to be
true. Hence, if a program cannot show that an infant was born in the UK, it assumes
that it was not.[1] More generally, researchers realized that several aspects of the law
which can not be dealt with standard (Fregean) logic had to be taken into account,
such as the need to handle exceptions, conflicting rules, vagueness, open texture (i.e.
the failure of natural languages to determine future usage of terms), counterfactual
conditionals and the possibility of rational disagreement. Some of these peculiarities
of legal reasoning were and are in fact still addressed by various non–classical logics
in an area sometimes called 'applied logic' [32].

Representation of law.   Among the logical formalisms used to represent law,
deontic logic provides formal tools for the clarification of the meaning of normative
terms such as 'may', 'must' and 'shall', which play a central role in specifying the
legal relations between agents (which can be human beings or machines). Therefore,
it has been used in the analysis of law [29] and in the area of automated contract
management. Other notions which play an important role in law and legal reasoning
were also analysed logically, such as the notion of power, as involved in sentences
such as "the president has the power to declare a state of emergency". Normative
systems propounded by Alchourrón and Bulygin were another influential approach
to represent legal systems [2]. A normative system is a set of norms, which are pairs
of the form ⟨*condition, consequence*⟩: if the condition holds then the consequent
must hold. Unlike formulas of deontic logic, they do not bear truth values.

Several structural features of legal regulations, such as the use of exception,
the use of hierarchies of legislation to resolve conflicts, cross references to other

---

[1]This thread of research in legal reasoning using logic programming is in fact still active, as
witnessed for example by the Japanese PROLEG system [36].

parts of the legislation, deeming provisions, conditions under which the legislation is applicable, and conditions for the validity of particular norms, led to the use of non-monotonic logics [34]. These are logics where the consequence relation is not monotonic, meaning that adding a formula to a theory does not necessarily produce an increase of its set of consequences. However, these formalisms proved inadequate to provide a general means of conflict resolution. These conflicts are sometimes due to conflicting interpretations of the law and get even more difficult to handle in the context of *stare decisis* (a legal principle of Common Law by which judges are obliged to respect the precedents established by prior decisions of higher jurisdictions). These difficulties led to a shift of focus from the modes of representation to the modes of reasoning.

Reasoning about law. Law and its practice are subject to different kinds of reasoning:

- *Case-based reasoning* uses considerations about precedent legal cases to show how they justify particular outcomes in a new case (following the *stare decisis* principle of the Anglo-American Common Law). The problem is then to map appropriately the precedent cases to the new case. Several models have been developped and logically formalized, notably by McCarty [31]. This problem involves the classification of the facts of a case under legal concepts and the interpretation of these legal concepts.

- *Practical and teleological reasoning* deals with the reasoning involved in the justification of choices that the arbiter has to make in some legal decisions. These justifications should be in line with the underlying purpose of the law. This involves to be able to derive normative consequences from the classification of facts and the interpretation of legal concepts.

- *Evidential reasoning* is the kind of reasoning that occurs when judges strive to establish the facts on the basis of evidences.

Different kinds of logical formalisms were developped to address these different kinds of reasonings, and in particular numerous works resorted to argumentation theory. However, legal reasoning mostly arises in the conduct of a dispute which is regulated by a particular procedure. The outcome of this dispute does not only depend on facts and a body of law, but also on the procedure itself: whether it is a criminal proceedings or a civil proceedings, to which party is assigned the burden of proof in this procedure, etc. A number of dialogue games models of legal procedure have been produced in the last 20 years [33].

All this said, a striking particularity of most of the works which have been pursued at the interface of logic and law in the last decades is that they were mostly driven by theoretical considerations and without much interaction with jurists and lawyers. Arguably, this work did not really catch the attention of the lawyers and jurists. In particular, they did not change the way they work or their actual practice of the law, except maybe for the adoption of large and online databases such as LexisNexis or Legifrance[2] (based on standards for legal documents such as Legal-RuleXML) and the use of knowledge management systems [17, 16, 1]. This theoretical work did not seem for jurists to answer an actual need and it was somehow remote from their daily preoccupations, although the researchers could sense the potential and the important applicability of their work in the practice of the law.

1.2 Current problems in the application of law in France

The joint work reported in this article stems from actual problems in the application of law in France that were expressed to us by jurists and magistrates of the French Cour de cassation.[3] These problems are in fact not specific to France. The application of law is plagued with a series of problems which are difficult to overcome with the standard and present methods employed by jurists [30, 11].[4] First, the increasing diversity and complexity of legal texts and jurisprudence makes the work of jurists (and lawyers) very difficult to pursue nowadays. This complexity appears not only at the local or national level but is sometimes worsened by its interaction with the European level, and sometimes even the international level. Second, legal texts and jurisprudence are changing at a high pace in some areas and it is difficult for jurists (and lawyers) to cope and keep up-to-date with the current legislation and regulations. Third, there is a lack of consistency in the application of law, depending on the geographical part in which trials take place, on the local customs and sometimes the personality of the judges, and more generally on the specific

---

[2]See www.lexisnexis.com and www.legifrance.gouv.fr.

[3]The main role of the Cour de cassation is to check that the law is applied correctly and uniformly in France mainly from a legal and procedural point of view, the determination of facts being left to the 'tribunaux de grande instance et d'instance' and 'cour d'appel'. Stemming from the 'justice retenue' dispensed by the French kings from the 13th century on and formerly called 'Conseil des parties' and then 'Tribunal de cassation' during the French revolution, the Cour de cassation is one of the oldest juridical institution in France. It "is the highest Court in the French judiciary. [...] the Court of Cassation is thus required to find whether the rules of law have been correctly applied by the lower courts based on the facts. [...] If the decision of the lower court is quashed [cassée in French], the case has consequently to be heard again." (www.courdecassation.fr/about_the_court_9256.html)

[4]In some countries subject to common law, like the United Kingdom, such problems are worsen by the fact that some legal decisions are not made by professional jurists [11].

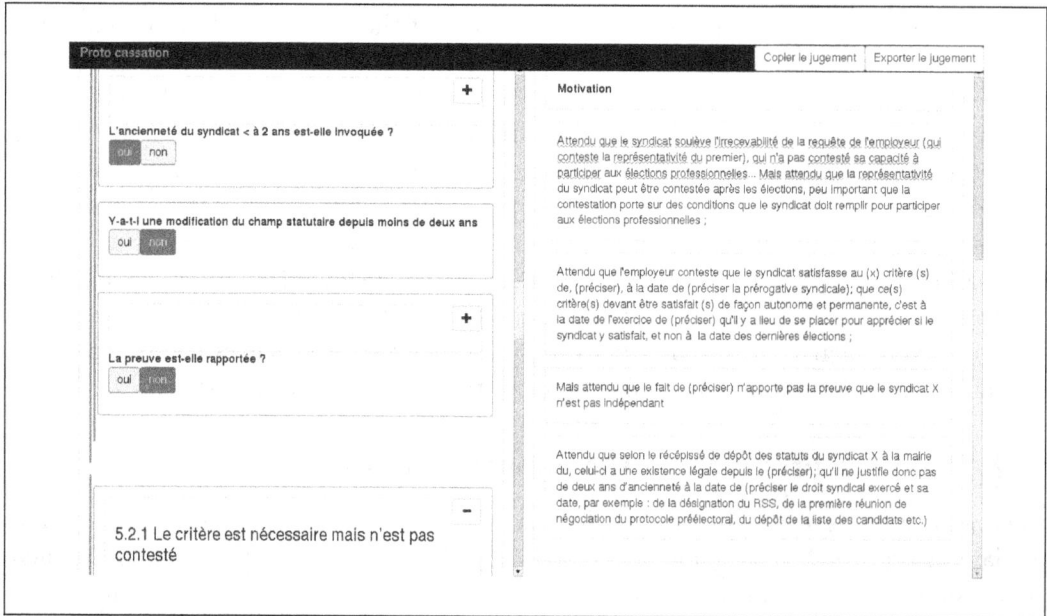

Figure 1: Screenshot of the graphical user interface.

political or social context in which a legal decision is taken.

The third problem is not novel and was already raised before the French revolution of 1789 by philosophers such as Voltaire, Diderot or Rousseau and the lawyer Linguet for example. It led to the vast enterprise of codification of the law. During the French revolution, the Assemblée voted for a Code penal inspired by ideas of Montesquieu and Beccaria and a series of 'revolutionary' projects led by Cambacérès were submitted. One had to wait for Napoleon and more peaceful times for the promulgation of the first comprehensive Code civil and other codes. The Napoleonian codification had a lasting impact over France and many other countries which adopted (partly) the French codes or took inspiration from them. However, as already noted by Thireau in 2009, "with the inflation of legal texts and regulations, the time of large codifications seems over" [39, p. 335]. Altogether, these three problems call for a new kind of solution.

1.3 A new kind of solution: a software assistant

The solution that we propose is to make use of a software, whose ultimate role is to help judges write a judgement and take better and well-informed decisions thanks to a series of questions to which she/he has to answer. These questions would be backed up by the corresponding legal texts and jurisprudence.

This software assistant would indeed be a solution to the problems mentioned above. First, it would unify and uniformize the application of law in France: the kind of reasoning proposed by the software to sort out a given (type of) litigation could then be controlled and it could also be the same in every jurisdiction of France. Second, like any software, it could take into account the evolution of legal texts and jurisprudence to update the different kinds of reasoning and therefore cope with the increasing complexity of law. Third, its easy access to a large and up-to-date knowledge base comprising the current legal texts and jurisprudence would increase the chance for the judge to make well–informed decisions.

The graphical user interface (GUI) of this software is depicted in Figure 1. On the left hand side of the GUI, a guide for reasoning consisting of a series of questions is displayed. These questions and their underlying reasoning are backed up by legal texts and jurisprudence to which the user can have access whenever she/he wants. On the right hand side of the GUI, a judgement is produced automatically as the user answers the questions. The user can modify at any time the judgement produced and can also have an alternative graphical representation of the web of questions to which she/he has to answer on the left hand side.

1.4   Structure of the article

In Section 2, we recall the basics of propositional logic and BDD. In Section 3, we extend propositional logic with the examination operator $!\varphi$ and we provide a semantics for this operator based on Multi-OBDD. In Section 4, we consider as case study the problems of determining whether an association of employees in a firm can indeed be considered ('qualified') as a trade union. In Section 5, we show how the various algorithms that have been designed for OBDD can be used to solve and address specific kinds of legal issues that arise in the practice of the law. In Section 6, we describe the prototype that we have implemented during our project and provide a link with restricted access to try it out. We conclude in Section 7.

2   Propositional logic and BDD

In this section, we recall the basics of propositional logic and Binary Decision Diagrams (BDD for short, [20, 21]) and we show how they are related to each other. BDD provide a graphical semantics for formulas of propositional logic in which the truth–values of their atoms can be assigned in a specific order. This feature will play a role in the legal context since it will allow us to represent the *procedural* aspect of the practice of law (during a trial especially).

$\mathcal{I}(\varphi)$	$\mathcal{I}(\psi)$	$\mathcal{I}(\neg\varphi)$	$\mathcal{I}(\varphi \wedge \psi)$	$\mathcal{I}(\varphi \to \psi)$	$\mathcal{I}(\varphi \vee \psi)$
T	T	F	T	T	T
T	F	F	F	F	T
F	T	T	F	T	T
F	F	T	F	T	F

Figure 2: Semantics of propositional connectives.

2.1 Propositional logic

In the sequel, $\mathbb{P}$ is a set of *atoms* (propositional letters) denoted $p, q, r, \ldots$ and T and F are two symbols called *truth values* standing for True and False.

Definition 1 (Propositional language $\mathcal{L}$). The language $\mathcal{L}$ is the set that contains $\mathbb{P}$ and such that

- if $\varphi, \psi \in \mathcal{L}$, then $\neg\varphi, (\varphi \wedge \psi), (\varphi \vee \psi), (\varphi \to \psi) \in \mathcal{L}$;

- $\mathcal{L}$ contains no more formulas.

We introduce the following abbreviations: $\varphi \leftrightarrow \psi \triangleq (\varphi \to \psi) \wedge (\psi \to \varphi), \top \triangleq (p \vee \neg p), \bot \triangleq (p \wedge \neg p)$ for some $p \in \mathbb{P}$. The formula $\varphi[p \backslash \psi]$ denotes the formula φ where the atom p is uniformly substituted with ψ.

The intuitive reading of the formulas is as follows: $\neg\varphi$: "φ does not hold"; $\varphi \wedge \psi$: "φ holds and ψ holds"; $\varphi \vee \psi$: "φ holds or ψ holds"; $\varphi \to \psi$: "If φ holds then ψ holds".

Definition 2 (Interpretation). A *total (partial) interpretation* is a total (resp. partial) function $\mathcal{I} : \mathbb{P} \mapsto \{T, F\}$ that assigns one of the *truth values* T or F to *every* (resp. *some* of the) atom(s) in $\mathbb{P}$. The set of total interpretations is denoted $\mathcal{C}$ and the set of partial interpretations is denoted $\mathcal{C}^p$. Note that $\mathcal{C} \subseteq \mathcal{C}^p$. If $\mathcal{I} \in \mathcal{C}^p$, then $Ext(\mathcal{I})$ is the set of *total* interpretations extending the interpretation $\mathcal{I}$, that is, for all $\mathcal{I}' \in Ext(\mathcal{I})$, for all $p \in \mathbb{P}$ such that $\mathcal{I}(p)$ is defined, we have that $\mathcal{I}(p) = \mathcal{I}'(p)$.

We can extend the domain of an interpretation function from the set of atoms to the set of all formulas of $\mathcal{L}$. This extension is inductively defined by the truth table given in Figure 2. If E is a set of interpretations, we say that a formula φ of $\mathcal{L}$ is *valid on* E when for all $\mathcal{I} \in E$, we have that $\mathcal{I}(\varphi) = T$. When $E = \mathcal{C}$, we simply say that φ is *valid*.

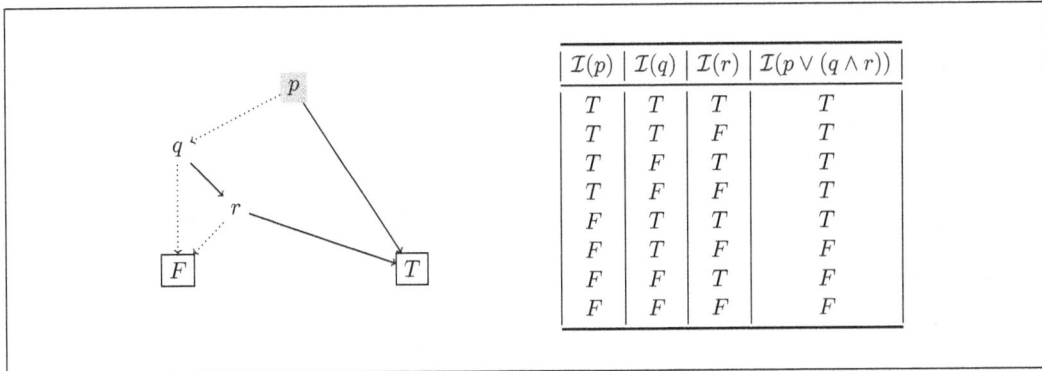

$\mathcal{I}(p)$	$\mathcal{I}(q)$	$\mathcal{I}(r)$	$\mathcal{I}(p \vee (q \wedge r))$
T	T	T	T
T	T	F	T
T	F	T	T
T	F	F	T
F	T	T	T
F	T	F	F
F	F	T	F
F	F	F	F

Figure 3: OBDD and truth table of the formula $(p \vee (q \wedge r))$.

2.2 Binary Decision Diagrams (BDD)

Binary decision diagrams are graphical data structures for representing compactly the semantics of formulas of propositional logics. They have been widely used in the industry for the verification of computer hardware. Our presentation is adapted from the book of Ben-Ari [14], based on the articles on BDDs by Bryant [20, 21].

Definition 3 (Binary Decision Diagram, BDD). A *binary decision diagram (BDD)* is a directed acyclic graph with a unique root, which is also called the *entry point*. Each leaf is labeled with one of the truth values T or F. Each interior node is labeled with an atom and has two outgoing edges: one, the *false edge*, is denoted by a dotted line, while the other, the *true edge*, is denoted by a solid line. No atom appears more than once in a branch from the root to a leaf.

During a trial, some questions have to be examined in a certain temporal order. This temporal order does not play a role from a logical point of view, in the sense that the truth value of a given statement will not depend on the order in which its arguments are examined. However, this temporal order plays a role from a procedural point of view when the judge constructs its judgment while answering the different questions.

This ordering is made explicit in *ordered* binary decision diagrams (OBDD): we can canonically associate to each OBDD an ordering corresponding to the order in which the different questions should be examined by the judge. However, these orderings of the different branches of the OBDD should be somehow 'compatible'.

Definition 4 (Compatible orderings). Let $<^1, \ldots, <^n$ be orderings on $\mathbb{P}$, that is, for each $i \in \{1, \ldots, n\}$, $<^i$ is a total relation on a subset $\mathbb{P}^i \subseteq \mathbb{P}$. We say that $<^1, \ldots, <^n$

Input: A BDD *bdd*.

Output: A reduced BDD denoted `Reduce`(*bdd*).

Perform a recursive traversal of the BDD:

- If *bdd* has more than two leaves T and F, remove duplicate leaves. Direct all edges that pointed to a removed leaf to the remaining respective leaf.

- Perform the following steps as long as possible:

  1. If all outgoing edges of a node labeled p point at the same node labeled q, delete this node for p and direct p's incoming edges to q.

  2. If two nodes labeled p (distinct from roots) are the roots of identical sub-BDDs, delete one sub-BDD and direct its incoming edges to the other node.

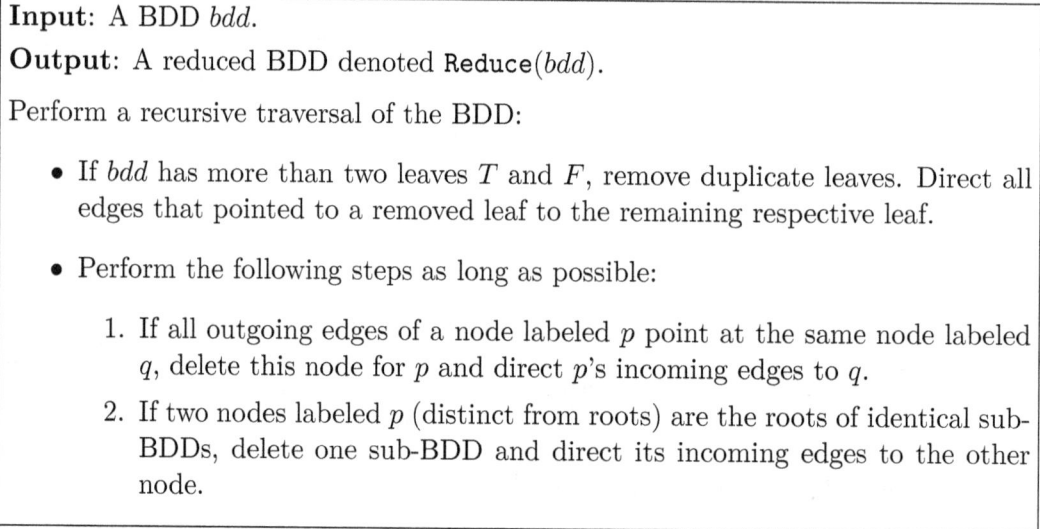

Figure 4: Schematic algorithm `Reduce`.

are *compatible orderings* when for all $i \neq j$, there are no atoms $p, p' \in \mathbb{P}^i \cap \mathbb{P}^j$ such that $p <^i p'$ while $p' <^j p$.

Definition 5 (Ordered Binary Decision Diagrams, OBDD). An *ordered binary decision diagram (OBDD)* is a BDD such that the orderings of atoms $<^1, \ldots, <^n$ defined by the branches stemming from the root are compatible. The *ordering associated to the OBDD* is $<^1 \cup \ldots \cup <^n$.

Example 1. Figure 3 shows an OBDD and the truth table of the formula $(p \vee (q \wedge r))$. The OBDD representation is more compact and avoids redundancy. One can indeed notice that the four rows of the truth table where p is true make the formula $(p \vee (q \wedge r))$ true, regardless of the truth values of q and r. This redundancy is represented in the OBDD by setting a true edge from the entry point p to the leaf labeled T.

Definition 6 (Operation `Reduce` on BDD). The operation `Reduce` on BDD is defined in Figure 4.

Example 2. Figure 5 shows the application of a sequence of `Reduce` on an OBDD equivalent to the OBDD of Figure 3. First, we merge all the leaves labeled by F in one single leaf and we merge all the leaves labeled by T in one single leaf. Then we 'bypass' the r-node and the two r-nodes on the right because all outgoing edges lead to the same node. In the last step, we 'bypass' the q-node on the right.

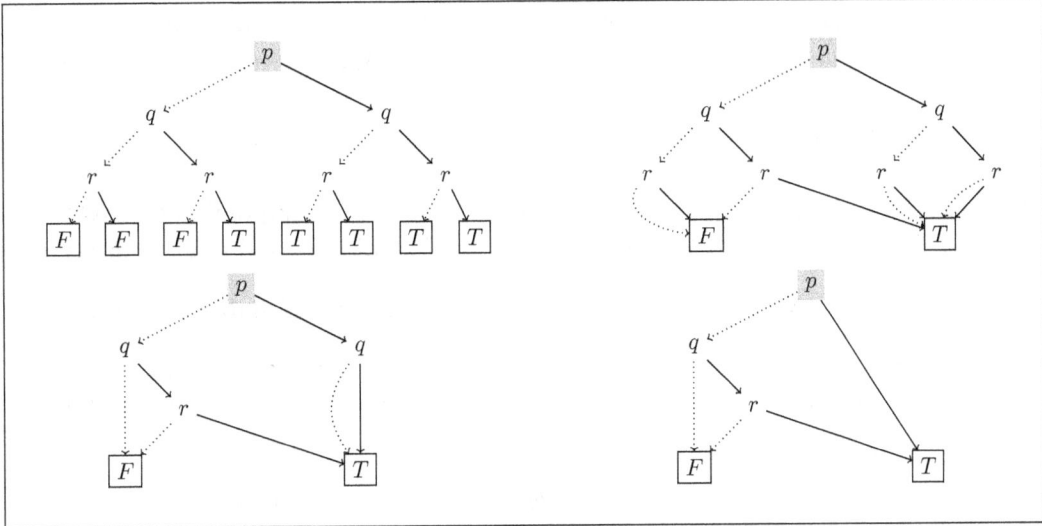

Figure 5: Step-by-step of algorithm `Reduce`.

Definition 7 (Operation `Apply` on OBDD). The operation `Apply` on OBDD is defined in Figure 6.

Instead of the set-theoretical semantics of propositional logic based on the notion of interpretation, we can provide a semantics to propositional logic in terms of OBDD. The meaning of a propositional formula is completely determined by the OBDD associated to that formula, which is itself built inductively with the `Apply` Algorithm of Figure 6. The soundness of `Apply` is ensured by the following theorem:

Theorem 1 (Shannon expansion). *For all formulas $\varphi, \psi \in \mathcal{L}_M$, for all $\star \in \{\wedge, \vee, \rightarrow\}$, the following formula is valid:*

$$\varphi \star \psi \leftrightarrow (p \wedge (\varphi[p\backslash\top] \star \psi[p\backslash\top])) \vee (\neg p \wedge (\varphi[p\backslash\bot] \star \psi[p\backslash\bot]))$$

For example, $\varphi \wedge \psi \leftrightarrow (p \wedge (\varphi[p\backslash\top] \wedge \psi[p\backslash\top])) \vee (\neg p \wedge (\varphi[p\backslash\bot] \wedge \psi[p\backslash\bot]))$.

Definition 8 (OBDD associated to a formula and an ordering). Let $\chi \in \mathcal{L}$ and let $<$ be an ordering on the set of atoms occuring in χ. The OBDD associated to χ and $<$, written $obdd_\chi$, is defined inductively on χ as follows:

- $obdd_\top$ is the OBDD consisting of a single node labeled T;

- $obdd_\bot$ is the OBDD consisting of a single node labeled F;

- $obdd_p$ is the following OBDD:

Input: MOBDDs *mobdd* and *mobdd'*, an ordering $<$ on all atoms of *mobdd* and *mobdd'* compatible with their associated orderings. A truth–functional connective $\star$.

Output: A MOBDD denoted $\mathtt{Apply}(mobdd, mobdd', \star, <)$.

1. Take the OBDDs *obdd* and *obdd'* generated by the entry points of *mobdd* and *mobdd'*. Let p and p' be the labels of the entry points of *obdd* and *obdd'* respectively.

   - If *obdd* and *obdd'* are both leaves, define the leaf $mobdd_0$ labeled by $p \star p'$.

   - If $p = p'$, define the OBDD $mobdd_0$ whose entry point is labeled by p, whose left sub-OBDD is $\mathtt{Apply}(obdd_l, obdd'_l, \star, <)$ and whose right sub-OBDD is $\mathtt{Apply}(obdd_r, obdd'_r, \star, <)$.

   - if $p < p'$, define the OBDD $mobdd_0$ whose entry point is labeled by p, whose left sub-OBDD is $\mathtt{Apply}(obdd_l, obdd', \star, <)$ and whose right sub-OBDD is $\mathtt{Apply}(obdd_r, obdd', \star, <)$.

     Otherwise, define the OBDD $mobdd_0$ whose entry point is p', whose left sub-OBDD is $\mathtt{Apply}(obdd, obdd'_l, \star, <)$ and whose right sub-OBDD is $\mathtt{Apply}(obdd, obdd'_r, \star, <)$.

   where $obdd_l, obdd'_l$ (resp. $obdd_r, obdd'_r$) are the left (resp. right) sub-BDDs of *obdd* and *obdd'*.

2. Return $obdd_0$ [with the disjoint union of the OBDDs generated by the other roots of *mobdd* and *mobdd'*].

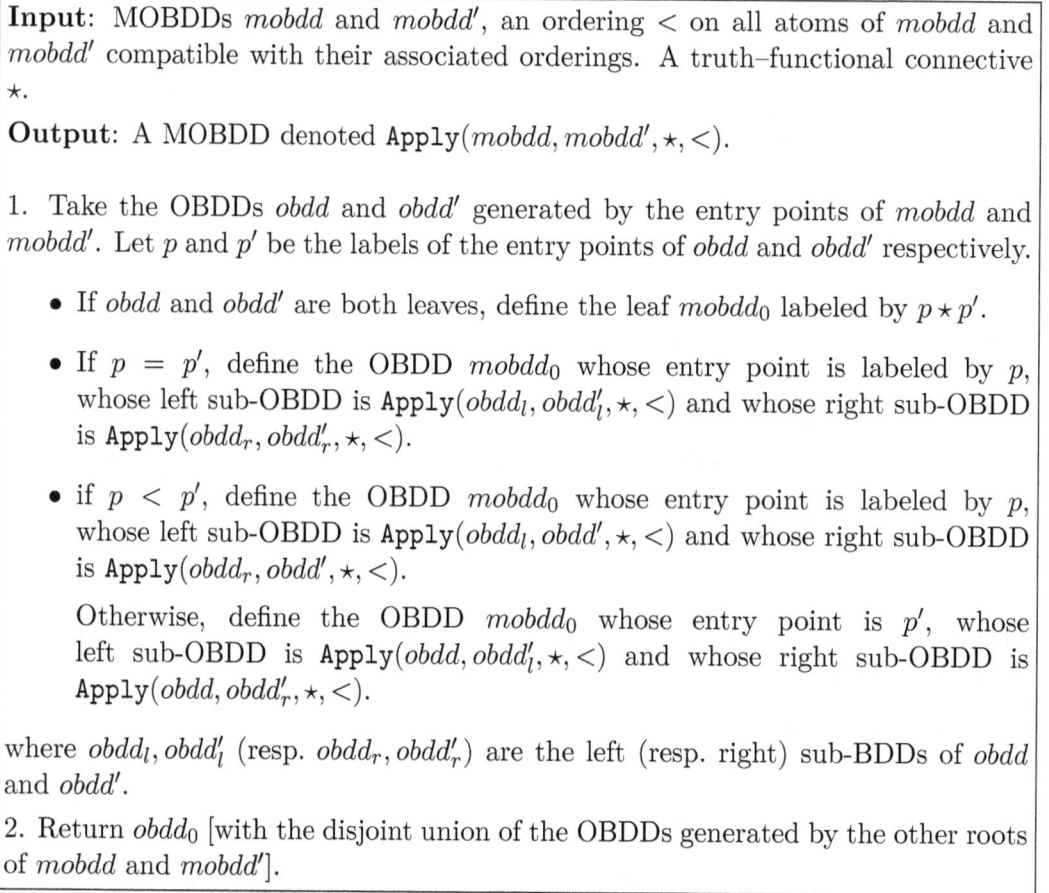

Figure 6: Schematic algorithm $\mathtt{Apply}$ for OBDD [and MOBDD].

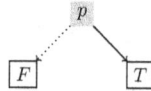

- if $\chi = \varphi \star \psi$, then $obdd_\chi \triangleq \mathtt{Apply}(obdd_\varphi, obdd_\psi, \star, <)$.

- if $\chi = \neg\varphi$, then $obdd_\chi$ is $obdd_\varphi$ where the labels T or F of the leafs have been switched.

Example 3. Let us consider the formula $\varphi \triangleq p \to ((q \wedge \neg r) \vee \neg s)$. The syntactic decomposition tree of formula φ is represented at the top of Figure 7. From this decomposition tree, we can apply the induction process of Definition 8 to obtain

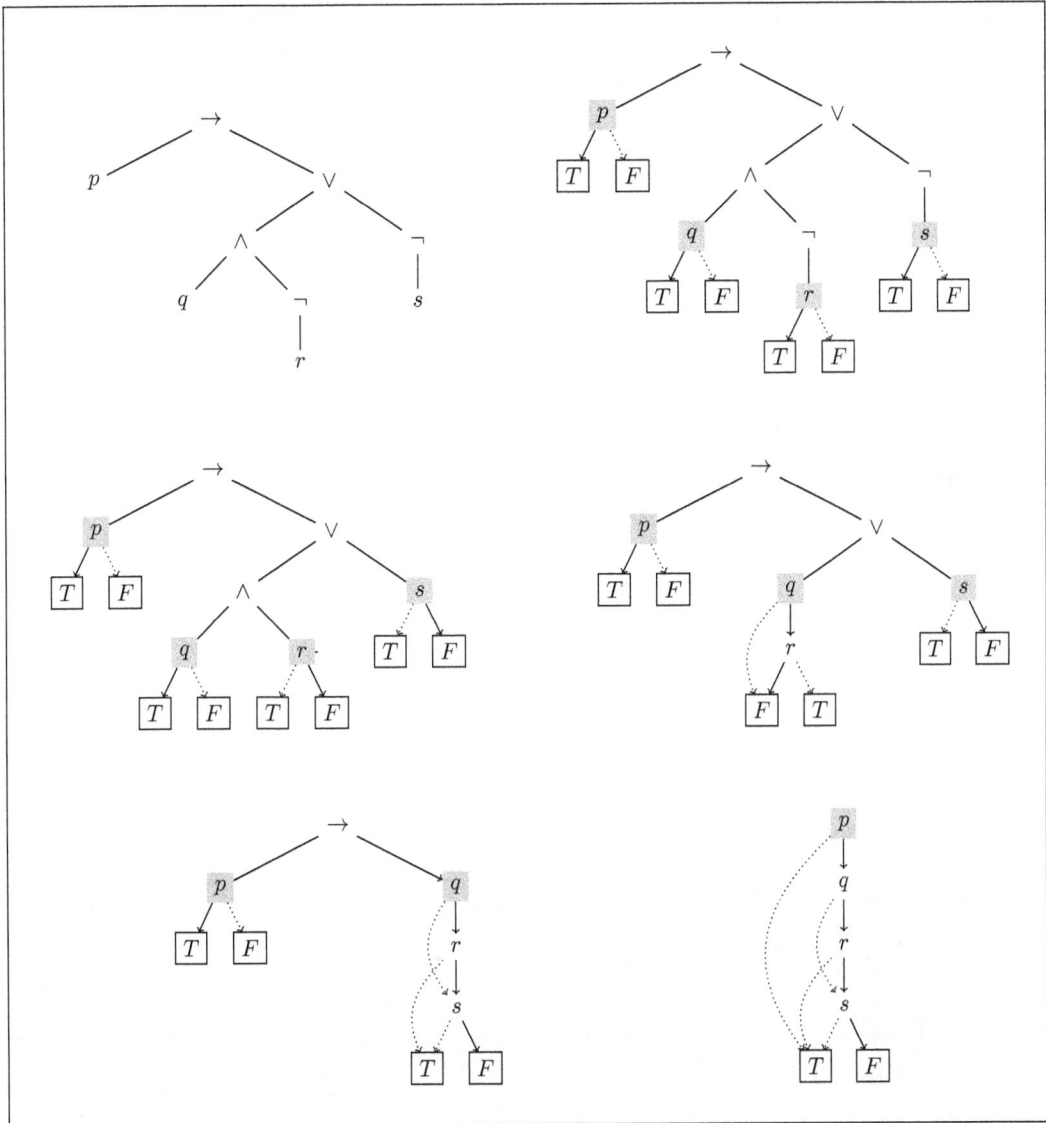

Figure 7: Successive applications of **Apply** with the ordering $p < q < r < s$: from the syntactic tree of $\varphi \triangleq p \to ((q \land \neg r) \lor \neg s)$ (*top left*) to *obdd*$_\varphi$ (*bottom right*).

Input: A BDD *obdd* for a formula φ; an interpretation $\mathcal{I} \in \mathcal{C}^p$ (partial or total).
Output: A BDD denoted $\texttt{Restrict}(\mathcal{I}, obdd)$.

Perform a recursive traversal of the BDD:

- If the root of *obdd* is a leaf, return the leaf.

- If the root of *obdd* is labeled p and $\mathcal{I}(p)$ is defined, return the sub-BDD reached by its true edge if $\mathcal{I}(p) = T$ and the sub-BDD reached by its false edge if $\mathcal{I}(p) = F$.

- Otherwise (the root of *obdd* is labeled p and $\mathcal{I}(p)$ is *not* defined), apply the algorithm to the left and right sub-BDD, and return the BDD whose root is p and whose left and right sub-BDD are those returned by the recursive calls.

Figure 8: Schematic algorithm $\texttt{Restrict}$.

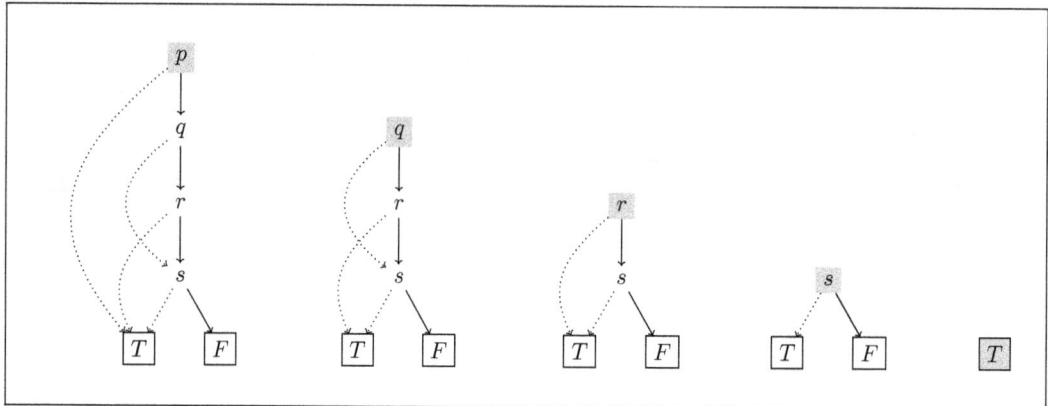

Figure 9: Successive applications of $\texttt{Restrict}$ to $obdd_\varphi$ of Figure 7 with the interpretations $\mathcal{I}_1$, $\mathcal{I}_2$, $\mathcal{I}_3$ and $\mathcal{I}_4$.

the OBDD $obdd_\varphi$. This process is represented in Figure 7, it consists in applying iteratively algorithm $\texttt{Apply}$ of Figure 6.

Definition 9 (Operation $\texttt{Restrict}$). The operation $\texttt{Restrict}$ on BDD and interpretations is defined in Figure 8.

Example 4. Figure 9 shows the successive application of $\texttt{Restrict}$ on the OBDD $obdd_\varphi$ of Figure 7, with the interpretations $\mathcal{I}_1(p) = T$ and $\mathcal{I}_1(t)$ undefined for all

$t \neq p$, $\mathcal{I}_2(q) = T$ and $\mathcal{I}_2(t)$ undefined for all $t \neq q$, $\mathcal{I}_3(r) = T$ and $\mathcal{I}_3(t)$ undefined for all $t \neq r$, $\mathcal{I}_4(s) = F$ and $\mathcal{I}_4(t)$ undefined for all $t \neq s$.

The key properties of the algorithms `Restrict` and `Reduce` are highlighted by the following results.

Proposition 1. *Let $obdd_\varphi$ be an OBDD associated to a formula $\varphi \in \mathcal{L}$ (and an ordering) and let $\mathcal{I} \in \mathcal{C}^p$ be an interpretation. Then, `Restrict`$(\mathcal{I}, obdd_\varphi)$ returns an OBDD associated to a formula ψ such that $\varphi \leftrightarrow \psi$ is valid on the set of interpretations extending $\mathcal{I}$, $Ext(\mathcal{I})$.*

Theorem 2. *The algorithm `Reduce` constructs an OBDD if the original BDD is ordered. For a given ordering of atoms, the reduced OBDDs for logically equivalent formulas are structurally identical.*

All operations on OBDD, `Apply`, `Reduce`, and `Restrict`, have a polynomial algorithmic complexity with the size of the OBDD they operate on. The size of the result of `Reduce` strongly depends on the atom ordering. It is exponential in the worst case. Some formulas even have an exponential size OBDD for all orderings. However, OBDD have shown to be of great pratical value, and have become a standard solution for dealing with industrial size propositional formulas, *e.g.* in the areas of digital system design, verification and testing [10].

3   The examination operator

Sometimes during a trial, the judge must necessarily examine or raise a specific issue. For example, the complainant can attack the legitimity of a trade union presenting a candidate to the professionnal elections of a firm. The complainant could argue that this association of employees cannot really be qualified as a syndicate because it is not senior enough. In that case, even if the trade union turns out to be senior enough (older than 2 years), the judge *must* nevertheless examine *all* the other criteria (different from seniority) that make the association of employees qualify as a trade union (even if she/he does not decide to raise the issue of the other criteria during the trial).

Definition 10 (Language $\mathcal{L}_M$). The language $\mathcal{L}_M$ is the set that contains $\mathbb{P} \cup \{\top, \bot\}$ and such that

- if $\varphi, \psi \in \mathcal{L}_M$, then $!\varphi, \neg\varphi, (\varphi \wedge \psi), (\varphi \vee \psi), (\varphi \rightarrow \psi) \in \mathcal{L}_M$;

- $\mathcal{L}_M$ contains no more formulas.

We use the same abbreviations and notations as in Definition 1.

The intuitive reading of the formula $!\varphi$ is as follows: "φ is examined and it holds". For example, the formula $q \wedge !p$ holds if, and only if, p is examined and p and q both hold.

We must provide a semantics to our extended language $\mathcal{L}_M$, in particular to the examination operator $!\varphi$. The Algorithm `Apply` provides already a semantics to every formula of the propositional language in terms of OBDD. Indeed, it suffices to apply it inductively to every subformula of a given formula φ (from the atoms up to the formula φ) to obtain an OBDD that captures the meaning of φ. Thus, we extend this algorithm to include the examination operator as well. With this aim in view, we introduce an extended form of BDD with several entry points. This extension is called a *Multi-BDD* and, contrary to BDD, it can have several roots.

Definition 11 (Multi-BDD, Multi-OBBB). A *Multi-BDD* (MBDD for short) is a BDD with possibly multiple roots $r_0, \ldots, r_k$. The root r_0 is distinguished and called the *entry point* of the MBDD. A *Multi-OBDD* (MOBDD for short) is a MBDD such that the orderings of atoms $<^0, \ldots, , <^k$ defined by the branches stemming from the roots $r_0, \ldots, r_k$ are compatible. A MBDD or MOBDD is *elementary* when it consists only of leaves.

Example 5. Figure 13 shows a MOBDD. The construction of this MOBDD is detailed in the next section.

Definition 12 (MOBDD associated to a formula and an ordering). Let $\chi \in \mathcal{L}_M$ and let $<$ be an ordering on the set of atoms occuring in χ. The MOBDD associated to χ (and $<$), written $mobdd_\chi$, is defined inductively on χ as follows.

- $mobdd_\top \triangleq obdd_\top$, $mobdd_\bot \triangleq obdd_\bot$, $mobdd_p \triangleq obdd_p$ of Definition 8.

- if $\chi = (\varphi \star \psi)$, then

    - if $\varphi \neq !\varphi'$ and $\psi \neq !\psi'$ for any $\varphi', \psi' \in \mathcal{L}_M$, then
      $mobdd_\chi \triangleq \texttt{Apply}(mobdd_\varphi, mobdd_\psi, \star, <)$ (as defined in Figure 6);
    - if $\varphi = !\varphi'$ and $\psi \neq !\psi'$ for some $\varphi' \in \mathcal{L}_M$ and any $\psi' \in \mathcal{L}_M$ (or $\varphi \neq !\varphi'$ and $\psi = !\psi'$ for some $\psi' \in \mathcal{L}_M$ and any $\varphi' \in \mathcal{L}_M$), then
      $mobdd_\chi \triangleq mobdd_{\varphi'} \sqcup \texttt{Apply}(mobdd_{\varphi'}, mobdd_\psi, \star, <)$
      (resp. $mobdd_\chi \triangleq mobdd_{\varphi'} \sqcup \texttt{Apply}(mobdd_\varphi, mobdd_{\psi'}, \star, <)$);
    - if $\varphi = !\varphi'$ and $\psi = !\psi'$ (for some $\varphi', \psi' \in \mathcal{L}_M$), then
      $mobdd_\chi \triangleq mobdd_{\varphi'} \sqcup mobdd_{\psi'} \sqcup \texttt{Apply}(mobdd_{\varphi'}, mobdd_{\psi'}, \star, <)$.

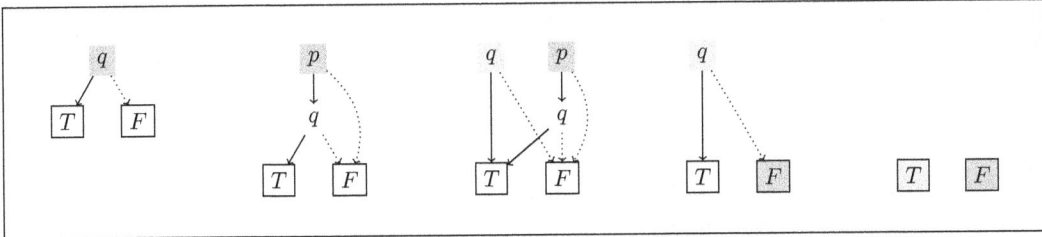

Figure 10: $obdd_q$ (*first*), $obdd_{p \wedge q}$ (*second*), $mobdd_{(p \wedge !q, <)}$ reduced (*third*), `Restrict`$\left(\mathcal{I}, mobdd_{(p \wedge !q, <)}\right)$ reduced, with $\mathcal{I}(p) = F$ and $\mathcal{I}(q)$ undefined (*fourth*), `Restrict`$\left(\mathcal{I}, mobdd_{(p \wedge !q, <)}\right)$ with $\mathcal{I}(p) = F$ and $\mathcal{I}(q) = T$ (*fifth*) with ordering $p < q$.

- if $\chi = \neg\varphi$, then $mobdd_\chi$ is $mobdd_\varphi$ where the labels T or F of the leafs have been switched.

It is important to notice that in order to reduce a MOBDD to an elementary MOBDD, the magistrate must evaluate *every* formula associated to a root of the MOBDD. The proposition below shows that our definition of MOBDD does capture that requirement.

Proposition 2. *Let $\varphi \in \mathcal{L}_M$ with subformula $!\psi$ and let $\mathcal{I} \in \mathcal{C}^p$ be a partial interpretation. If $\mathcal{I}(\psi) \notin \{T, F\}$, then* `Reduce`*$($*`Restrict`*$(\mathcal{I}, mobdd_\varphi))$ is not elementary.*

Example 6. In Figure 10, we represent the MOBDD associated to the formula $p \wedge !q$ and ordering $p < q$ (disjoint union of first and second OBDD, reduced in the third graph). Roots are in gray and the entry points are darker. This formula is true if, and only if, p and q are both true and q is examined. Hence, if p is given the value F (by the judge), then, even if the truth value of the formula $p \wedge !q$ is determined, the MOBDD is still not elementary. Indeed, we must *examine* the 'question' q and give a truth value to q. Then, once the judge has examined question 'q', we reach the fifth MOBDD, which is elementary.

4 Case study: French trade unions

Our case study deals with French professional election in a firm [38]. Among the problems to be decided upon, one is to determine whether an association of employees is really qualified as a trade union so that this association can propose employees to come forward as candidates at the professional elections of the firm. Law introduces four criteria that have to be fulfilled so that an association can indeed be qualified as a trade union. These four criteria have to hold altogether, and during a trial, the judge must necessarily examine all of them. They are the following:

1. the association of employees should respect the 'Republican values';

2. the association of employees should be 'Independent' (from the directorate of the firm for example);

3. the association of employees should be 'Senior' enough (minimum 2 years of existence);

4. the association of employees should be within the appropriate 'Geographical and professional range'.

Hence, we introduce the formula R (resp. I, S and G) which stands respectively for "the criteria of the *Republican values* (resp. *Independence, Seniority, Geographical and professional range*) is fulfilled". The four criteria must hold for an association of employees to be legally qualified as trade union in a firm, and they must all be examined by the judge, even if they are not contested. Therefore, the following formula must be true: $(!R \land (!I \land (!G \land !S)))$.

4.1 The criterias of 'republican values' and 'independence'

To determine whether the criteria of 'Republican values' holds, the judge has to answer a number of questions. To formulate these questions, we introduce the following set of atoms:

$$\mathbb{P}_R \triangleq \{Lit_R, OldJug_R, NewElt_R, Proof_{\neg R}\}$$

These atoms stand for the following propositions:

- Lit_R: "The plaintiff contests the criteria of 'Republican values' ";

- $OldJug_R$: "An old judgement dealing with the criteria of 'Republican values' already established that the association of employees fulfills that criteria";

- $NewElt_R$: "New elements have been brought to the fore that oblige to reconsider the old judgement";

- $Proof_{\neg R}$: "The plaintiff presents the proof that the criteria of 'Republican values' is not fulfilled".

Then, we can give the formula in the language $\mathcal{L}_M$ that determine in which case the criteria of 'Republican values' holds. It is the following:

$$R \triangleq Lit_R \rightarrow ((OldJug_R \wedge \neg NewElt_R) \vee \neg Proof_{\neg R})$$

The above formula reads as follows: "if the plaintiff contests that the criteria of 'Republican values' is satisfied, then either there is an old judgement which already established that this criteria was fulfilled and no new elements have been brought to the fore which oblige to reconsider this old judgement, or there is no old judgement and the plaintiff has not been able to prove that the criteria is not fulfilled". However, in the procedure that the judge must follow during a trial, he must first determine whether or not there was an old judgment (that already established that this criteria was fulfilled) before asking the plaintiff to provide a proof that the criteria is not fulfilled. This procedural reasoning is captured by our OBDDs. In the left OBDD of Figure 11, the judge first has to check that there was an old judgement establishing the criteria. In the right OBDD of Figure 11, he first has to ask the plaintiff to provide a proof that the criteria is not fulfilled (without wondering whether an old judgement was already established or not).

Dealing with the criteria of 'Independence' is completely similar. Hence, we introduce the following set of atoms:

$$\mathbb{P}_I \triangleq \{Lit_I, OldJug_I, NewElt_I, Proof_{\neg I}\}$$

Their interpretation is the same as for the criteria of Republican values, except that the term "Republican values" has to be replaced by "Independence" everywhere. So, the formula I of the language $\mathcal{L}_M$ which determines in which case I holds is the following:

$$I \triangleq Lit_I \rightarrow ((OldJug_I \wedge \neg NewElt_I) \vee \neg Proof_{\neg I})$$

Its intuitive interpretation is the same as for the previous criteria.

4.2   The criteria of 'geographical and professional range'

For the criteria of 'Geographical and professional range', we introduce the following set of atoms:

$$\mathbb{P}_G \triangleq \{Lit_G, Decide_G, Proof_{\neg G}\}$$

These atoms stand for the following propositions:

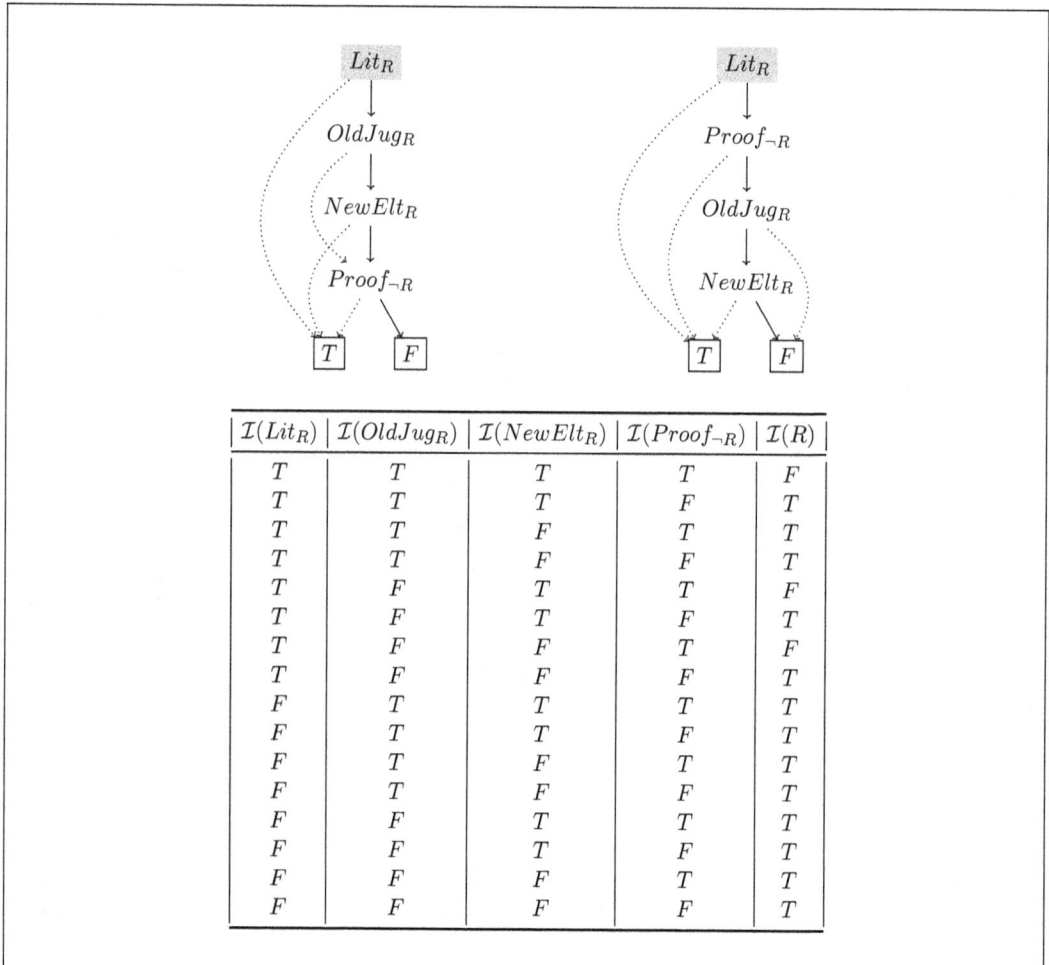

$\mathcal{I}(Lit_R)$	$\mathcal{I}(OldJug_R)$	$\mathcal{I}(NewElt_R)$	$\mathcal{I}(Proof_{\neg R})$	$\mathcal{I}(R)$
T	T	T	T	F
T	T	T	F	T
T	T	F	T	T
T	T	F	F	T
T	F	T	T	F
T	F	T	F	T
T	F	F	T	F
T	F	F	F	T
F	T	T	T	T
F	T	T	F	T
F	T	F	T	T
F	T	F	F	T
F	F	T	T	T
F	F	T	F	T
F	F	F	T	T
F	F	F	F	T

Figure 11: Three logically equivalent representations. OBDD associated to $R = Lit_R \rightarrow ((OldJug_R \wedge \neg NewElt_R) \vee \neg Proof_{\neg R})$ and $Lit_R < OldJug_R < NewElt_R < Proof_{\neg R}$ (*top left*), $Lit_R < Proof_{\neg R} < OldJug_R < NewElt_R$ (*top right*). Truth table of R (*bottom*).

- Lit_G: "The plaintiff contests the criteria of 'Geographical and professional range' ";

- $Decide_G$: "The judge decides to examine the criteria of 'Geographical and professional range' ";

- $Proof_{\neg G}$: "The plaintiff presents the proof that the criteria of 'Geographical and professional range' is not fulfilled".

The formula G of the language $\mathcal{L}_M$ determines in which case the criteria of 'Geographical and professional range' holds:

$$G \triangleq (Lit_G \vee Decide_G) \rightarrow \neg Proof_{\neg G}$$

The formula G reads as follows: "if the plaintiff contests that the criteria of 'Geographical and professional range' is fulfilled or if the judge decides to consider this criteria, then the plaintiff presents the proof that the criteria is not fulfilled". The OBDD associated to the formula G is depicted in Figure 12, it is the third OBDD from the left. In that last example, even if the criteria was not contested by the plaintiff, the judge can decide to solicit or not the plaintiff on that issue.

Logically equivalent representations. In Figure 11, we provide three equivalent and alternative representations of the semantics of the formula R: the truth table of R and two OBDDs associated to R that only differ on their associated orderings. The construction of the top left OBDD using the algorithm `Apply` is given in Figure 7 (with the appropriate atom substitutions). During a trial, the judge answers the questions corresponding to the nodes of the OBDD in the order specified by the ordering of the OBDD. However, this order could be changed automatically thanks to the algorithms dedicated to OBDDs, if needed. He could also 'navigate' in the binary decision diagram to explore it without having to answer any question node. Each time the judge answers a question, the OBDD is updated to a simpler OBDD. This update is performed by the `Restrict` algorithm.

Example 7. A series of questions answering and updates is given in Figure 9 (with the appropriate atoms substitutions). First, the judge answers 'yes' to question p (alias Lit_R) because the plaintiff contests the criteria. The software tool then applies the `Restrict` algorithm on the first OBDD of Figure 9 with the interpretation $\mathcal{I}_1$, yielding the second OBDD of Figure 9. Then, the judge answers 'yes' again to the question q (alias $OldJug_R$) because an old judgement dealing with the criteria has indeed already established its validity. The software tool then applies the `Restrict` algorithm to the second underlying OBDD of Figure 9 with $\mathcal{I}_2$, yielding the third OBDD of Figure 9. Then, he answers 'yes' to r (alias $NewElt_R$) because new elements have been brought to the fore that oblige to reconsider the old judgement, yielding the fourth OBDD of Figure 9 by application of `Restrict` with $\mathcal{I}_3$. Finally, he answers 'no' to s (alias $Proof_{\neg R}$) because these new elements do not invalidate the old judgement establishing the validity of the criteria. The software tool finally reaches the last elementary OBDD of Figure 9 by application of `Restrict` with $\mathcal{I}_4$, stating that the criteria of 'republican values' is fulfilled in that particular litigation.

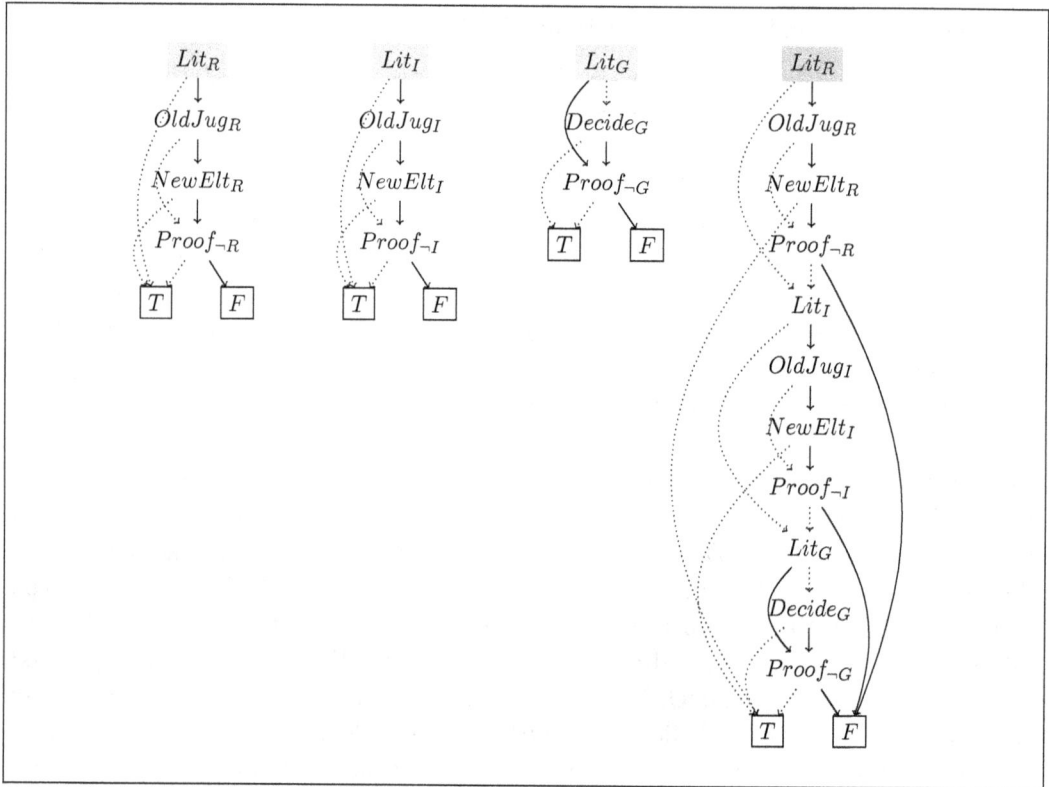

Figure 12: MOBDD of $(!R \wedge (!I \wedge !G))$ with ordering $Lit_R < OldJug_R < NewElt_R < Proof_{\neg R} < Lit_I < OldJug_I < NewElt_I < Proof_{\neg I} < Lit_G < OldJug_G < NewElt_G < Proof_{\neg G}$. The four OBDD are associated to R, I, G and $(R \wedge (I \wedge G))$, the entry point is in dark gray.

4.3 A MOBDD for qualifying as a trade union

We have not considered the criteria of 'Seniority' until now and we will not deal with it in order to ease the presentation and because it is more complex to represent than the other criteria, in the sense that many more atoms (questions) are needed to deal with it. The MOBDD of $(!R \wedge (!I \wedge !G))$ with ordering $Lit_R < OldJug_R < NewElt_R < Proof_{\neg R} < Lit_I < OldJug_I < NewElt_I < Proof_{\neg I} < Lit_G < OldJug_G < NewElt_G < Proof_{\neg G}$ is represented in Figure 12. It is simply the disjoint union of the three OBDD associated to R, I, G and $(R \wedge (I \wedge G))$. Then, it can be reduced equivalently thanks to the Algorithm **Reduce** of Figure 4 to the MOBDD of Figure 13. Note that in this MOBDD, all the leaf nodes have been merged into two nodes ($obdd_\top$ and $obdd_\bot$).

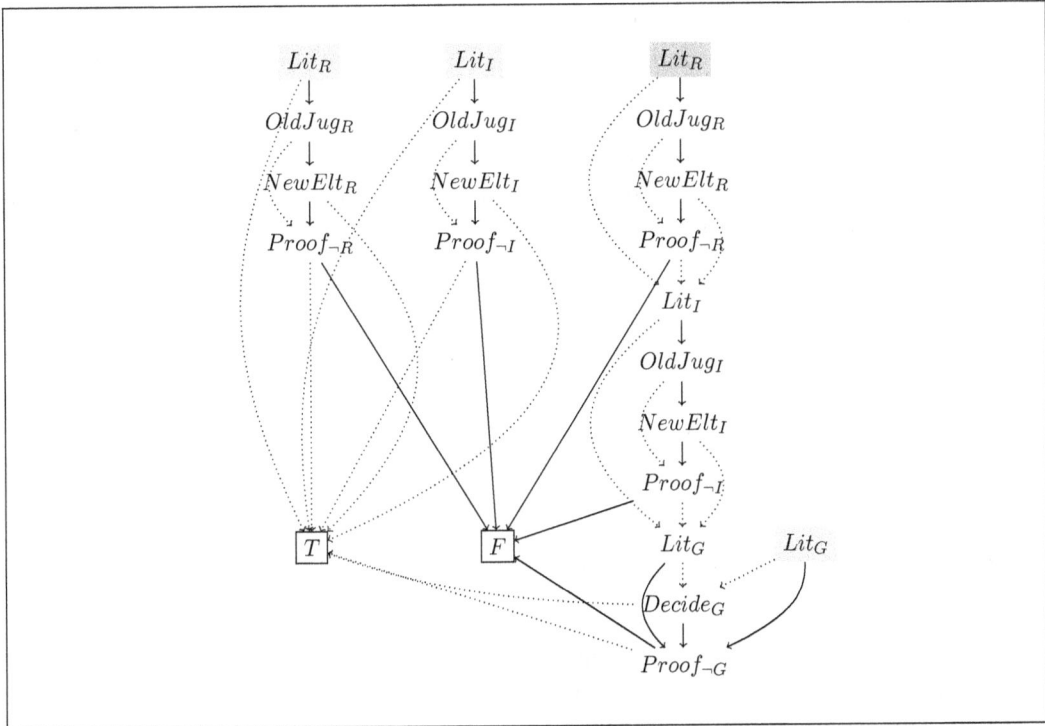

Figure 13: MOBDD of $(!R \wedge (!I \wedge !G))$ of Figure 12 reduced.

5   Applications of BDD algorithms to legal reasoning

Because our solution is based on BDD, we inherit from the vast amount of work for BDD a number of algorithms and software applications that can play an important role in legal reasoning. Even if they were not initially intended to be used in the legal domain, these algorithms and software applications turn out to be really relevant for solving specific problems or answer specific queries of the judge. We list below some of these algorithms (some of them have already been considered above) and show how they can be used by a judge during, before or after a trial. We start with the algorithms of this article:

- Algorithm **Restrict**. This algorithm can be used when the judge answer questions: each question answered corresponds in fact to an application of the algorithm **Restrict**. After each answer, the MBDD is instanciated and the node corresponding to the question disappears (see Example 7).

- Algorithm **Reduce**. This algorithm can be used to determine whether two kinds of legal reasoning represented by two different MBDD are in fact equivalent: in case the MBDD returned by this algorithm is the same in both cases, then they are indeed equivalent (see Theorem 2).

- Algorithm `Apply`. This algorithm can be used to construct a MOBDD corresponding to a formula expressed in our language $\mathcal{L}_M$. It can also be used when we want to combine two kinds of legal reasoning that deal with different but complementary issues that have already been represented by two MOBDD.

Other algorithms based on BDD dealing with quantification over propositional atoms can be used to solve the following tasks:

- Determine whether the answer to a specific question will allow the judge to conclude about a litigation or a specific subproblem without having to examine all the other questions exhaustively.

- Determine whether a question is redundant and can thus be removed from the MBDD.

These algorithms are only a few among the large amount of algorithms for BDD which are available. Many other algorithms can be used or designed or derived to solve specific legal reasoning tasks.

6   Our prototype

In this section, we introduce the prototype that was developed during our project. Our prototype does not use any BDD software library: we realized that the graphs elaborated in collaboration with the jurists were in fact BDD at a rather well-advanced stage of the project. The global architecture of the prototype is given in Figure 14. The latest version of our prototype is available at the following address:

`http://cassation.gforge.inria.fr/prototype-2015-06-08`

To try it out, please contact one of the authors to obtain the access codes.

In Section 6.1, we describe the front-end graphical user interface. The development process was iterative (inspired by the agile method) and Section 6.2 explains how the graphs were designed interactively by the jurists and computer scientists for creating the input graph. In Section 6.3, we introduce the architecture of the graph generation from `.doc` and `.dia` files using Microsoft^{TM} Word and the software Dia. In Section 6.4, we describe the architecture of the front-end that takes the graph as input.

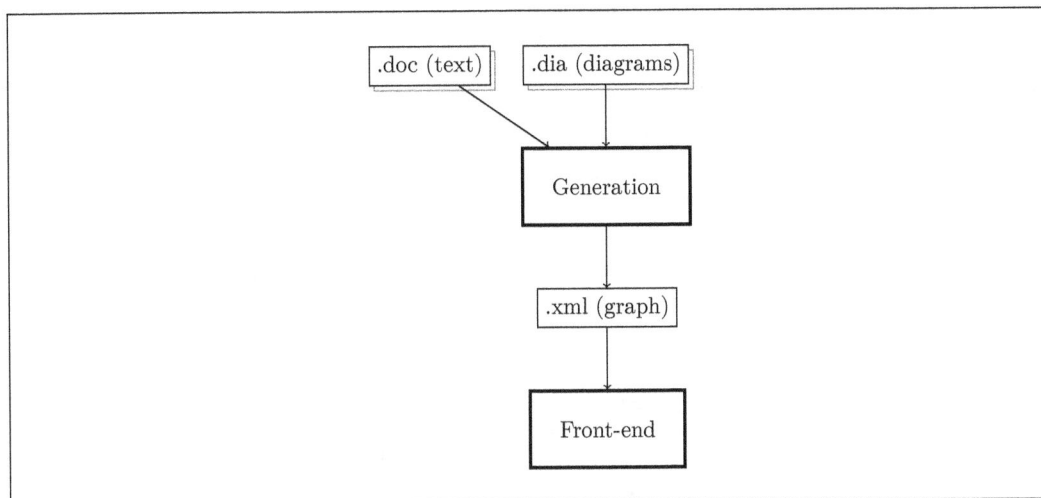

Figure 14: Global architecture

6.1 Graphical user interface

The graphical user interface is represented in Figure 1 and the front-end is split up into two parts. The left part proposes questions to be answered and the documentation for helping the judge to make decisions. While the judge answers questions, the right part shows the produced judgment text.

6.2 Iterative design method

During the project, computer scientists and jurists needed to find out a common 'language'. Jurists first resorted to graphical representations of their reasonings under the form of binary decision trees, which are more natural than the formulas and truth tables of logic. At some point, to ease communication, we decided to adopt a common code and use a software editor called Dia (`http://dia-installer.de/`). The jurists had quite some flexibility and autonomy. For instance, they could decide to adopt new symbols for new concepts whenever they felt that it was needed. It was very instructive, both for computer scientists and jurists, to discuss and exchange ideas based on these diagrams during several meetings. We agreed on a set of graphical conventions. The textual documentation was already created by jurists with informal structural conventions. It consisted in a collection of `.doc` files and we decided to keep this format.

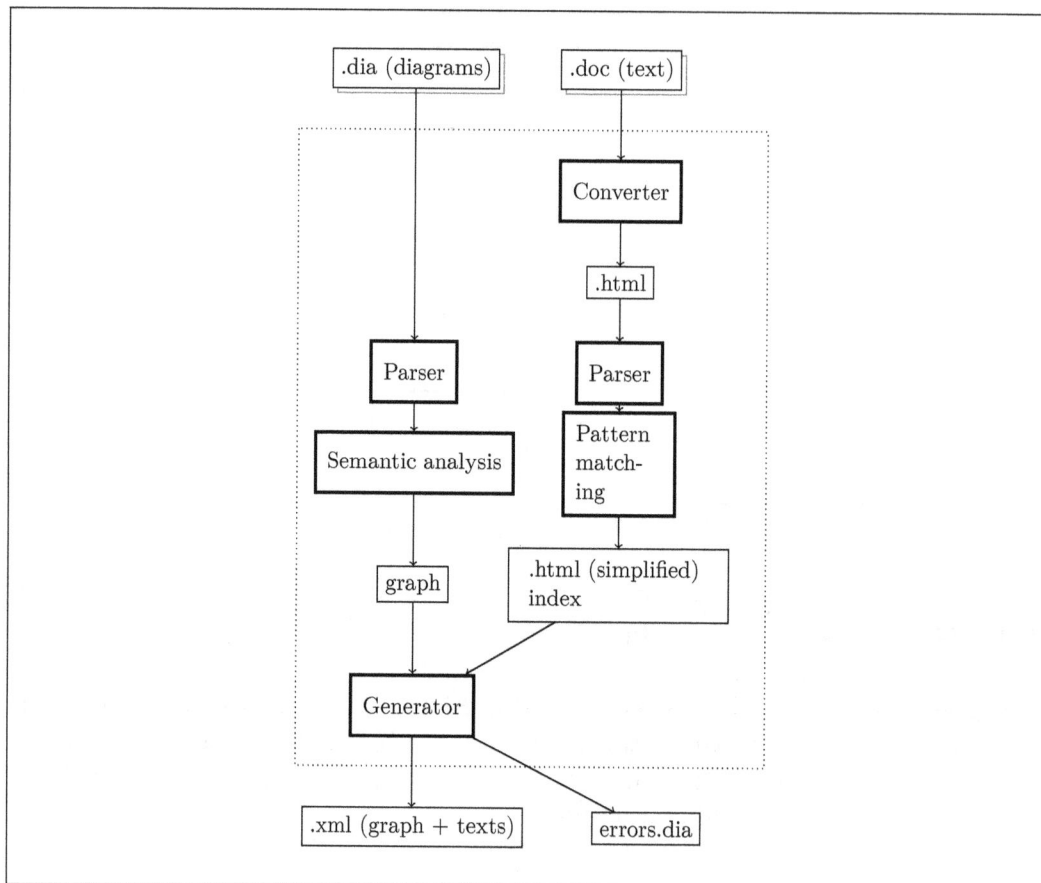

Figure 15: Graph generation architecture.

6.3 Graph generation architecture

Figure 15 represents the graph generation architecture. The input is divided in two parts: diagrams in `.dia` and the documentation in `.doc` files. On the one hand, BDD are represented by diagrams saved in `.dia` files. We used some graphical conventions to encode the various elements (questions, answers, connectives). The diagrams are parsed and a semantic analysis generates a graph from diagrams by using the graphical conventions. On the other hand, `.doc` files are converted in `.html` files, which are then parsed. We identify the different paragraphs in `.html` files and we produce a single simplified `.html` file and an index identifies the sections and the judgment paragraphs.

The last step generates a `.xml` file which merges the graph and the textual documentation (sections and judgment paragraphs). Errors are reported graphically in a `.dia` file.

6.4   Front-end architecture

The front-end is implemented in Javascript and is fully executed in the web browser. The front-end is based on a model-view-controller architecture. The model is dynamically generated from the `.xml` file given as input. The generated `.xml` file contains all the data for displaying questions and judgment paragraphs in the front-end. We can export the produced judgment output as a text file.

7   Conclusion

One could argue that our proposed solution is not really suitable because it is merely based on propositional logic and does not integrate non-classical reasoning such as deontic, causal or defeasible reasoning, which has often been claimed to be more appropriate to deal with legal reasoning (see the various references in Section 1.1 and [13]).[5] As such, our work is only a first step and it is quite possible to extend it to other kinds of reasoning hardly amenable to propositional reasoning. In fact, even if we have not addressed in this article standard problems in AI and law such as those related to 'contrary to duties' or exceptions, we did encounter such kinds of exceptional reasoning in the course of our project. As it turns out, exceptions could also be dealt with BDD and we actually introduced with that aim in view in the course of our project an operator called "sinon" ("otherwise" in English) with a semantics based on BDD. Our solution for handling exceptions based on BDD turned out to be also very appealing to jurists.

This research was carried out from the outset hand in hand by both jurists and computer scientists, with regular and numerous communications between both parties. The graphical representation that we made up and shared for our common work and formalization turned out in the end to correspond to BDD. This type of representation was more along the lines of the actual practice of the jurists. In particular, the procedural and temporal aspect of BDD was in fact very close to their actual practice as lawyers or jurists: in a BDD, one has to answer one after the other questions corresponding to the nodes of the BDD. This procedural and

---

[5]Quite different approaches based on neural networks and machine learning have been proposed to justify *a posteriori* the reasoning of the judge, and also predict it, such as for example the work of Borges and Bourcier [19]. However, these other kinds of models could not really be used to solve the problems that concerned us here because they do not really provide means to help judges to form their judgements: the crucial reasoning part is absent from these models.

temporal aspect of the legal practice cannot be captured by the usual syntactic representation of propositional formulas. We made an experiment with law students of the Ecole Normale Supérieure of Rennes to determine whether they prefer to write the kind of meta-regulations that we use in the software tool with formulas of (propositional) logic or with a graph-based representation like BDD. We designed a kind of experimental protocol to figure this out. Even if the results were hard to interpret because they had no prior teaching in logic or graph theory, it turns out that they had somehow more facility to use the graph-based representation than the formula-based representation.

Propositional logic is not intended to model the current state of legal texts and regulations. It only serves here as a theoretical basis for the *rewriting* of most of legal texts and regulations together with their dual representation in terms of BDD.[6] In fact, this graphical representation in terms of BDD could also be extended to more complex kinds of legal reasoning, as mentionned above. Hence, criticisms regarding the adequacy of propositional logic for legal reasoning do not really apply to our work, especially in this specific context of the development of a judgement editor. This rewriting of legal texts and regulations in logical terms can be viewed as a new type of codification (see Section 1.2). This new codification would then provide theoretical basis for the development of software tools that could be used by jurists and lawyers, and probably change their actual practice of the law. If our proposal is adopted, the current legal texts and regulations would have to be all rewritten and adapted in order to fit the format based on BDD propounded in this article, as we did during our project with the case study of the "French trade unions" (see Section 4). The rewriting and adaptation phase could start with small fragments of the law and it could then be expanded step by step to all parts of the law.

From a theoretical point of view, we believe that our solution is the most promising and realistic approach to answer the needs and problems summarized in Section 1.2. Indeed, it is based on methods and techniques of logic that are very well understood, worked out and applied and therefore provides a rigorous and solid foundation to subsequent technological developments. Our approach has the advantage to provide a strong control over our representation of the legal reasoning and over the changes that we may want to make to this reasoning. Moreover, BDD are very well–studied and their associated algorithms are able to scale–up to a large number

---

[6]In [6, 8, 7], we propounded another rewriting of legal texts and regulations in logical terms in order to deal with problems of privacy. The proposed reformulation was different from a logical point of view, somehow more specific, because it was meant for other legal purposes in a particular context, namely to check that the privacy policies declared by a company or organization are compliant with respect to the privacy regulations of a given legislation and to check that the company or organization does enforce its declared privacy policy over the internet.

of nodes. Thus, using BDD is clearly a realistic solution to deal with the complexity of the law and the large amount of texts and jurisprudences. Finally, our approach is very flexible and can take into account the dynamism and unpredictable changes of the law. Indeed, because it is based on propositional logic, the various changes in the law, such as promulgation, abrogation and annulment can be modeled naturally as update operations in propositional logic. Historically, the well-known AGM theory of belief change [3] was propounded by three researchers whose one of them, Alchourrón, was a jurist: AGM theory from his point of view was supposed to model and deal with changes in the law, viewed as a theory of propositional logic, such as promulgation and abrogation in an abstract way.[7] As it turns out, dealing with dynamism and change has been at the core of most of the recent developments in logic and artificial intelligence in the last decades and many extensions of propositional logic with dynamic operators have been introduced (see for instance [23, 40, 41, 9]). Hence, the dynamic character of the law could be dealt within the software by importing, adapting and implementing the various methods and techniques which have been developed in logic for dealing with dynamism and change.

Our case study based on the "French trade unions" (Section 4) shows that our approach is feasible and can be extended to any kind of regulations, even if a tremendous amount of work would need to be carried out by the jurists (in collaboration with computer scientists and logicians) to rewrite and adapt all the existing texts and regulations in order to define corresponding BDD. Our case study has been implemented in a prototype in the course of our project (see Section 6). This prototype was tried out by four judges of 'tribunaux d'instance' in France accustomed to our case study. They were all rather impressed and satisfied with the prototype tool. This said, the software tool introduced here is only the first part of a larger project since this software tool would only *use* the BDD representing the legal reasoning underlying specific litigations. But these BDD would first need to be defined and created by a legal expert or a legislator. The second tool that complements the software described in this article still has to be specified precisely. Its role would be to create and edit these BDD that capture any other case studies. In particular, this second software tool should provide mechanisms for modifying the graphs that underly the BDD. From a theoretical point of view, graph modifications and change is also an area of research that is currently very active in logic [5, 9, 24, 4, 12]. These theoretical works could provide algorithms and associated software tools for checking and verifying that a particular change or modification of the BDD has indeed been made and that these changes do correspond to the idea and the intention of the user/legislator who triggered them.

---

[7]Governatori & Al. [26, 25] attempt to model abrogation and annulment more realistically.

Finally, if such kinds of softwares would ever be developed and used by judges, this would entail as a prerequisite that the jurists be trained and taught some rudiments of logic, especially the law experts who would have to rewrite and adapt the legal texts and regulations to create and edit the corresponding BDD. This would also call for the enactment of regulations to determine in which kinds of legal context and litigations these softwares can and should be used. Indeed, the novelty of such softwares and their impact on society would raise a number of ethical issues that would need to be harnessed by the law. This said, we want to stress that this work will not replace by any means judges by 'machines', nor will suppress the responsibility that judges endorse when they make decisions. In this article, we only propose some theoretical foundations that could lead to the development of a software tool to help judges to make fair, well-informed and maybe sometimes better decisions. This software tool would also provide and recall judges some well–structured information about the relevant legal texts and jurisprudence that they should not omit, together with some crucial information about the legal procedure that should be followed in order to deal with a specific litigation. If such a software tool were available some day to judges, they should in any case remain fully responsible of their decisions and they should not have the possibility to discharge the responsibility of their decisions to that software tool.

Acknowledgments. The work reported in this article was carried out in the context of a collaborative project with the Cour de cassation. The coordinator of this project was Guillaume Aucher.[8] We thank Anthony Baire, Annie Foret, Jean-Baptiste Lenhof and François Schwarzentruber for their participation. We thank François Schwarzentruber for developing the very first prototype and Anthony Baire for developing the subsequent prototypes. We thank Marie-Pierre Lanoue, Daniel Tardif, Eloi Buat-Menard, Laurence Pecaut-Rivolier, Llio Humphreys, Guido Boella as well as Jean-Paul Jean, Damien Pons and Ronan Guerlot for their advices and participation and for contributing to make this project happen. We thank Aude Bubbe, Laurence Pecaut-Rivolier, Aurélie Police and Françoise Simond for trying out our prototype. We thank Olivier Ridoux for carefully reading the article and for proposing to use multi–BDD instead of three–valued OBDD.

References

[1] AJANI, G., BOELLA, G., CARO, L. D., ROBALDO, L., HUMPHREYS, L., PRADUROUX,

---

[8]Guillaume Aucher was interviewed at the outset of the project by Jean-Michel Prima: `http://emergences.inria.fr/lettres2013/newsletter-n28/L28-OUTILDECISION`.

S., ROSSI, P., AND VIOLATO, A. The European taxonomy syllabus: A multi-lingual, multi-level ontology framework to untangle the web of european legal terminologyuropean taxonomy syllabus: A multi-lingual, multi-level ontology framework to untangle the web of european legal terminology. *Applied Ontology 11*, 4 (2016), 325–375.

[2] ALCHOURRÓN, C. E., AND BULYGIN, E. *Normative systems*. Springer-Verlag, Wien, New York, 1971.

[3] ALCHOURRÓN, C. E., GÄRDENFORS, P., AND MAKINSON, D. On the logic of theory change: Partial meet contraction and revision functions. *J. Symb. Log. 50*, 2 (1985), 510–530.

[4] ARECES, C., FERVARI, R., AND HOFFMANN, G. Relation-changing modal operators. *Logic Journal of the IGPL 23*, 4 (2015), 601–627.

[5] AUCHER, G., BALBIANI, P., CERRO, L. F. D., AND HERZIG, A. Global and local graph modifiers. In *Methods for Modalities 5 (M4M-5)* (Cachan, France, 2007), ENTCS, Elsevier.

[6] AUCHER, G., BARREAU-SALIOU, C., BOELLA, G., BLANDIN-OBERNESSER, A., GAMBS, S., PIOLLE, G., AND VAN DER TORRE, L. The Coprelobri project : the logical approach to privacy. In *2e Atelier Protection de la Vie Privée (APVP 2011)* (Sorèze, France, June 2011).

[7] AUCHER, G., BOELLA, G., AND VAN DER TORRE, L. Privacy policies with modal logic: the dynamic turn. In *Deontic Logic in Computer Science (DEON 2010)* (2010), G. Governatori and G. Sartor, Eds.

[8] AUCHER, G., BOELLA, G., AND VAN DER TORRE, L. A dynamic logic for privacy compliance. *Journal of artificial intelligence and law 19*, 2–3 (2011), 187–231.

[9] AUCHER, G., VAN BENTHEM, J., AND GROSSI, D. Modal logics of sabotage revisited. *J. Log. Comput. 28*, 2 (2018), 269–303.

[10] BAIER, C., AND KATOEN, J.-P. *Principles of model checking*. MIT press, 2008.

[11] BAINBRIDGE, D. Case: Computer assisted sentencing in magistrates' courts. In *BILETA Conference* (1990).

[12] BALBIANI, P., ECHAHED, R., AND HERZIG, A. A dynamic logic for termgraph rewriting. In *ICGT* (2010), pp. 59–74.

[13] BEIERLE, C., FREUND, B., KERN-ISBERNER, G., AND THIMM, M. Using defeasible logic programming for argumentation-based decision support in private law. In *COMMA* (2010), vol. 216 of *Frontiers in Artificial Intelligence and Applications*, IOS Press, pp. 87–98.

[14] BEN-ARI, M. *Mathematical logic for computer science*. Springer Science & Business Media, 2012.

[15] BENCH-CAPON, T., AND PRAKKEN, H. Introducing the logic and law corner. *Journal of logic and computation 18*, 1 (2008), 1–12.

[16] BOELLA, G., CARO, L. D., HUMPHREYS, L., ROBALDO, L., ROSSI, P., AND VAN DER TORRE, L. Eunomos, a legal document and knowledge management system for the web to provide relevant, reliable and up-to-date information on the law. *Artif. Intell.*

Law 24, 3 (2016), 245–283.

[17] BOELLA, G., HUMPHREYS, L., MARTIN, M., ROSSI, P., AND VAN DER TORRE, L. Eunomos, a legal document and knowledge management system to build legal services. In *International Workshop on AI Approaches to the Complexity of Legal Systems* (2011), Springer, pp. 131–146.

[18] BONGIOVANNI, G., POSTEMA, G., ROTOLO, A., SARTOR, G., VALENTINI, C., AND WALTON, D., Eds. *Handbook of Legal Reasoning and Argumentation*. Springer, 2018.

[19] BORGES, F., BORGES, R., AND BOURCIER, D. A connectionist model to justify the reasoning of the judge. In *Legal Knowledge and Information Systems. Jurix 2002: The Fifteenth Annual Conference* (Amsterdam, 2002), T. Bench-Capon, A. Daskalopulu, and R. Winkels, Eds., IOS Press, pp. 113–122.

[20] BRYANT, R. E. Graph-based algorithms for boolean function manipulation. *Computers, IEEE Transactions on 100*, 8 (1986), 677–691.

[21] BRYANT, R. E. Symbolic boolean manipulation with ordered binary-decision diagrams. *ACM Comput. Surv. 24*, 3 (1992), 293–318.

[22] GABBAY, D., HORTY, J., PARENT, X., VAN DER MEYDEN, R., AND VAN DER TORRE, L., Eds. *Handbook of deontic logic and normative systems*. College Publication, 2013.

[23] GÄRDENFORS, P. *Knowledge in Flux (Modeling the Dynamics of Epistemic States)*. Bradford/MIT Press, Cambridge, Massachusetts, 1988.

[24] GIRARD, P., SELIGMAN, J., AND LIU, F. General dynamic dynamic logic. In *Advances in Modal Logic* (2012), pp. 239–260.

[25] GOVERNATORI, G., PADMANABHAN, V., ROTOLO, A., AND SATTAR, A. A defeasible logic for modelling policy-based intentions and motivational attitudes. *Logic Journal of the IGPL 17*, 3 (2009), 227–265.

[26] GOVERNATORI, G., AND ROTOLO, A. Changing legal systems: legal abrogations and annulments in defeasible logic. *Logic Journal of the IGPL 18*, 1 (2010), 157–194.

[27] GROSSI, D., AND ROTOLO, A. *A New Survey of Active Directions in Modern Logic*, vol. 30 of *Studies in Logic*. College Publications London, 2011, ch. Logic in the Law: A Concise Overview, pp. 251–274.

[28] HALPERN, J., HARPER, R., IMMERMAN, N., KOLAITIS, P., VARDI, M., AND VIANU, V. On the unusual effectiveness of logic in computer science. *The Bulletin of Symbolic Logic 7*, 2 (2001), 213–236.

[29] JONES, A., AND SERGOT, M. Deontic logic in the representation of law: Towards a methodology. *Artificial Intelligence and Law 1*, 1 (1992), 45–64.

[30] LEITH, P. The rise and fall of the legal expert system. *European Journal of Law and Technology 1*, 1 (2010).

[31] MCCARTY, L. A language for legal discourse i. basic features. In *Proceedings of ICAIL* (1989), ACM, pp. 180–189.

[32] MOSS, L. S. Applied logic: A manifesto. In *Mathematical problems from applied logic I*. Springer, 2006, pp. 317–343.

[33] PRAKKEN, H. Formal systems for persuasion dialogue. *The Knowledge Engineering*

Review 21, 2 (2006), 163–188.

[34] PRAKKEN, H., AND SARTOR, G. The role of logic in computational models of legal argument: a critical survey. *Computational Logic: Logic Programming and Beyond* (2002), 175–188.

[35] RISSLAND, E. L. *A companion to cognitive science*. Wiley-Blackwell, 1999, ch. Legal Reasoning, pp. 722–733.

[36] SATOH, K., ASAI, K., KOGAWA, T., KUBOTA, M., NAKAMURA, M., NISHIGAI, Y., SHIRAKAWA, K., AND TAKANO, C. Proleg: An implementation of the presupposed ultimate fact theory of japanese civil code by prolog technology. In *New Frontiers in Artificial Intelligence: JSAI-isAI 2010 Workshops* (2012), LNAI 6797, Springer, pp. 153–164.

[37] SERGOT, M. J., SADRI, F., KOWALSKI, R., KRIWACZEK, F., HAMMOND, P., AND CORY, H. The british nationality act as a logic program. *Communications of the ACM* 29, 5 (1986), 370–386.

[38] STRUILLOU, Y., MORIN, M.-L., AND PÉCAUT-RIVOLIER, L. *Le guide des élections professionnelles 2016-2017 et des désignations de représentants syndicaux dans l'entreprise*. No. 3. Dalloz / Guides Dalloz, 11 2015.

[39] THIREAU, J.-L. *Introduction Historique au Droit*. Champs Université. Flammarion, 2009.

[40] VAN BENTHEM, J. *Exploring logical dynamics*. CSLI publications Stanford, 1996.

[41] VAN BENTHEM, J. *Logical Dynamics of Information and Interaction*. Cambridge University Press, 2011.

[42] WAGNER, P. *Machine en Logique*. Presses Universitaires de France – PUF, 1998.

Received 16 December 2018

TIME AND COMPENSATION MECHANISMS IN CHECKING LEGAL COMPLIANCE

GUIDO GOVERNATORI
CSIRO's Data61, Dutton Park, Australia
guido.governatori@data61.csiro.au

ANTONINO ROTOLO
CIRSFID, University of Bologna, Italy
antonino.rotolo@unibo.it

Abstract

In this paper we extend the logic of violation proposed by [23] with time, more precisely, we temporalise that logic. The resulting system allows us to capture many subtleties of the concept of legal compliance. In particular, the formal characterisation of compliance can handle different types of legal obligation and different temporal constraints over them. The logic is also able to represent, and reason about, chains of reparative obligations, since in many cases the fulfillment of these types of obligation still amounts to legally acceptable situations.

1 Introduction

Developments in open MASs(Multi-Agent Systems) have pointed out that normative concepts can play a crucial role in modelling agents' interaction [38, 14, 3]. Like in human societies, desirable properties of MAS can be ensured if the interaction of artificial agents adopts institutional models whose goal is to regiment agents' behaviour through normative systems in supporting coordination, cooperation and decision-making. However, to keep agents autonomous it is often suggested that norms should not simply work as hard constraints, but rather as soft constraints [7]. In this sense, norms should not limit in advance agents' behaviour, but would instead provide standards which can be violated, even though any violations should result in sanctions or other normative effects applying to non-compliant agents. The detection of violations and the design of agents' compliance can amount to a relatively affordable operation when we have to check whether agents comply with simple normative systems. However, things are tremendously harder when we deal with realistic, large and articulated systems of norms such as the law. To the best of our knowledge, no systematic investigation has been so far proposed in this regard in the MAS field.

Among other things, the complexities behind the concept of legal compliance are due to the following reasons:

Reparative Obligations Legal norms often specify obligatory actions to be taken in case of their violation. Obligations in force after some other obligations have been violated correspond in general to contrary-to-duty obligations (CTDs) (see [11] for an overview). A peculiar subclass of CTDs is particularly relevant for the law: the so-called reparative obligations. For instance, in contract and in tort law reparative obligations protect individual legitimate interests by imposing actions that compensate any damages following from non-compliance [19, 20]. These constructions affect the formal characterisation of legal compliance since they identify situations that are not ideal, but still legally acceptable. Consider the following example (where norms have as usual a conditional structure: if the antecedents are jointly the case, then the consequent is obligatory):

$$Invoice \Rightarrow OBLPayBy7days$$
$$OBLPayBy7days, \neg PayBy7days \Rightarrow OBLPay5\%Interest$$
$$OBLPay5\%Interest, \neg Pay5\%Interest \Rightarrow OBLPay10\%Interest$$

What about if a customer violates both the obligation to pay by seven days after having received the invoice for her purchase, and the obligation to pay the 5% of interest of the due amount, but she pays the total amount plus the 10% of interest? In the legal perspective (which aims at protecting the rights of the vendor), the customer is compliant.

If so, these constructions can give rise to very complex rule dependencies, because we can have that the violation of a single rule can activate other (reparative) rules, which, in case of their violation, refer to other rules, and so forth [24]. If we take the above legal norms in isolation, the depicted situation is non-compliant, since two applicable legal norms are violated. However, if we compensate for the violations, then we are still in a "legal" situation.

Obligation and Time The law makes use of different types of obligations (see Section 2) also depending on how legal effects are temporally qualified. A first basic distinction is between those legal obligations which persist over time unless some other and subsequent events terminate them (e.g., "If one causes damage, one has to provide compensation"), and those that hold at a specific time on the condition that the norm preconditions hold and with a specific temporal relationship between such preconditions and the obligation (e.g., "If one is in a public building, one is forbidden to smoke").

Concerning the concept of compliance, it is worth noting that we may have obligations requiring:

1. to be always fulfilled during a specified time interval;

2. that a particular condition must occur at least once before a certain deadline and such that the obligations may, or may not, persist after this deadline if they are not complied with;

3. that something is done at a precise time [21].

Things are definitely harder when these types of obligations occur in chains of reparative obligations. For example, if the primary obligation is persistent and states to pay before tomorrow, and the secondary (reparative) obligation is to pay a fine in three days after the violation of the primary obligation, we are compliant not only when we pay by tomorrow, but also when we do not meet this deadline and pay both the due amount and the fine on the day after tomorrow.

Formal Requirements for Legal Compliance From a logical point of view, a formal characterisation of the concept of legal compliance requires to address the following related research tasks: (a) We need a logic able to handle different types of legal obligation and different temporal constraints over them; (b) This logic should be able to represent, and reason about, chains of reparative obligations. In particular, we need a procedure for making hidden conditions and reparative chains explicit; without this, we do not know whether a certain situation is legally acceptable; (c) We have to embed into the logic aspects of time, such as persistence and deadlines.

The paper is organised as follows: in Section 2 we informally discuss the types of obligation we will handle in the proposed framework. Then, in Section 3 we formally introduce the language of the framework, a set of inference rules to infer additional norms/rules from the rules given in a (defeasible) theory (meant to correspond to a normative system) and to remove rules subsumed by others in a given theory. In Section 4 we provide the proof theory for inferring what obligations are in force and when they are in force from a given theory. In Section 5 we provide a normalisation procedure to be used to identify, given a defeasible theory what are the rules that

817

have to be used to determine what are the obligations in force, and whether they have been complied with. Finally, in Section 6 we provide some hints about future directions of research and we shortly discuss some related works.

2 The Many Faces of Obligations

We can distinguish *achievement* from *maintenance obligations* [10, 21, 30]. For an *achievement obligation*, a certain condition must occur at least once before a deadline:

Example 1. *Customers must pay within 7 days, after receiving the invoice.*

The deadline refers to an obligation triggered by receipt of the invoice. After that, the customer is obliged to pay. The fulfilment of the obligation by its deadline terminates the persistence of the obligation.

For *maintenance obligations*, a certain condition must obtain during all instants before the deadline:

Example 2. *After opening a bank account, customers must keep a positive balance for 30 days.*

In Example 2 the deadline only signals that the obligation is terminated: a violation occurs when the obliged state does not obtain at some time before the deadline.

Finally, *punctual obligations* only apply to single instants:

Example 3. *When banks proceed with any wire transfer, they must transmit a message, via SWIFT, to the receiving bank requesting that the payment is made according to the instructions given.*

Punctual obligations apply to single instants; they can be thought of as maintenance obligations in force in time intervals where the endpoints are equal. Typically punctual obligations must occur at the same time of their triggering conditions.

Norms can be associated with an explicit sanction. For example,

Example 4. *Customers must pay within 7 days, after receiving the invoice. Otherwise, 10% of interest must be paid within 10 days.*

Example 5. *After opening a bank account, customers must keep a positive balance for 30 days. Otherwise, their account must be immediately blocked.*

A sanction is often implemented through a separate obligation, which is triggered by a detected violation. Thus, different types of obligations can be combined in chains of reparative obligations: in Example 4, the violation of the primary achievement obligation is supposed to be repaired by another achievement obligation; in Example 5, the violation of a primary maintenance obligation is compensated by a punctual obligation.

We introduced in [23] the non-boolean connective $\otimes$: a formula like $a \otimes b$ means that a is obligatory, but if the obligation a is not fulfilled, then the obligation b is activated and becomes in force until it is satisfied or violated. However, the violation condition of an obligation varies depending on the types of obligations used. In the remainder, we will extend the approach of [23, 24] by adding temporal qualifications to cover these cases.

Notice that the classification of obligation types given in [30], which is a complete classification over the temporal, compliance and violation dimensions, provides a more fine grained distinctions of the different types of normative requirements; however, for space reasons, we restrict our analysis to the cases presented in this section.

3   Temporalised Violation Logic

To start with, we consider a logic whose language is defined as follows:

Definition 1 (Language). *Let $\mathcal{T} = (t_1, t_2, \dots)$ be a discrete linear order of instants of time, $Atm = \{a, b, \dots\}$ be a set of atomic propositions, and O be a deontic operator.*

- *A literal is either an atomic proposition or the negation of an atomic proposition, that is: $Lit = Atm \cup \{\neg l : l \in Atm\}$.*

- *If $l \in Lit$ and $t \in \mathcal{T}$, then l^t is a temporal literal; $\top$ and $\bot$ are temporal literals. $TLit$ denotes the set of temporal literals.*

- *If l^t is a temporal literal, then Ol^t and $\neg Ol^t$ are deontic literals. The set of deontic literals is denoted by $DLit$.*

- *If a^{t_a} and b^{t_b} are temporal literals, $t \in \mathcal{T}$, and $t_a \leq t$, then $a^{t_a} \otimes_t^x b^{t_b}$ (for $x \in \{p, m, a\}$) is an $\otimes$-chain.*

- *If α is an $\otimes$-chain, a^{t_a} is a temporal literal and $t \in \mathcal{T}$, then $\alpha \otimes_t^x a^{t_a}$ (for $x \in \{p, m, a\}$) is an $\otimes$-chain.*

- *Let α be either a temporal literal, or an $\otimes$-chain, $t \in \mathcal{T}$, then $\bot$, $\alpha \otimes \bot$ and $\alpha \otimes_t \bot$ are deontic expressions. Nothing else is a deontic expression. The set of deontic expressions is denoted by $DExp$.*

Let us explain the intuitive meaning of the various elements of the language. The meaning of a temporal literal a^t is that proposition a holds at time t. The deontic literal Oa^t means that we have the obligation that a holds at time t. The meaning of $\top$ and $\bot$ is that $\top$ is a proposition that is always complied with (or in other terms, it is impossible to violate) and $\bot$, on the other hand, is a proposition that is always violated (or it is impossible to comply with). According to the intended meaning, it is useless in the present context to temporalise them. $\otimes$ is a binary operator to express complex normative positions. More specifically, the meaning of a deontic expression like

$$\alpha \otimes^x_{t_a} a^{t_a} \otimes^y_{t'_a} b^{t_b}$$

is that the violation of a triggers a normative position whose content is b^{t_b}. What counts as a violation of a^{t_a} depends on the parameter x, encoding the type of obligation whose content is a, and the two temporal parameters t_a (the time when the obligation enters in force) and t'_a (the deadline to fulfil the obligation). The nature of the normative position whose content is b^{t_b} depends on $\otimes^y$. The type of obligation whose content is a^{t_a} is determined by x. If $x = p$, then we have a punctual obligation (in this case we require that $t_a = t'_a$), and this means that to comply with this prescription a must hold at time t_a. If $x = a$, then we have an achievement obligation; in this case, a is obligatory from t_a to t'_a, and the obligation is fulfilled if a holds for at least one instant of time in the interval $[t_a, t'_a]$. Finally, if $x = m$, similarly to the previous case, a is obligatory in the interval $[t_a, t'_a]$, but, in this case, to comply with the prescription, a must hold for all the instants in the interval. As we have said, the $\otimes$ operator introduces normative positions in response to a violation of the formula on the left of the operator; thus this is a contrary-to-duty operator. An important application of contrary-to-duties is that a contrary-to-duty can be used to encode a sanction or compensation or reparation for a violation. The focus of this paper is mostly on this type of contrary-to-duties. What about $DExp$? The meaning of a $DExp$, in particular of $\bot$ at the end of them, is that we have reached a situation that cannot be compensated for, This means that the penultimate element of a deontic expression identifies the 'last chance' to be compliant. After that, the deontic expression results in a situation that cannot be complied with anymore.

Definition 2 (Rules/norms[1]). *A rule*

$$r : \Gamma \hookrightarrow \alpha$$

is an expression where r is a unique rule label, $\Gamma \subseteq TLit \cup DLit$, $\hookrightarrow \in \{\Rightarrow^x, \rightsquigarrow\}$, $\alpha \in DExp$.

- *If $\hookrightarrow$ is $\Rightarrow^x$, the rule is a* defeasible rule*;*
- *If $\hookrightarrow$ is $\rightsquigarrow$, the rule is a* defeater*.*

For defeasible rules $x \in \{a, m, p\}$, and:

- *If $x = a$ the rule is an* achievement rule*;*
- *If $x = m$ the rule is a* maintenance rule*;*
- *If $x = p$ the rule is a* punctual rule*.*

For defeaters $\alpha \in TLit$.

A rule is a relationship between a set of premises and a conclusion; thus we use several types of rules to describe different types of relationships. We use the distinction of the types of rules (defeasible and defeater) for the strength of the relationship between the premises and the conclusion. The superscript x indicates the mode of a rule. The mode of a rule tells us what kind of conclusion we can obtain from the rule. In the context, the mode identifies the type of obligation we can derive. The idea is that from a rule of mode a, an achievement rule, we derive an achievement obligation.

A defeasible rule is a rule where when the body holds then typically the conclusion holds too unless there are other rules/norms overriding it. For example, when you receive an invoice, you have the obligation to pay for it:

$$r_1 : invoice^t \Rightarrow^a pay^t \tag{1}$$

The meaning of the above rule is that if you received an invoice at time t, then you have the obligation to pay for it, starting from time t.[2]

Defeaters are the weakest rules. They cannot be used to derive obligations, but they can be used to prevent the derivation of an obligation. Hence, they can be used

---

[1]In the reminder, we will interchangeably use both the terms 'norm' and 'rule', but we will prefer 'norm' whenever the usage of the term 'rule' may be confused with 'inference rule'.

[2]We assume the usual inter-definability between obligations and prohibition, thus $O\neg \equiv F$, and $F\neg \equiv O$.

to describe exceptions to obligations, and in this perspective they can be used to terminate existing obligations. For this reason, the arrow $\leadsto$ is not labeled by either a, m, nor p. Continuing the previous example, paying for the invoice terminates the obligation to pay for it:

$$r_2 : paid^t \leadsto pay^t \tag{2}$$

Rule r_2 says that if you pay at time t then, from time t on, there is no longer the obligation to pay. Notice that the defeater does not introduce the prohibition to pay again.

Definition 3 (Defeasible Theory). *A Defeasible Theory is a structure $(F, R, \succ)$, where*

- *F, the set of facts, is a set of temporal literals;*

- *R is a set of rules; and*

- *$\succ$, the superiority relation, is a binary relation over R.*

A theory corresponds to a normative system, i.e., a set of norms, where every norm is modelled by rules. The superiority relation is used for conflicting rules, i.e., rules whose conclusions are complementary literals, in case both rules fire. Notice that we do not impose any restriction on the superiority relation, which is a binary relation that just determines the relative strength of two rules. For example, if we consider the two rules in (1) and (2), given an invoice, and that the invoice has been paid the two rules alone cannot allow us to conclude anything due to the sceptical nature of Defeasible Logic. But if we further establish that $r_2 \succ r_1$, then the second rule prevails, and we will conclude that we are permitted not to pay.

Definition 4. *Given an $\otimes$-chain α, the* length *of α is the number of elements in it. Given an $\otimes$-chain $\alpha \otimes_t^x b^{t_b}$, the* index *of b^{t_b} is n iff the length of $\alpha \otimes_t^x b^{t_b}$ is n. We also say that b appears at index n in $\alpha \otimes_t^x b^{t_b}$.*

Definition 5 (Notation). *Given a rule $r \colon \Gamma \hookrightarrow \alpha$, we use $A(r) = \Gamma$ to indicate the antecedent or body of the rule, and $C(r) = \alpha$ for the consequent or conclusion or head of r. Given a set or rules R:*

- *$R_\Rightarrow$ is the set of defeasible rules in R;*

- *$R_\leadsto$ is the set of defeaters in R;*

- *R^a is the set of achievement rules in R;*

- *R^m is the set of maintenance rules in R;*

- R^p *is the set of punctual rules in* R*;*

- $R[a^t]$ *is the set of rules whose head contains* a^t*.*

- $R[a^t, k]$ *is the set of rules where* a^t *is at index* k *in the head of the rules.*

To simplify and uniform the notation we can combine the above notations, and we use subscripts and superscripts before the indication relative of the head. Thus, for example, $R_{\leadsto}[p^{10}]$ is the set of defeaters whose head is the temporal literal p^{10}, and the rule

$$r: a_1^{t_1} \ldots, a_n^{t_n} \Rightarrow^p a^{10} \otimes_{10}^m b^{20}$$

is in $R_{\Rightarrow}^m[b^{20}]$, as well as in $R^p[a^{10}]$ and $R_{\Rightarrow}[b^{20}, 2]$.

Finally, notice that we will sometimes abuse the notation and omit (a) the timestamp t_l in the temporal literal l^{t_l} whenever it is irrelevant to refer to it in the specific context, (b) the mode x in the rule arrow $\Rightarrow^x$ when x can be instantiated with any of a, m or p, (c) x and y in $\otimes_y^x$ when x and y can be instantiated, respectively, with any of a, m, p and with any time instants.

Properties of the $\otimes$ operator   When we have a deontic expression $\alpha = a_1 \otimes \cdots \otimes a_n$ we do not have information about the type of obligation for the first element. This information is provided when we use the expression in a rule. In this section we are going to investigate properties of $\otimes$, in particular when two (sub-)sequences of deontic expression are equivalent and thus we can replace them preserving the meaning of the whole expression (or rule). To simplify the notation, we introduce the following conventions.

Definition 6. *Let* $r: \Gamma \Rightarrow^x \alpha$ *be a rule, then* $^x\alpha$ *is an* $\otimes$*-sequence. The empty sequence is an* $\otimes$*-sequence. If* $\alpha \otimes_{t_\alpha}^x a^{t_a} \otimes_{t'_a}^y \beta \otimes_{t_\beta}^z \gamma$ *is an* $\otimes$*-sequence, where* α, β, γ *are* $\otimes$*-sequences, then* $^x a^{t_a} \otimes_{t'_a}^y \beta$ *is an* $\otimes$*-sequence.*

Given a rule $r: \Gamma \hookrightarrow^x \alpha \otimes_t^y \beta$*,* α *can be the empty* $\otimes$*-sequence, and if so, then the rule reduces to* $r: \Gamma \Rightarrow^y \beta$*.*

From now on, we will refer to $\otimes$-sequences simply as sequences and we will provide properties for sequences to be used in rules.

The first property we want to list is the commutativity of the $\otimes$ operator.

$$\alpha \otimes_t^x (\beta \otimes_{t'}^y \gamma) \equiv (\alpha \otimes_t^x \beta) \otimes_{t'}^y \gamma. \tag{3}$$

We extend the language with $\top$ and $\bot$. Given their meaning, those two propositions can be defined in terms of the following sequence and equivalence[3]

$$^p a^0 \otimes_0^p \neg a^0 \equiv \top \qquad \bot \equiv \neg \top. \tag{4}$$

The two new propositions are useful to define reduction rules for deontic expressions. Let us start with equivalences for $\top$.

$$\top \otimes \alpha \equiv \top. \tag{5}$$

This equivalence says that a violation of $\top$ can be compensated by α; however, $\top$ is a proposition that cannot be violated. Thus, the whole expression cannot be violated. What about when $\top$ appears as the last element of $\otimes$?

$$\alpha \otimes \top \equiv \top. \tag{6}$$

The meaning of $\alpha \otimes \top$ is that $\top$ is the compensation of α, thus the violation of α is sanctioned by $\top$. This means that the violation of α is always compensated for, thus we have a norm whose violation does not result in any effective sanction, thus violating α does not produce any effect. Hence, we have two possibilities: to reject (6) if we are interested to keep trace of violations, or to accept it if we want to investigate the effects of violations. In this paper we take the first option and we reject the equivalence of $\alpha \otimes \top$ and $\top$. Notice that reducing $\alpha \otimes \top$ to α would change the meaning, since this would mean that the violation of α cannot be repaired. To see this we move to the properties involving $\bot$.

$$^p a^{t_a} \otimes_{t_a}^x \bot \equiv a^{t_a}. \tag{7}$$

The above equivalence specifies that if $\bot$ is the compensation of a punctual obligation a at time t, then there is no compensation, since the compensation cannot be complied with. The effect of the rules is that we can eliminate $\bot$ from the deontic expression and we maintain the same meaning. Notice, however, that the same is not true for other types of obligations. For example, for $x \in \{a, m\}$, we cannot eliminate $\bot$ from rules like

$$\Gamma \Rightarrow^x a^t \otimes_{t'}^m \bot$$

since the resulting expression would be $\Gamma \Rightarrow^x a^t$ and we would miss the information about the deadline to comply with a. Nevertheless, the following equivalence states that $\bot$ can be safely eliminated if it is not the last element of a deontic expression, or when it is the 'compensation' of a maintenance obligation without deadline.

$$\alpha \otimes_{t_\alpha}^x \bot \otimes_t^y \beta \equiv \alpha \otimes_{t_\alpha}^y \beta \qquad {}^m a^{t_a} \otimes \bot \equiv {}^m a^{t_a} \tag{8}$$

---

[3]In case one wants the temporalised version, $\top^t \equiv {}^p a^t \otimes_t^p \neg a^t$, and $\bot^t \equiv \neg \top^t$.

To complete the description for the properties for $\bot$, we need to specify when we can generate a new rule introducing $\bot$ from two other rules.

$$\frac{\Gamma \Rightarrow^x \alpha \otimes_{t_\alpha}^y a^t \otimes_{t_a} \bot \qquad \Delta \hookrightarrow \neg a^{t'} \otimes_{t''} \bot}{\Gamma, \Delta \Rightarrow^x \alpha \otimes_{t_\alpha}^y a^t \otimes_{t'-1} \bot} t < t' \text{ and } y \in \{a, m\}. \qquad (9)$$

The meaning of the above inference rule is that if we have a norm determining the termination of an obligation, then we can encode the obligation, the time when the obligation comes to force and the time when the norm terminates its normative effect. The idea behind a norm like $a^t \Rightarrow^x b^{t'}$ is the obligation b enters into force from time t'. Here we assume the intuition developed in [27] that a 'new' rule takes precedence over a conclusion obtained in the past and carrying over to the current moment by persistence. Thus if we have a rule $c^{t_c} \Rightarrow \neg b^{t''}$ with $t'' > t'$ the rule for $\neg b^{t''}$ effectively terminates the force of the obligation b. Consider the following instance of the rule

$$\frac{r_1 : a^5 \Rightarrow^m b^{10} \otimes_{15} \bot \qquad c^{12} \Rightarrow^a \neg b^{12} \otimes_{20} \bot}{a^5, c^{12} \Rightarrow^m b^{10} \otimes_{11} \bot}.$$

In this case r_1 puts the obligation of b in force in the interval from 10 to 15, and r_2 enforces $\neg b$ from 12 to 20, thus when both conditions to apply, the effective time when the obligation of b is in force is from 10 to 11 (after that the obligation $\neg b$ enters into force).

The $\otimes$ operator, introduced in [23], is a substructural operator corresponding to the comma on the right hand side of a sequent in sequent system. In a classical sequent system both the left hand side and right hand side of a sequent are set of formulas, thus the order of the formulas does not matter, and properties like contraction and duplication hold. In [23] we established the equivalence

$$\alpha \otimes a \otimes \beta \otimes a \otimes \gamma \equiv \alpha \otimes a \otimes \beta \otimes \gamma.$$

This states that if a literal occurs multiple times, we can remove all but the first occurrence. The different types of obligation and times make thing more complicated. Thus we turn our attention to study conditions under which we have contraction for the various (combination of) $\otimes$ operators we have.

Tables 1 and 2 give the conditions to remove duplicates of the same atom. Consider for example, the instance $^p a^{10} \otimes^m a^0 \otimes_{20} \bot$ of the reduction Punctual-Maintenance (P-M) in Table 1, where the primary obligation is to have a at time 10, and whose compensation is to maintain a from 0 to 20. To trigger the secondary obligation we should have the violation of the primary obligation. This means that $\sim a$ holds at 10, but this implies that it is not possible to maintain a from 0 to 20,

P-P	$^p a^t \otimes^x_t \beta \otimes^p_{t_\beta} a^{t'} \otimes^y_{t'} \gamma \equiv^p a^t \otimes^x_t \beta \otimes^p_{t_\beta} \bot \otimes^y \gamma$	$t = t'$
P-A	$^p a^t \otimes^x_t \beta \otimes^a_{t_\beta} a^{t_s} \otimes^y_{t_e} \gamma \equiv^p a^t \otimes^x_t \beta \otimes^a_{t_\beta} \bot \otimes^y \gamma$	$t = t_s = t_e$
P-M	$^p a^t \otimes^x_t \beta \otimes^m_{t_\beta} a^{t_s} \otimes^y_{t_e} \gamma \equiv^p a^t \otimes^x_t \beta \otimes^m_{t_\beta} \bot \otimes^y \gamma$	$t \in [t_s, t_e]$
A-P	$^a a^{t_s} \otimes^x_{t_e} \beta \otimes^p_{t_\beta} a^{t'} \otimes^y_{t'} \gamma \equiv^a a^{t_s} \otimes^x_{t_e} \beta \otimes^p_{t_\beta} \bot \otimes^y \gamma$	$t' = t_s = t_e$
A-A	$^a a^{t_s} \otimes^x_{t_e} \beta \otimes^a_{t_\beta} a^{t'_s} \otimes^y_{t'_e} \gamma \equiv^a a^{t_s} \otimes^x_{t_e} \beta \otimes^a_{t_\beta} \bot \otimes^y \gamma$	$[t'_s, t'_e] \subseteq [t_s, t_e]$
A-M	$^a a^{t_s} \otimes^x_{t_e} \beta \otimes^m_{t_\beta} a^{t'_s} \otimes^y_{t'_e} \gamma \equiv^a a^{t_s} \otimes^x_{t_e} \beta \otimes^m_{t_\beta} \bot \otimes^y \gamma$	$[t_s, t_e] \cap [t'_s, t'_e] \neq \emptyset$
M-P	$^m a^{t_s} \otimes^x_{t_e} \beta \otimes^p_{t_\beta} a^{t'} \otimes^y_{t'} \gamma \equiv^m a^{t_s} \otimes^x_{t_e} \beta \otimes^p_{t_\beta} \bot \otimes^y \gamma$	$t' = t_s = t_e$
M-A	$^m a^{t_s} \otimes^x_{t_e} \beta \otimes^a_{t_\beta} a^{t'_s} \otimes^y_{t'_e} \gamma \equiv^m a^{t_s} \otimes^x_{t_e} \beta \otimes^a_{t_\beta} \bot \otimes^y \gamma$	$t_s = t_e = t'_s = t'_e$
M-M	$^m a^{t_s} \otimes^x_{t_e} \beta \otimes^m_{t_\beta} a^{t'_s} \otimes^y_{t'_e} \gamma \equiv^m a^{t_s} \otimes^x_{t_e} \beta \otimes^m_{t_\beta} \bot \otimes^y \gamma$	$[t_s, t_e] \subseteq [t'_s, t'_e]$

Table 1: Reductions to $\bot$ (P, A, M, stand respectively for Punctual, Achievement, Maintenance)

P-P	$^p a^t \otimes^x_t \beta \otimes^p_{t_\beta} \sim a^{t'} \otimes^y_{t'} \gamma \equiv^p a^t \otimes^x_t \beta \otimes^p_{t_\beta} \top \otimes^y \gamma$	$t = t'$
P-A	$^p a^t \otimes^x_t \beta \otimes^a_{t_\beta} \sim a^{t_s} \otimes^y_{t_e} \gamma \equiv^p a^t \otimes^x_t \beta \otimes^a_{t_\beta} \top \otimes^y \gamma$	$t \in [t_s, t_e]$
P-M	$^p a^t \otimes^x_t \beta \otimes^m_{t_\beta} \sim a^{t_s} \otimes^y_{t_e} \gamma \equiv^p a^t \otimes^x_t \beta \otimes^m_{t_\beta} \top \otimes^y \gamma$	$t = t_s = t_e$
A-P	$^a a^{t_s} \otimes^x_{t_e} \beta \otimes^p_{t_\beta} \sim a^{t'} \otimes^y_{t'} \gamma \equiv^a a^{t_s} \otimes^x_{t_e} \beta \otimes^p_{t_\beta} \top \otimes^y \gamma$	$t' = t_s = t_e$
A-A	$^a a^{t_s} \otimes^x_{t_e} \beta \otimes^a_{t_\beta} \sim a^{t'_s} \otimes^y_{t'_e} \gamma \equiv^a a^{t_s} \otimes^x_{t_e} \beta \otimes^a_{t_\beta} \top \otimes^y \gamma$	$[t_s, t_e] \subseteq [t'_s, t'_e]$
A-M	$^a a^{t_s} \otimes^x_{t_e} \beta \otimes^m_{t_\beta} \sim a^{t'_s} \otimes^y_{t'_e} \gamma \equiv^a a^{t_s} \otimes^x_{t_e} \beta \otimes^m_{t_\beta} \top \otimes^y \gamma$	$[t'_s, t'_e] \subseteq [t_s, t_e]$
M-P	$^m a^{t_s} \otimes^x_{t_e} \beta \otimes^p_{t_\beta} \sim a^{t'} \otimes^y_{t'} \gamma \equiv^m a^{t_s} \otimes^x_{t_e} \beta \otimes^p_{t_\beta} \top \otimes^y \gamma$	$t' = t_s = t_e$
M-A	$^m a^{t_s} \otimes^x_{t_e} \beta \otimes^a_{t_\beta} \sim a^{t'_s} \otimes^y_{t'_e} \gamma \equiv^m a^{t_s} \otimes^x_{t_e} \beta \otimes^a_{t_\beta} \top \otimes^y \gamma$	$[t_s, t_e] \subseteq [t'_s, t'_e]$
M-M	$^m a^{t_s} \otimes^x_{t_e} \beta \otimes^m_{t_\beta} \sim a^{t'_s} \otimes^y_{t'_e} \gamma \equiv^m a^{t_s} \otimes^x_{t_e} \beta \otimes^m_{t_\beta} \top \otimes^y \gamma$	$t_s = t_e = t'_s = t_e$

Table 2: Reductions to $\top$

thus it is not possible to compensate the violation of the primary obligation. Notice that in several cases the reductions are possible only when the intervals are just single instants.

Introduction Rules   Besides the properties given so far the full meaning of the $\otimes$ operator is given by the rules to introduce (and modify) the operator. The general idea of the introduction rules is to determine the conditions under which a norm is violated. If these conditions imply a particular obligation then, then this obligation can be seen as a compensation of the norm the conditions violate.

$$\frac{\Gamma \Rightarrow^x \alpha \otimes^p_{t_\alpha} b^{t_b} \otimes^Y_{t_b} \gamma \quad \Delta, \neg b^{t_b} \hookrightarrow^z \delta}{\Gamma, \Delta \Rightarrow^x \alpha \otimes^p_{t_\alpha} b^{t_b} \otimes^z_{t_b} \delta} \otimes I_p.$$

The punctual obligation $O^p b^{t_b}$ (implied by the first sequent) holds only at time t_b thus the only instant when the obligation can be violated is exactly t_b.

Rule $\otimes I_p$ is the standard rule to introduce a (novel) compensation or CTD (see [23] for further discussion about it).

$$\frac{\Gamma \Rightarrow^x \alpha \otimes^m_{t_\alpha} b^{t_s} \otimes^y_{t_e} \beta \quad \Delta, \Theta \Rightarrow^z \delta}{\Gamma, \Delta \Rightarrow^x \alpha \otimes^m_{t_\alpha} b^{t'_s} \otimes^z_{t'_e} \delta} \otimes I_m \text{ where } \Theta = \{\sim b^{t'} : t_s < t'_s \le t' \le t'_e \le t_e\}.$$

The introduction rule for $\otimes^m$ defines a slice of the interval where a specific compensation of the violation holds. This conditions requires a rule whose antecedent contains the complement of a maintenance obligation in the head of the other rule, such that the literal is temporalised with the last n consecutive instants. For example given the rules

$$a^{10} \Rightarrow^m b^{10} \otimes_{20} \bot \qquad c^{15}, \neg b^{17}, \neg b^{18}, \neg b^{19} \Rightarrow^p d^{20} \otimes_{20} \bot$$

we can derive the new rule

$$a^{10}, c^{15} \Rightarrow^m b^{17} \otimes^p_{19} d^{20} \otimes_{20} \bot.$$

The conditions to derive a new compensation rule for an achievement obligation are more complicated. As we have seen from the previous two cases, the structure of the introduction rules is that the negation of a consequent of a norm is a member of the antecedent of another norm (with the appropriate time). This ensures that the antecedent of the norm is a breach of the other one. The idea is the same for achievement obligations, but now detecting a violation is more complex.

$$\frac{\Gamma \Rightarrow^x \alpha \otimes^a_{t_\alpha} a^{t_a^s} \otimes^x_{t_a^e} \beta \quad \Delta, Oa^{t_a}, \sim a^{t'_a} \Rightarrow^z \delta \quad \{\Delta, \sim a^{t''_a} \Rightarrow^z \delta\}_{\forall t''_a : t_a^s < t'_a \le t''_a \le t_a^e}}{\Gamma, \Delta \Rightarrow^x \alpha \otimes^a_{t_\alpha} a^{t_a^s} \otimes^z_{t'_a} \delta} \otimes I_a.$$

The idea behind the introduction of a compensation for achievement obligation is that we have to determine that the obligation has not been fulfilled at a time before the deadline and for all instant greater or equal to it the complement is required. Essentially, the $\otimes I_a$ amounts to shortening the deadline for an achievement obligation.

$$\frac{a^1 \Rightarrow^a b^5 \otimes_{10} \bot \quad Ob^8, \neg b^8 \Rightarrow^p c^{15} \otimes_{15} \bot \quad \neg b^9 \Rightarrow^p c^{15} \otimes_{15} \bot \quad \neg b^{10} \Rightarrow^p c^{15} \otimes_{15} \bot}{a^1 \Rightarrow^a b^5 \otimes^p_8 c^{15} \otimes_{15} \bot}.$$

The first norm initially sets the deadline by when b has to be achieved to 10. The last n norms, in this case $n = 2$, have as premises the opposite of an obligation of the first norm covering the last n instant of the force period of the obligation and the same conclusion. This means that refraining to fulfill the obligation in the last n instants results in the same consequence. The last part is to assess that we have a violation. This is achieved by the second norm; here, we have the obligation in the antecedent (an achievement obligation is no longer in force in two cases: we are after the deadline or the content of the obligation has been achieved), thus the condition Ob^8 and $\neg b^8$ is to ensure that the obligation is still in force at the time, and the combination of the norms ensures that from now on not fulfilling the obligation results in the same compensation.

Subsumption The inference rules combine premises in such a way as the deontic content of at least one of them is included by the conclusion. Consequently, some original rules are no longer needed. To deal with this issue we introduce the notion of subsumption. A norm subsumes a second when the behaviour of the second norm (its compliance condition) is implied by the first one. Here below is an example illustrating this idea.

Example 6. *Consider the following norms:*

$$r \colon Invoice^t \Rightarrow^a Pay^t \otimes^p_{t+6} PayInterest^{t+7} \otimes_{t+7} \bot,$$
$$r' \colon Invoice^t, OPay^{t+6}, \neg Pay^{t+6} \Rightarrow^a PayInterest^{t+7} \otimes_{t+8} \bot.$$

The first norm says that after the seller sends the invoice, the buyer has the achievement obligation to pay within 7 days, otherwise immediately after the violation the buyer has to pay the principal plus the interest (punctual obligation to pay at $t+7$). According to the second norm, given the same set of circumstances Invoice at time t, if we have still the obligation on the seventh day after the invoice receipt date and the payment is not made yet, we have the achievement obligation to pay the interest by the eighth day. However, (a) the primary obligation of r' obtains when we have a violation of the primary obligation of r; (b) after the primary obligation of r is violated, complying with its secondary obligation entails complying with the primary obligation of r' (but not vice versa); (c) hence, r is more general than r', and so the latter can be discarded.

In what follows, Definition 10 characterizes the concept of subsumption that we have informally illustrated in Example 6. Since we need to check whether the compliance of a norm guarantees the compliance of another norm (the subsumed one), we provide below the following auxiliary definitions to establish (a) Definition 7: the modes with which the compliance conditions for one obligation covers the compliance conditions of another one; (b) Definition 8: when the compliance conditions of an $\otimes$-chain cover the compliance conditions of another $\otimes$-chain; (c) Definition 9: the conditions under which a literal belonging to an $\otimes$-chains is violated (indeed, subsumption allows to remove the norms whose applicability conditions require to violate another norm, while these conditions are encoded in the $\otimes$-chain of the subsuming norm).

Definition 7. *Let* $X, Y \in \{a, m, p\}$. *Then,* $Y \sqsubseteq X$ *iff*

(i) *if* $Y = a$, *then* $X \in \{a, m, p\}$;

(ii) *if* $Y = m$, *then* $X = m$;

(iii) *if* $Y = p$, *then* $X \in \{p, m\}$.

Definition 8. *Let*

$$\gamma = {}^{x_1}c_1{}^{t_{c_1}}_{t'_{c_1}} \otimes^{x_2} c_2{}^{t_{c_2}}_{t'_{c_2}} \otimes^{x_3}_{t'_{c_2}} \cdots \otimes^{x_j}_{t'_{c_{j-1}}} c_j{}^{t_{c_j}} \qquad \beta = {}^{y_1}b_1{}^{t_{b_1}} \otimes^{y_2} b_2{}^{t_{b_2}}_{t'_{b_1}} \otimes^{y_3}_{t'_{b_2}} \cdots \otimes^{y_k}_{t'_{b_{k-1}}} b_k{}^{t_{b_k}}$$

be $\otimes$-*chains. The* $\otimes$-*chain* γ *d-includes the* $\otimes$-*chain* β *iff*

1. $j = k$,

2. $c_i = b_i$,

3. $y_i \sqsubseteq x_i$;

4. (a) *if* $y_i = a$, *then* $t'_{c_i} \geq t_{b_i}$ *when* $x_i = m$, *otherwise* $t_{c_i} = t_{b_i}$ *and* $t'_{c_i} \leq t'_{b_i}$;

   (b) *if either* $y_i = m$ *or* $y_i = p$, *then* $t_{c_i} \leq t_{b_i}$ *and* $t'_{c_i} \geq t'_{b_i}$

where $1 \leq i \leq j, k$.

Definition 9. *Let* ${}^{x_1}c_1{}^{t_{c_1}}_{t'_{c_1}} \otimes^{x_2} c_2{}^{t_{c_2}}_{t'_{c_1}} \otimes^{x_3}_{t'_{c_2}} \cdots \otimes^{x_j}_{t'_{c_{j-1}}} c_j{}^{t_{c_j}}$ *be any* $\otimes$-*chain. For any* c_i, *where* $1 \leq i \leq j$, *a set* X *violates* c_i *iff*

1. *if* $x_i = a$, *then* $X = \{Oc_i^{t'_{c_i}}, \sim c_i^{t'_{c_i}}\}$;

2. *if* $x_i = m$ *or* $x_i = p$, *then* $X \subseteq \{\sim c_i^t | t_{c_i} \leq t \leq t'_{c_i}\}$.

Definition 10. *Let* $r_1 : \Gamma \Rightarrow \alpha \otimes \beta \otimes \gamma$ *and* $r_2 : \Delta \Rightarrow \delta$ *be two rules, where* α, β, γ, *and* δ *are* $\otimes$-*chains such that* $\gamma = {}^{z_1}c_1{}^{t_{c_1}}_{t'_{c_1}} \otimes^{z_2} c_2{}^{t_{c_2}}_{t'_{c_2}} \otimes^{z_3}_{t'_{c_2}} \cdots \otimes^{z_l}_{t'_{c_{l-1}}} c_l{}^{t_{c_l}}$.

Then r_1 *subsumes* r_2 *iff*

1. $\Gamma = \Delta$ and α d-includes δ; or

2. $\Gamma \cup X = \Delta$, where X violates all elements in α, and β d-includes δ; or

3. $\Gamma \cup Y = \Delta$, where Y violates all elements in β, and $\alpha \otimes^{z_1} c_1^{t_{c_1}} \otimes^{z_2}_{t'_{c_1}} c_2^{t_{c_2}} \otimes^{z_3}_{t'_{c_2}}$
   $\cdots \otimes^{z_n}_{t'_{c_{n-1}}} c_n^{t_{c_n}}$ d-includes δ, where $n \leq l$.

4 Proof Conditions

We introduce the conditions that allow us to determine whether an obligation is in force at time t (and the type of obligation as well). The problem reduces to determine whether a (temporalised) literal follows from a theory, in other terms whether we can derive the (temporalised) literal. In addition the conditions allow us to establish whether a theory has been complied with. As we discussed in the previous sections our language focuses on obligations as well as compensatory obligations. Thus compliance amount to check that violated and not compensated occurred (thus it is possible to have norms that have been violated, but they have been compensated for).

In Definition 1 we stated that a deontic expression extends an $\otimes$-chain with $\bot$ at the end. Thus effectively the penultimate element of a deontic expression identifies the 'last chance' to be compliant. After that the deontic expression results in a situation that cannot be complied with anymore. Hence, checking whether a theory is not compliant amounts to deriving $\bot$.

Definition 11. *A tagged literal is an expression $\#l$, where $\# \in \{+\partial, -\partial, +\partial^p, -\partial^p, +\partial^a, -\partial^a, +\partial^m, -\partial^m\}$.*

Definition 12. *A proof P is a sequence $P(1) \ldots P(n)$ of tagged literals satisfying the proof conditions given in Definitions 15, 16, 17 and 18. Each $P(i)$, $1 \leq i \leq n$ is called a line of the proof. Given a proof P, $P(1..n)$ denotes the first n lines of the proof.*

Definition 13. *A rule r is applicable at index i in a proof P at line $P(n+1)$ iff[4]*

1. $\forall a \in A(r)$:

   (a) *if $a \in TLit$, then $a \in F$, and*

---

[4]In the following, if $^{x_1} c_1^{t_{c_1}} \otimes^{x_2}_{t'_{c_1}} \cdots \otimes^{x_j}_{t'_{c_{j-1}}} c_j^{t_{c_j}} \otimes^{x_{j+1}}_{t'_j} \cdots \otimes_{t'_n} \bot$ is an $\otimes$-chain of length $n+1$, $mode(c_j) = x_j$, $start(c_j) = t_{c_j}$, and $end(c_j) = t'_{c_j}$.

(b)   i. *if $a = Ol^t$, then $+\partial l^t \in P(1..n)$,*

ii. *if $a = \neg Ol^t$, then $-\partial l^t \in P(1..n)$; and*

2. $\forall c_j \in C(r), 1 \leq j \leq i$:

(a) *$\forall t, start(c_j) \leq t \leq end(c_j) +\partial c_j^t \in P(1..n)$ and*

i. *if $mode(c_j) = p$, then $c_j^t \notin F$ or $\sim c_j^t \in F$, $start(c_j) = t$,*

ii. *if $mode(c_j) = a$, then $\forall t, start(c_j) \leq t \leq end(c_j)$, $c_j^t \notin F$ or $\sim c_j^t \in F$,*

iii. *if $mode(c_j) = m$, then $\exists t, start(c_j) \leq t \leq end(c_j)$, $c_j^t \notin F$ or $\sim c_j^t \in F$.*

Definition 14. *A rule r is discarded at index i in a proof P at line $P(n+1)$ iff*

1. $\exists a \in A(r)$:

(a) *if $a \in TLit$, then $a \in F$; or*

i. *if $a = Ol^t$, then $-\partial l^t \in P(1..n)$,*

ii. *if $a = \neg Ol^t$, then $+\partial l^t \in P(1..n)$; or*

2. $\exists c_j \in C(r), 1 \leq j \leq i$:

(a) *$\exists t, start(c_j) \leq t \leq end(c_j) -\partial c_j^t \in P(1..n)$ or*

i. *if $mode(c_j) = p$, then $c_j^t \in F$, $start(c_j) = t$,*

ii. *if $mode(c_j) = a$, then $\exists t, start(c_j) \leq t \leq end(c_j)$, $c_j^t \in F$,*

iii. *if $mode(c_j) = m$, then $\forall t, start(c_j) \leq t \leq end(c_j)$, $c_j^t \in F$.*

The intuition behind the definitions of applicable and discarded given above is as follows: Let us start from the conditions for a rule to be applicable at index i at line $P(n+1)$. First off all we have to ensure that all the elements of the body or antecedent of the rule are already provable. If the element is a plain literal, then it must be given as a fact, if it is a deontic literal, then if it is a positive deontic modality, we have to have that we have already proved it positively in the previous line of the derivation, if it is a negated deontic modality, it must be proved with $-\partial$ (or in other words it has been rejected, meaning that we failed to prove that it is obligatory). For the consequent of a rule, remember that the reading of the chain $^x a^t \otimes_{t'}^y b$ is that the obligation of b is in force when we have the violation of a. This means that Oa^{t^*} has to be in force ($+\partial a^{t^*}$) for the duration of the interval associated to a and $\neg a^{t^*}$ must hold (i.e., $a^{t^*} \in F$ or alternatively there is no evidence that a holds, $a^{t^*} \notin F$) at the appropriate times. The intuition for discarded is similar, for the antecedent we have to determine that we cannot trigger the rule: thus some of the antecedents are rejected. For the consequent, we check that at least one of the element preceding the element at index i is not violated, meaning that it is not obligatory; or if it is, it has been complied with.

In the proof conditions below we will simply use applicable/discarded at index i, instead of applicable/discarded at index i in the proof P at line $P(n+1)$.

All proof tags presented in the paper will be defined according the principle of strong negation [4]. According to it, the pairs of tag $+\#$ and $-\#$ are the strong negations of each other, where the strong negation is a function replacing/exchanging: $\forall$ and $\exists$, conjunctions and disjunctions, and 'applicable' and 'discarded'. For space reasons, we provide the definition of both the positive and negative proof tags for punctual obligation (i.e., $+\partial^p$ and $-\partial^p$), and only the positive definition of the proof tags for achievement and maintenance obligations; the corresponding negative proof tags can be derived using the above mentioned principle.

Definition 15 (Proof Conditions for $\pm\partial^p$).

If $P(n+1) = +\partial^p p^t$ then
(1) $\exists r \in R^p_\Rightarrow[p^t, i]$ r is applicable at index i and
(2) $\forall s \in R[\sim p^t, j]$, either
    (2.1) s is discarded at index j or
    (2.2) $\exists w \in R[p^t, k]$ such that w is applicable at k and $w \succ s$.

If $P(n+1) = -\partial^p p^t$ then
(1) $\forall r \in R^p_\Rightarrow[p^t, i]$ either r is discarded at i, or
(2) $\exists s \in R[\sim p^t, j]$ such that
    (2.1) r is applicable at index j and
    (2.2) $\forall w \in R[p^t, k]$ either w is discarded at k or $s \not\succ w$.

The proof conditions above are essentially a simple combination of the condition for $\otimes$ given in [19] and those for punctual obligation of [27]. To prove $+\partial^p a^t$, there must be a rule for a^t such that all the antecedents have to be provable, and for all elements preceding a^t in the head, we have to ensure that a violation occurred. This means that we have to examine the mode of the conclusions at indexes lower that the index of a^t, and then for a punctual obligation we have to see that the content of the obligation did not happen at t. We have two cases: the first is that we do not have a^t in the set of facts, and second we have the opposite, i.e., we have $\sim a^t$. For an achievement obligation we have to check that for all instants in the interval the same condition as that for a punctual obligation is satisfied, while for a maintenance obligation, a violation occurs when the condition holds for at least one instant of time in the interval. Condition (2.1) and (2.2) are the usual conditions of

Defeasible Logic, that is: we have to verify that rules for the opposite either do not fire (2.1), they are not applicable, or (2.2) they are defeated by applicable rules for the conclusion we want to prove.

Definition 16 (Proof Conditions for $\pm\partial^a$).

If $P(n+1) = +\partial^a p^t$ then
(1) $\exists r \in R^a_\Rightarrow[p^t, i]$ r is applicable at index i and
(2) $\forall s \in R[\sim p^t, j]$, either
    (2.1) s is discarded at index j or
    (2.2) $\exists w \in R[p^t, k]$ such that w is applicable at k and $w \succ s$; or
(3) $\exists x \in R^a_\Rightarrow[p^{t'}, i]$, $t' < t$, $end(p^{t'}) \geq t$ and
    (3.1) x is applicable at index i, and
    (3.2) $\forall y \in R[\sim p^{t''}, j]$, $t' \leq t'' < t$ either
        (3.2.1) y is discarded at j or
        (3.2.3) $\exists z \in R[p^{t''}, k]$, z is applicable at k and $z \succ y$; and
    (3.3) $\forall t''', t'' < t''' \leq t$, $p^{t'''} \notin F$.

The conditions for $+\partial^a p^t$ are similar to those for punctual obligations. The difference is that we have to consider persistence, clause (3). This means that we could have derived the obligation in the past, let us say at time t', and the obligation has not been terminated since them. We have two ways to terminate it: there is a rule for the opposite that is applicable between t and t' (3.2) see [27], or the obligation has been already fulfilled (3.3).

Definition 17 (Proof Conditions for $\pm\partial^m$).

If $P(n+1) = +\partial^m p^t$ then
(1) $\exists r \in R^m_\Rightarrow[p^t, i]$ r is applicable at index i and
(2) $\forall s \in R[\sim p^t, j]$, either
    (2.1) s is discarded at index j or
    (2.2) $\exists w \in R[p^t, k]$ such that w is applicable at k and $w \succ s$; or
(3) $\exists x \in R^m_\Rightarrow[p^{t'}, i]$, $t' < t$, $end(p^{t'}) \geq t$ and
    (3.1) x is applicable at index i, and
    (3.2) $\forall y \in R[\sim p^{t''}, j]$, $t' \leq t'' < t$ either
        (3.2.1) y is discarded at j or
        (3.2.3) $\exists z \in R[p^{t''}, k]$, z is applicable at k and $z \succ y$.

The conditions for maintenance obligations are the same as those for achievement obligation with the difference that fulfilling the obligation does not terminate it.

Definition 18 (Proof Condition for $\pm\partial$). *If $P(n+1) = +\partial p^t$, then either $+\partial^p p^t \in P(1..n)$, or $+\partial^a p^t \in P(1..n)$, or $+\partial^m p^t \in P(1..n)$.*

If $P(n+1) = -\partial p^t$, then $-\partial^p p^t \in P(1..n)$, and $-\partial^a p^t \in P(1..n)$, and $-\partial^m p^t \in P(1..n)$.

Definition 19. *Given a theory D, the universe of D (U^D) is the set of all the atoms occurring in D. The extension E^D of D is a structure (∂^+, ∂^-), where, for $X \in \{p, a, m\}$:*

$$\partial_D^+ = \{l^t : D \vdash +\partial^X l^t\} \qquad \partial_D^- = \{l^t : D \vdash -\partial^X l^t\}.$$

Example 7. *Consider the following theory:*

$$F = \{Invoice^t, \neg Pay^t, \neg Pay^{t+1}, PayInterest^{t+2}, Defective^t\}$$
$$R = \{r_1: Invoice^t \Rightarrow^a Pay^t \otimes_{t+1} \bot$$
$$\qquad r_2: Invoice^t, OPay^{t+1}, \neg Pay^{t+1} \Rightarrow^a PayInterest^{t+2} \otimes_{t+3} \bot,$$
$$\qquad r_3: Defective^t \rightsquigarrow \neg Pay^t\}$$
$$\succ = \{r_1 \succ r_3\}.$$

The first two norms basically describe the same situation of Example 6: the only difference is that here we have not yet applied any introduction rule for $\otimes$. r_3 states that, if the delivered good is defective, the customer is allowed not to pay. The facts trigger r_1, thus we derive the obligation to pay by $t+1$ (starting from t): also r_3 is triggered but is weaker than r_1. The obligation to pay is however not fulfilled by F. Since $\neg Pay^t \in F$, we obtain $OPay^{t+1}$ from r_1, which contributes to trigger r_2, thus obtaining the obligation to pay the interest by $t+3$ (starting from $t+2$). Since the obligation to pay by $t+1$ is not fulfilled, the extension of the theory D contains $\bot$: r_1 was not complied with.

5 Checking Compliance

If we work on the idea that a set of facts may fulfill a set of norms even when some of these norms are violated (but such violations are always compensated), then the following definition of compliance does not suffice:

Definition 20 (Theory compliance). *A Defeasible Theory D is compliant iff $\bot \notin \partial_D^+$.*

Definition 20 is very simple and exploits the basic properties of any temporalized obligations: since all $\otimes$-chains have $\bot$ as their last element, they have an ultimate deadline beyond which we derive $\bot$: this amounts to saying that after that deadline we state that it is impossible to compensate. Since the proof conditions for our logic establish that an obligation in an $\otimes$-chain is derived only if the previous obligations in that chain are violated, if we have $\bot$ in the positive extension of a theory, this means that there is at least one obligation whose violation cannot be compensated. For instance, if we consider Example 7, according to Definition 20 the theory D is not compliant because the theory extension contains $\bot$. However, such a theory should be considered compliant, since norm r_2, which provides a compensation for the violation of r_1, is indeed fulfilled.

Normalisation Process The inference rules $(\otimes I_p)$, $(\otimes I_m)$, and $(\otimes I_a)$ provide a method for representing the norms in a format that can be used to check the compliance of a theory. In fact, they allow for making explicit the hidden reparative relation between obligations. Once applied, the redundant rules can be removed. For instance, in Example 7 above, we could apply $(\otimes I_a)$ to r_1 and r_2 and obtain the new rule

$$r_3 : Invoice^t \Rightarrow^a Pay^t \otimes^a_{t+1} PayInterest^{t+2} \otimes_{t+3} \bot.$$

Once r_3 is obtained, since r_2 is subsumed by r_3, then r_2 is deontically redundant and can be removed from the theory.

Formally, this process is called normalisation of a theory. Before presenting the process, some auxiliary notions are needed: (a) Definition 21 identifies all the instances of inference rules we can obtain from a theory; (b) since such instances allow to introduce new norms, we should establish when these norms can inherit the same strength qualifications (via $\succ$) of previous norms; we should also remove redundant norms and norm priorities (Definitions 22 and 23); (c) Definition 24 introduces the deductive closure of a theory under the inference conditions for $\otimes$.

Definition 21. *Let $D = (F, R \succ)$ be any defeasible theory. Any instance I of the inference rules $(\otimes I_p)$, $(\otimes I_m)$, and $(\otimes I_a)$ is based on D if each of the premises r_i and r_j of I is either (a) in R (in which case, the instance is* rooted*), or (b) is the conclusion of another instance of the inference rules $(\otimes I_p)$, $(\otimes I_m)$, and $(\otimes I_a)$ based on D.*

The instances of the inference rules $(\otimes I_p)$, $(\otimes I_m)$, and $(\otimes I_a)$ based on D are also called D-$\otimes$-instances.

Definition 22. *Let $D = (F, R \succ)$ be any defeasible theory. The superiority relation $\succ^\infty = \cup^\infty_{i=1} \succ_i$ is recursively defined as follows:*

- *$\succ_0 = \succ \cup \{(j,k) | j$ (or k) is the conclusion of a rooted D-$\otimes$-instance such that $k \in R$ (or $j \in R$) and, for any $i \in R$, $(i, k) \in \succ$ (or $(j, i) \in \succ$) \};$*

- $\succ_{i+1} = \succ_i \cup \{(j,k) \mid j \ (or \ k) \ is \ the \ conclusion \ of \ a \ D\text{-}\otimes\text{-instance such that}$
  $(i,k) \in \succ_i \ (or \ (j,i) \in \succ_i) \}$.

The relation $\succ^\infty$ is called the D-saturation of $\succ$.

Definition 23. *Let $D = (F, R \succ)$ be any defeasible theory. Let $\mathcal{S}$ be an operation over D defined as follows: if $\Pi = \{r \mid r \in R, \exists r' \in R : r' \ subsumes \ r\}$, then*

$$\mathcal{S}(D) = \begin{cases} D' & where \ D' = (F, R', \succ') \ such \ that \\ & R' = R - \Pi \ and \\ & \succ' = \succ^\infty - \{(x,y) \in \succ \mid either \ x \in \Pi \ or \ y \in \Pi\} \\ D & otherwise \end{cases} \quad (10)$$

Definition 24. *If $D = (F, R \succ)$ is any defeasible theory, let $\vdash_\otimes$ be the consequence relation defined by the inference rules $(\otimes I_p)$, $(\otimes I_m)$, and $(\otimes I_a)$. The closure $(D, \vdash_\otimes)$ of D under $\vdash_\otimes$ is a theory $D' = (F, R', \succ')$ where (a) R' is the smallest set containing all elements of R and the conclusions of all D-$\otimes$-instances; (b) $\succ'$ is the D-saturation of $\succ$.*

Definition 25 (Theory normalisation). *The normalisation D^∞ of a theory D is a theory recursively obtained as follows: (a) $D_0 = D$, (b) $D_{i+1} = \mathcal{S}(D_i, \vdash_\otimes)$.*

Example 8. *The inference rules and the rule removal via subsumption must be done several times in the appropriate order. The normalised theory is the fixed-point of the above constructions. At each step of the the procedure we have to first apply the inference rules for $\otimes$ and then the subsumption: suppose we have a theory containing the following three norms*

$r_1: f^{t_f} \Rightarrow^p a^{t_a} \otimes^p_{t_a} g^{t_g} \otimes_{t_g} \bot,$ $\qquad r_2: e^{t_e} \Rightarrow^p a^{t_a} \otimes^p_{t_a} b^{t_b} \otimes^p_{t_b} c^{t_c} \otimes^p_{t_c} d^{t_d} \otimes_{t_d} \bot,$
$r_3: e^{t_e}, \neg a^{t_a}, \neg b^{t_b} \Rightarrow^p c^{t_c} \otimes_{t_c} \bot.$

The normalisation process would consist here in a single cycle leading to apply (i) $(\otimes I_p)$ to r_1 and r_3, thus producing

$$r_4: e^{t_e}, f^{t_f}, \neg b^{t_b} \Rightarrow^p a^{t_a} \otimes^p_{t_a} c^{t_c} \otimes_{t_c} \bot;$$

(ii) subsumption and remove r_3. Notice that also r_2 subsumes r_3. However, if we apply subsumption first on this basis we have to delete r_3 and r_4 would be no longer derivable from r_1 and r_3 alone.

After a theory is normalised, Definition 20 can be safely applied, as all redundant rules are removed and all hidden reparative connections between obligations are made explicit.

Before proving some significant properties of the normalisation process, let us state some preliminary lemmas and introduce some auxiliary concepts:

Lemma 1. *For any defeasible theory D, its normalisation D^∞ is also a defeasible theory.*

The proof of this first lemma is straightforward and is omitted.

Lemma 2. *Any instance of the inference rules $(\otimes I_p)$, $(\otimes I_m)$, and $(\otimes I_a)$ is such that at least one premise is subsumed by the conclusion.*

Proof. Consider the inference rule $(\otimes I_p)$ and let us introduce labels to denote the premises and the conclusion:

$$\frac{p_1 : \Gamma \Rightarrow^x \alpha \otimes_{t_\alpha}^p b^{t_b} \otimes_{t_b}^Y \gamma \qquad p_2 : \Delta, \neg b^{t_b} \hookrightarrow^z \delta}{c : \Gamma, \Delta \Rightarrow^x \alpha \otimes_{t_\alpha}^p b^{t_b} \otimes_{t_b}^z \delta} \otimes I_p.$$

The schema guarantees that, in all instances, c subsumes p_2 since, $\Gamma \cup \Delta \cup \{\neg b^{t_b}\}$ violates b^{t_b} in the head of c, thus satisfying condition 3 in Definition 10.

Consider $(\otimes I_m)$:

$$\frac{p_1' : \Gamma \Rightarrow^x \alpha \otimes_{t_\alpha}^m b^{t_s} \otimes_{t_e}^y \beta \qquad p_2' : \Delta, \Theta \Rightarrow^z \delta}{c' : \Gamma, \Delta \Rightarrow^x \alpha \otimes_{t_\alpha}^m b^{t_s'} \otimes_{t_e'}^z \delta} \otimes I_m$$

where $\Theta = \{\sim b^{t'} : t_s < t_s' \leq t' \leq t_e' \leq t_e\}$. Since Θ contains $\sim b^{t'}$, namely, $\sim b$ holds at least one time between t_s' and t_e', then $\Gamma \cup \Delta \cup \Theta$ violates b^{t_s} in the head of c', thus satisfying condition 3 in Definition 10.

Finally, let us examine $(\otimes I_a)$:

$$\frac{p_1'' : \Gamma \Rightarrow^x \alpha \otimes_{t_\alpha}^a a^{t_a^s} \otimes_{t_e^a}^x \beta \quad p_2'' : \Delta, O a^{t_a'}, \sim a^{t_a'} \Rightarrow^z \delta \quad p_{3,\ldots,n}'' : \{\Delta, \sim a^{t_a''} \Rightarrow^z \delta\}}{c'' : \Gamma, \Delta \Rightarrow^x \alpha \otimes_{t_\alpha}^a a^{t_a^s} \otimes_{t_a'}^z \delta} \otimes I_a$$

where $\forall t_a'' : t_a^s < t_a' \leq t_a'' \leq t_a^e$. Consider in the body of p_2'' the set of antecedents $\{O a^{t_a'}, \sim a^{t_a'}\}$: this set violates $a^{t_a^s}$ in the head of c'', thus satisfying condition 3 in Definition 10. Also, $\sim a^{t_a''}$ in the antecedents of $p_3'', \ldots p_n''$ violates $a^{t_a^s}$ in the head of c'', thus satisfying, too, condition 3 in Definition 10. $\square$

Definition 26. *Let $D = (F, R \succ)$ be any defeasible theory. We associate D with an operator Θ_D defined as follows[5]:*

$$\Theta_D(\mathbf{Rul}, \mathbf{Sup}) = (\mathbf{Rul'}, \mathbf{Sup'}) \qquad where$$
$$\mathbf{Rul'} = R \cup \{r | \exists I : I \text{ is a } \mathbf{D}\text{-}\otimes\text{-instance s.t. } r \text{ is the conclusion of } I$$
$$and \ \mathbf{D} = (F, R \cup \mathbf{Rul}, \succ \cup \mathbf{Sup})\}$$
$$\mathbf{Sup'} = \succ \cup \{(j, k) | \ j \ (or \ k) \text{ is the conclusion of a } \mathbf{D}\text{-}\otimes\text{-instance}$$
$$such that \ (i, k) \in \mathbf{Sup} \ (or \ (j, i) \in \mathbf{Sup})\}$$

The set of 2-tuples forms a complete lattice under pointwise containment ordering, where $\bot = (\emptyset, \emptyset)$ as its least element. The least upper bound operation is the pointwise union $\sqcup$. The sequence of of repeated applications of Θ_D to $\bot$, i.e. the Kleene sequence of Θ_D is defined as follows:

- $\Theta_D \uparrow 0 = \bot$;

- $\Theta_D \uparrow (n+1) = \Theta_D(\Theta_D \uparrow n)$;

- $\Theta_D \uparrow n = \bigcup_{x<n} \Theta_D \uparrow x$ if n is a limit ordinal.

Lemma 3. Θ_D *is monotonic and the Kleene sequence from $\bot$ is increasing. Hence, the limit $F = (\mathbf{Rul}_F, \mathbf{Sup}_F)$ of all finite elements in the sequence exists, and Θ_D has a least fixpoint $L = (\mathbf{Rul}_L, \mathbf{Sup}_L)$. Since any theory D is a finite, $F = L$.*

Proof. Let us prove by induction that Θ_D is pointwise monotonic. The other properties follow as standard results from set theory.

Inductive base. The inductive base is trivial, since we have $(\emptyset, \emptyset)$.

Inductive step. We have two cases: (i) $\mathbf{Rul}_{n-1} \subseteq \mathbf{Rul}_n$ and (ii) $\mathbf{Sup}_{n-1} \subseteq \mathbf{Sup}_n$.

Case (i). Let us examine why $r \in \mathbf{Rul}_n$. If $r \in R$, then $\forall x : r \in \mathbf{Rul}_x$ and so $r \in \mathbf{Rul}_{n+1}$. Otherwise, there is a $\mathbf{D}_{n-1}\text{-}\otimes$-instance I s.t. its premises $p_1, \ldots p_m \in \mathbf{Rul}_{n-1}$ and r is the conclusion of I. By inductive hypothesis, $p_1, \ldots p_m \in \mathbf{Rul}_n$, so $r \in \mathbf{Rul}_{n+1}$.

Case (ii). Let us examine why $(j, k) \in \mathbf{Sup}_n$. Again, if $(j, k) \in \succ$, then $\forall x : (j, k) \in \mathbf{Sup}_x$ and so $(j, k) \in \mathbf{Sup}_{n+1}$. Otherwise: (a) there is a $\mathbf{D}_{n-1}\text{-}\otimes$-instance I s.t. its conclusion is either j or k (suppose, without lack of generality, that the conclusion of I is j); (b) one of the premises p of I is such that $(p, k) \in \mathbf{Sup}_{n-1}$. By inductive hypothesis, (a) and (b) should hold also for $\mathbf{Sup}_n$, so $(j, k) \in \mathbf{Sup}_{n+1}$. $\square$

We can now prove the following result:

---

[5]This construction recalls the one used in [5] to define theory extensions.

Theorem 4. *The normalisation D^∞ of any defeasible theory D exists and is unique.*

Proof. The result follows from that fact any defeasible theory contains only finitely many rules and each rule has finitely many elements. Since the construction of D^∞ ensures to remove at each step the subsumed rules and, in each instance of $(\otimes I_p)$, $(\otimes I_m)$, and $(\otimes I_a)$ at least one premise is subsumed by the conclusion, then Lemmas 2 and 3 guarantee that $R^\infty \subseteq \mathbf{Rul}_L$. Analogously, since the definition of $\succ^\infty$ strictly depends on how R^∞ is built (all redundant rule priorities are removed when rules are removed via subsumption), then $\succ^\infty \subseteq \mathbf{Sup}_L$. Hence, $(F, R^\infty, \succ^\infty)$ exists and is unique. $\square$

Definition 27. *Let $D = (F, R, \succ)$ any defeasible theory. We say that $\succ$ is consistent iff $\forall x, y \in R : (x, y), (y, x) \notin \succ$.*

The following result holds:

Theorem 5. *For any defeasible theory $D = (F, R, \succ)$, if $\succ$ is consistent, then the normalisation $D^\infty = (F, R^\infty, \succ^\infty)$ is such that $\succ^\infty$ is consistent.*

Sketch. The proof is by induction on the construction of D^∞ and is straightforward: we omit the details. Just notice that, by construction, for each D_n leading to D^∞, (i) the operation in Definition 22 adds new elements in $\succ_n$ when a new rule in D_n inherits a superiority that applied to another rule which appeared as a premise of an instance of $(\otimes I_p)$, $(\otimes I_m)$, or $(\otimes I_a)$ in D_{n-1}, and (ii) the operation in Definition 23 removes redundant elements in $\succ_{n-1}$, because they applied to rules that are no longer in R_n. Hence, the only case where, given $(r_i, r_j) \in \succ_{n-1}$, we could add (r_j, r_i) to $\succ_n$ without removing (r_i, r_j) is when r_i (or, respectively, r_j) is the conclusion of at least two instances of $(\otimes I_p)$, $(\otimes I_m)$, or $(\otimes I_a)$, while r_j (or, respectively, r_i) is not the premise of those two instances. More precisely, we should have the following case: given $r_j \in R_n$

$$\frac{r_x \quad r_y}{r_i} \quad (r_y, r_j) \in \succ_n \qquad \frac{r_w \quad r_z}{r_i} \quad (r_j, r_z,) \in \succ_n .$$

However, a simple inspection of the definition of $(\otimes I_p)$, $(\otimes I_m)$, and $(\otimes I_a)$ shows that one rule r_i cannot be derived by different sets of premises and different inference rules. Hence, the only admissible case is when $r_w = r_x$ and $r_z = r_y$, but this would mean that $(r_j, r_i), (r_i, r_j) \in \succ_{n-1}$, contrary to the assumption in the proof (i.e., by inductive hypothesis with respect to the inductive step n: recall that we build D^∞ starting with a theory D where $\succ$ is assumed to be consistent). $\square$

6   Summary and Related Work

This paper extends the logic of violation proposed by [23] with time. This extension introduces a temporal dimension to the language saying when a norm produces its normative effects, or in other terms when the obligation (or, in general the normative position) corresponding to the normative effect of the norm is in force. An immediate consequence of the extended language is that it is possible to investigate the 'lifecycle' of obligations, and more precisely if there are deadlines to comply with an obligation. The extension is done to properly deal with the concept of legal compliance. To do this we argue that we have to handle different types of temporalised legal obligations and devise a normalisation procedure for making hidden conditions and reparative chains explicit. One open research issue is to investigate the complexity of this procedure, which requires, several times and in the appropriate order, to apply the inference rules for $\otimes$ and to remove redundant norms. Related to this is how to implement the framework we present in an efficient way. Computing the extension of the temporalised defeasible theory can be computed in time linear the size of a theory and the time points present in the theory [26] and so is the computation of the extension of a (non temporal) defeasible theory with compensation chains [22]. We expect that the complexity of computing the extension is polynomial. Consider the transformation outlined below that takes a rule

$$r: a_1, \ldots, a_n \Rightarrow^{x_1} c_1^{t_1} \otimes_{t_1'}^{x_2} c_2^{t_2} \otimes_{t_2'} \cdots \otimes_{t_{m-1}'}^{x_m} c_m^{t_m} \otimes_{t_m'} \bot \tag{11}$$

and produces following rules in the temporal defeasible logic of [26]

$$a_1, \ldots, a_n \Rightarrow^\tau \mathbf{r} \tag{12}$$

$$a_1, \ldots, a_n \Rightarrow^\tau c_1^{t_1, t_1'} \qquad a_1, \ldots, a_n \Rightarrow^\tau \sim \bar{c}_1^{t_1, t*} \tag{13}$$

$$\mathbf{r}, c_i^{t_i, t_i'} \Rightarrow_O^\pi c_i^{t_i} \qquad \mathbf{r}, c_i^{t_i, t_i'}, \mathbf{t}_{t_i'} \rightsquigarrow_O \sim c_i^{t_i'} \tag{14}$$

$$\Rightarrow \mathbf{naf} c_i^t \qquad c_i^t \Rightarrow \neg \mathbf{naf} c_i^t \qquad \text{for } t_i \le t \le t_i' \tag{15}$$

where $\mathbf{r}$, $\mathbf{naf} c_i$, $c_i^{t,t'}$, $\bar{c}_i^{t_i, t_i*}$ and $\mathbf{t}_t$ are new literals corresponding, respectively, to the rule r, the literal c_i, and the interval of force of c_i determined by r, that the complement of c_i is not in force in an interval starting from t_i and the clock event for the time instant t.

In addition we have the following rules depending on the mode of the literals in the head of the rule. If $x_i = m$, then, for $1 < i \le m$ and $t_i \le t \le t_i'$ we have

$$\mathbf{r}, c_i^{t_i, t_i'}, O c_i^t, \mathbf{naf} c_i^t \Rightarrow^\tau c_{i+1}^{t_{i+1}, t_{i+1}'} \tag{16}$$

$$\mathbf{r}, c_i^{t_i, t_i'}, O c_i^t, \mathbf{naf} c_i^t \Rightarrow^\tau \sim \bar{c}_{i+1}^{t_{i+1}, t*} \tag{17}$$

otherwise, $1 < i \leq m$, we have

$$\mathbf{r}, c_i^{t_i, t_i'}, Oc_i^{t_i}, \mathbf{naf} c_i^{t_i}, \ldots Oc_i^{t_i'}, \mathbf{naf} c_i^{t_i'} \Rightarrow^\tau c_{i+1}^{t_{i+1}, t_{i+1}'} \tag{18}$$

$$\mathbf{r}, c_i^{t_i, t_i'}, Oc_i^{t_i}, \mathbf{naf} c_i^{t_i}, \ldots Oc_i^{t_i'}, \mathbf{naf} c_i^{t_i'} \Rightarrow^\tau \sim \bar{c}_{i+1}^{t_{i+1}, t*} \tag{19}$$

The intuition of the transformation above is as follows: rule (12) indicates that the rule r is applicable; then the rules in (13) indicates that if the rules is applicable the obligation of c_1 (the first element in the chain in the head fo the rule) is in force from t_1 to t_1', but this means that the complement of c_1 ($\bar{c}_1$) cannot be in force in any interval starting from t_1. Next, the rules in (14) specify that if the rule is applicable and the interval of force holds then, the obligation of c_i enters in force (at the start of the interval), and, since the first rule is persistent, the rule remains in force until it is terminated (explicitly by a rule for the opposite) or by its deadline. The rule for the deadline is given by the second rules, and the deadline is indicated by the clock literal $\mathbf{t}_{t_i'}$. The rules in (15) are meant to represent negation as failure for literal c_i. Notice that the rules in (13) are specific for the first element of a chain, for successive elements in the chain we have to use (16) and (17) in case of the i-th element in a chain refers to a maintenance obligation, and (18) and (19) for the other types of obligation. The difference from the case for the first element of a chain is that we have to determine (1) that obligation for the previous element holds and (2) that the obligation has been violated. For the violation we use the negation as failure (for one instant in the interval for maintenance, for all elements in the interval for achievement). The idea of using negation as failure corresponds to have the condition of $-\partial c_i$ in the proof conditions.

It is clear that the transformation above is linear in the number of rules, literals appearing in the rules, and time instants, thus the computation of extension of a theory is polynomial. However, the complexity of the normalisation procedure is not know (though, the upper bound is limited by the number of permutations of the literals in the head of the rules). Nevertheless, from a practical point of view we do not expect the normalisation process to be a concern, given that in real normative systems the number of possible combinations that are feasible is limited (and in general it is specified in normative systems what are the penalties for specific obligations).

The literature on norm compliance in MAS is large (see, e.g., [8, 16, 33, 17, 1, 18, 28, 6, 32, 13]). However, to the best of our knowledge no work in the field has so far attempted to model *legal* compliance pertaining to realistic systems where complex norm-enforcement mechanisms such as reparative chains are combined with a rich ontology of obligations as the one described here. In the literature on deontic logic, besides a few exceptions like [9], the research has mostly devoted extensive, but separate, efforts to the role of time for dealing with CTDs (since the seminal [39]) and on logical systems for modeling the concept deontic preference and CTDs (for an overview, [37]). This paper combines the two perspectives: in this sense, it also inherits from [23] the advantage of avoiding the most well-known CTD paradoxes. In this sense, [9] shares with our paper the same general view, but time is captured there at the semantic level and the language does not explicitly handle timestamps. Another approach similar to our using timestamps, and somehow inspired by Event Calculus is [35] with construction to check if norms have been complied with or violated, however such an approach does not consider reparation clauses. Similarly, [15] provides a survey of approaches to solve conflicts among norms, including approaches based on time and temporal logics, but the majority of such approaches do not consider compensatory norms.

Combination of time and norms are not novel, as many combinations of temporal (or tense) logic and deontic logic have been investigated. However, temporal logic cannot handle specific times (or timestamps). Typically these logics can express the temporal relationships between events (represented by propositions), or the relationships between states. A possible solution to obviate this is to consider hybrid logics using nominals to capture nominals [36]. A nominal represents a proposition true only in one possible world. A temporal nominal represents a particular instant of time. In most temporal logic it is possible to model branching of time, and the meaning of nominals is not clear in this kind of situations (is the world corresponding to a nominal the same in all the branches, or we have different copies of the same instant of time?). On the other hand timestamps (and events) have been used in the Event Calculus. Event Calculus has been used to model the interaction between norms and time (see, e.g., [34, 12]). However, Event Calculus is a dialect of first-order logic and Herrestad [31] has shown that these types of logic are not suitable to model normative reasoning in presence of violations and reparative clause. In addition systems to model temporal norms based on (standard) Event Calculus are not able to handle delays between the trigger and when the obligations enter in force (and similar temporal aspects) [2]. A deontic extension of Event Calculus sharing many features with the present work, apart the normalisation process, and the handling of conflicts provided by the defeasible logic, and addressing the shortcomings of other Event Calculus approaches has been developed in [29].

Acknowledgements

The paper is an extended and revised version of [25] presented at CLIMA XII. This work was partially supported by EU H2020 research and innovation programme under the Marie Skłodowska-Curie grant agreement No. 690974 for the project MIREL: MIning and REasoning with Legal texts.

We thank the anonymous referees for their valuable comments that help improving the paper.

References

[1] M. Alberti, M. Gavanelli, E. Lamma, F. Chesani, P. Mello, and P. Torroni. Compliance verification of agent interaction: a logic-based software tool. *Applied Artificial Intelligence*, 20(2-4):133–157, 2006.

[2] W. Alrawagfeh. Norm Representation and Reasoning: A Formalization in Event Calculus. In G. Boella, E. Elkind, B. Savarimuthu, F. Dignum, and M. Purvis, editors, *Proceedings of the 16th International Conference on Principles and Practice of Multi-Agent Systems (PRIMA 2013)*, pages 5–20. Springer, 2013.

[3] G. Andrighetto, G. Governatori, P. Noriega, and L. W. N. van der Torre, editors. *Normative Multi-Agent Systems*, volume 4 of *Dagstuhl Follow-Ups*. Schloss Dagstuhl - Leibniz-Zentrum fuer Informatik, 2013.

[4] G. Antoniou, D. Billington, G. Governatori, and M. Maher. A flexible framework for defeasible logics. In *Proc. AAAI-2000*. AAAI Press, 2000.

[5] G. Antoniou, D. Billington, G. Governatori, and M. Maher. Embedding defeasible logic into logic programming. *Theory and Practice of Logic Programming*, 6(6):703–735, 2006.

[6] G. Boella, J. Broersen, and L. van der Torre. Reasoning about constitutive norms, counts-as conditionals, institutions, deadlines and violations. In *PRIMA*. Springer, 2008.

[7] G. Boella and L. van der Torre. Fulfilling or violating obligations in multiagent systems. In *Procs. IAT04*, 2004.

[8] E. Bou, M. López-Sánchez, and J. A. Rodríguez-Aguilar. Adaptation of autonomic electronic institutions through norms and institutional agents. In *Proc. ESAW'06*. Springer, 2006.

[9] J. Broersen and L. van der Torre. Conditional norms and dyadic obligations in time. In *Proc. ECAI 2008*. IOS Press, 2008.

[10] J. M. Broersen. Strategic deontic temporal logic as a reduction to atl, with an application to chisholm's scenario. In L. Goble and J. C. Meyer, editors, *Proc DEON 2006*, volume 4048 of *Lecture Notes in Computer Science*, pages 53–68. Springer, 2006.

[11] J. Carmo and A. Jones. Deontic logic and contrary to duties. In D. Gabbay and F. Guenther, editors, *Handbook of Philosophical Logic, 2nd Edition*. Kluwer, 2002.

[12] F. Chesani, P. Mello, M. Montali, and P. Torroni. Representing and Monitoring Social Commitments using the Event Calculus. *Autonomous Agents and Multi-Agent Systems*, 27(1):85–130, 2013.

[13] M. Dastani, G. Governatori, A. Rotolo, and L. van der Torre. Programming cognitive agents in defeasible logic. In G. Sutcliffe and A. Voronkov, editors, *12th International Conference on Logic for Programming, Artificial Intelligence, and Reasoning*, volume 3835 of *LNAI*, pages 621–636, Heidelberg, 2005. Springer.

[14] M. Dastani, D. Grossi, J.-J. C. Meyer, and N. Tinnemeier. Normative multi-agent programs and their logics. In R. Bordini, M. Dastani, J. Dix, and A. E. Fallah-Seghrouchni, editors, *Programming Multi-Agent Systems*, number 08361 in Dagstuhl Seminar Proceedings, Dagstuhl, Germany, 2008. Schloss Dagstuhl - Leibniz-Zentrum fuer Informatik, Germany.

[15] J. S. dos Santos, J. de Oliveira Zahn, E. A. Silvestre, V. T. da Silva, and W. W. Vasconcelos. Detection and resolution of normative conflicts in multi-agent systems: a literature survey. *Autonomous Agents and Multi-Agent Systems*, 31(6):1236–1282, 2017.

[16] M. Esteva, B. Rosell, J. A. Rodríguez-Aguilar, and J. L. Arcos. Ameli: An agent-based middleware for electronic institutions. In *Proc. AAMAS 2004*. ACM, 2004.

[17] R. A. Flores and B. Chaib-draa. Modelling flexible social commitments and their enforcement. In *Proc. Engineering Societies in the Agents World V*. Springer, 2004.

[18] D. Gaertner, A. Garcia-Camino, P. Noriega, J.-A. Rodriguez-Aguilar, and W. Vasconcelos. Distributed norm management in regulated multiagent systems. In *Proc. AAMAS '07*. ACM, 2007.

[19] G. Governatori. Representing business contracts in RuleML. *International Journal of Cooperative Information Systems*, 14(2-3):181–216, 2005.

[20] G. Governatori. Thou shalt is not you will. In K. Atkinson, editor, *Proceedings of the Fifteenth International Conference on Artificial Intelligence and Law*, pages 63–68, New York, 2015. ACM.

[21] G. Governatori, J. Hulstijn, R. Riveret, and A. Rotolo. Characterising deadlines in temporal modal defeasible logic. In M. A. Orgun and J. Thornton, editors, *20th Australian Joint Conference on Artificial Intelligence*, volume 4830 of *Lecture Notes in Artificial Intelligence*, pages 486–496, Heidelberg, 2007. Springer.

[22] G. Governatori, F. Olivieri, A. Rotolo, and S. Scannapieco. Computing strong and weak permissions in defeasible logic. *Journal of Philosophical Logic*, 42(6):799–829, 2013.

[23] G. Governatori and A. Rotolo. Logic of violations: A Gentzen system for reasoning with contrary-to-duty obligations. *Australasian Journal of Logic*, 4:193–215, 2006.

[24] G. Governatori and A. Rotolo. A conceptually rich model of business process compliance. In S. Link and A. Ghose, editors, *7th Asia-Pacific Conference on Conceptual Modelling*, volume 110 of *CRPIT*, pages 3–12. ACS, 2010.

[25] G. Governatori and A. Rotolo. Justice delayed is justice denied: Logics for a temporal account of reparations and legal compliance. In J. Leite, P. Torroni, T. Ågotnes, G. Boella, and L. van der Torre, editors, *12th International Workshop on Computational Logic in Multi-Agent Systems*, volume 6814, pages 364–382, Heidelberg, 2011. Springer.

[26] G. Governatori and A. Rotolo. Computing temporal defeasible logic. In L. Morgenstern, P. S. Stefaneas, F. Lévy, A. Wyner, and A. Paschke, editors, *RuleML 2013*, volume 8035 of *Lecture Notes in Computer Science*, pages 114–128. Springer, 2013.

[27] G. Governatori, A. Rotolo, and G. Sartor. Temporalised normative positions in defeasible logic. In *10th International Conference on Artificial Intelligence and Law (ICAIL05)*, pages 25–34, 2005.

[28] D. Grossi, H. Aldewereld, and F. Dignum. Ubi lex, ibi poena: Designing norm enforcement in e-institutions. In *In Coordination, Organizations, Institutions, and Norms in Multi-Agent Systems II*. Springer, 2006.

[29] M. Hashmi, G. Governatori, and M. T. Wynn. Modeling Obligations with Event-Calculus. In *Proceedings of 8th International Web Rule Symposium (RuleML 2014)*, pages 296–310, Prague, Czech Republic, Aug. 2014. Springer.

[30] M. Hashmi, G. Governatori, and M. T. Wynn. Normative requirements for regulatory compliance: An abstract formal framework. *Information Systems Frontiers*, 18(3):429–455, 2016.

[31] H. Herrestad. Norms and formalization. In *ICAIL*, pages 175–184, 1991.

[32] J. F. Hübner, O. Boissier, and R. Bordini. From organisation specification to normative programming in multi-agent organisations. In *CLIMA XI*, 2010.

[33] F. López y López, M. Luck, and M. d'Inverno. Constraining autonomy through norms. In *Proc. AAMAS '02*. ACM, 2002.

[34] R. H. Marín and G. Sartor. Time and norms: a formalisation in the event-calculus. In *ICAIL*, pages 90–99, 1999.

[35] Z. Shams, M. D. Vos, J. Padget, and W. W. Vasconcelos. Practical reasoning with norms for autonomous software agents. *Eng. Appl. of AI*, 65:388–399, 2017.

[36] C. Smith, A. Rotolo, and G. Sartor. Temporal reasoning and mas. In *SNAMAS 2010*, 2010.

[37] J. Van Benthem, D. Grossi, and F. Liu. Deontics = betterness + priority. In *Proc. DEON'10*. Springer, 2010.

[38] L. van der Torre, G. Boella, and H. Verhagen, editors. *Normative Multi-agent Systems*, Special Issue of *JAAMAS*, vol. 17(1), 2008.

[39] J. van Eck. A system of temporally relative modal and deontic predicate logic and its philosophical applications. *Logique et Analyse*, 25:339–381, 1982.

Received 27 December 2018

Rights and Punishment: The Hohfeldian theory's applicability and morals in understanding criminal law

Réka Markovich *

Computer Science and Communications Research Unit, University of Luxembourg
Department of Business Law, Budapest University of Technology and Economics
Department of Logic, Eötvös Loránd University
`reka.markovich@uni.lu`

Abstract

It is often suggested that criminal law is a limitation of the general applicability of the Hohfeldian theory of rights and duties and their correlativity. The first part of this paper shows how a formalization of normative positions and a clear understanding of how rights work refuses this thesis. This part leads us to the notion of sanction. The second part of the paper presents an analysis of sanction in terms of rights and duties in order to resolve the seemingly paradoxical situation of the legal systems in which one has the right to escape from the prison.

1 The Hohfeldian Theory and Its Alleged Limitations

Wesley Newcomb Hohfeld's analysis on the different types of rights and duties (Fundamental Legal Conceptions as Applied in Judicial Reasoning, 1913, 1917) is highly influential and often discussed in analytical legal theory, and it is considered as a fundamental theory in AI&Law and normative multi-agent systems. Yet a century later, the formalization of this theory remains, in various ways, unresolved. The classical formalization developed by Stig Kanger and Lars Lindahl [8, 9, 10, 12] concentrating on computational features of duties has some well-known and documented limitations, for example in Makinson [14] and Sergot [18]. The theory of

---

*Support provided by the research project K-116191 of the Hungarian Scientific Research Fund is gratefully acknowledged. The research reported in this paper was supported by the Higher Education Excellence Program of the Ministry of Human Capacities in the frame of Artificial Intelligence research area of Budapest University of Technology and Economics (BME FIKP-MI/FM).

Hohfeld and its possible formal approaches have been commented on by many logicians and legal theorists since then. Some—famously Hart in [4], but also Lyons et al. [13], Sreenivasan [20], Kocourek [11]—argue against the general validity of the Hohfeldian model: they say it might be valid in (some areas of) private law where there are clearly (two) counterparties as the theory considers, but branches of law where we deal with undirected, general or absolute duties/prohibitions, or rights where there is no clear (one) other party obliged realize an obvious limitation of the theory's applicability. In the following I'm going to argue that if we understand properly the Hohfeldian system, we might incline to admit that whatever relational his system is, areas of law like criminal law—where we consider general obligations (prohibitions)—*do not* serve as a counterargument of the theory's general validity.

The well-known system of the correlative pairs of rights (upper line) and duties (lower line) Hohfeld built in [7] can be reconstructed in the following diagram:[1]

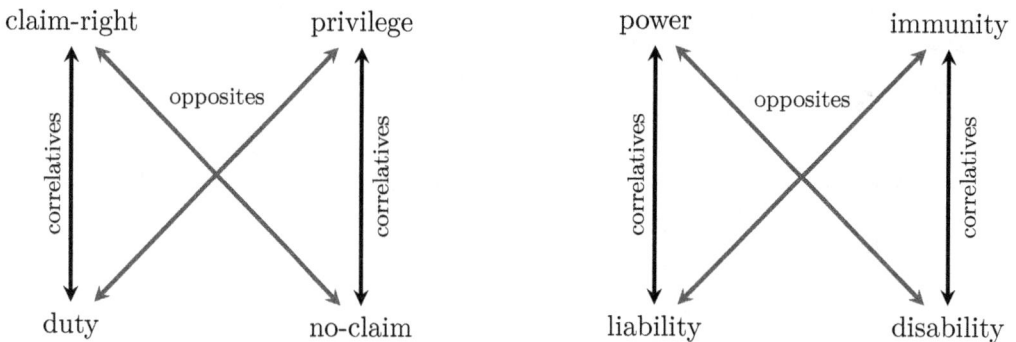

Hohfeld's reason to distinguish these, as he calls them, fundamental legal conceptions is that the word 'right' was overused in judicial reasoning. This overuse, though, leads to conceptual problems, as he puts it: "even if the difficulty related merely to inadequacy and ambiguity of terminology, its seriousness would nevertheless be worthy of definite recognition and persistent effort toward improvement."

Let's see a complex—and complete—example from the reception to represent the difference between the types of right behind the word 'right' and the necessity of their conceptual clarification. We find it in Szabó [21]: the sentence "Peter has a right to be in this house" can refer any of these types above depending where we say it: if this house is a building rent by Peter then his right is a claim-right against the

---

[1]It is worth to note already at this point that while Hohfeld was precise in using the word 'correlatives' to describe the relation within the pairs below, the word 'opposites' is covering different need to involve negation. See details in 2.1.

owner to ensure his occupancy (which means that the owner has the duty to do so); if this building is a public one then Peter's right is a privilege to be within it as being a citizen (which means other people have no claim-right against him *not* to do so); if Peter is a detective dashing in a house with an official warrant for search in his hand then he has a power (which means that the house's owner has a liability to this action of Peter, that is, Peter can impose a duty on her to let him in); but if Peter chains himself to the radiator in his own house stressing that he is exempt from the execution of the eviction, then what he refers to is his immunity (which also means a referring to the person's disability to conduct the eviction).[2] These normative positions have different computational properties, different consequences,[3] handling them and their difference properly both in legal theory and in logic representing law is crucial. From the viewpoint of deontic logic (or other logics used to formalize law) the virtue of this system is handling agency—this is the point of correlativity: according to Hohfeld, someone's right always involves someone else's duty, and the other way around. It is often discussed in legal theory whether this correlativity is general, i.e. true everywhere in law, see for instance Hart [4] and Lyons et al. [13].

In what follows, I introduce a formalism in order to show how the Hohfeldian theory can be used to describe rights and duties in criminal law. Please note that elements of and arguments for this formalism have already been presented in Markovich [15] and recently and most detailedly in Markovich [16]; the contribution of this paper, in the first part, is justifying the validity of the Hohfeldian theory in areas like criminal law using this formalism, and, in the second, showing what morals this theory and its proper understanding has about a specific notion of criminal law: the sanction or punishment.

2 Formalization of a "Resolutely Relational" Theory

If we consider the theory of normative positions presented by Kanger [8, 9, 10] as an attempt to formally represent the Hohfeldian rights and duties, missing the representation of legal relations's counterparties is a well-known limitation of it,[4] meanwhile, as Makinson stresses in [14], Hohfeld's theory is "resolutely relational". Makinson,

---

[2]Please note that it is not typical that the very same sentence can be interpreted in all the four different ways. It is the word 'right' which can be interpreted in the four different ways in different sentences, contexts.

[3]These different consequences is the central question of Markovich [16].

[4]It has been noted, for instance, in Hansson [3], in Makisnon [14], in Sergot [18]. Some considerations has been made, though, on whether the intended reading of Kangerian formulae can count as the representation of relationality, see in Makinson [14].

therefore, deems it necessary to introduce some explicit indexing of counterparties in the formal representation in order to properly capture the full relationality of rights relationships—even if it can be redundant sometimes in certain contexts. The notion of 'directed obligation' is explicitly introduced ten years later in Herrestad and Krogh [5] to the notion of duty (being the correlative pair of claim-right) in order to distinguish it from the general notion (standard) deontic logic uses. The necessity of this introduction can be shown easily with a simple formal setup which is used in rephrasing the classical formalizations in Makinson [14] and Sergot [18], that is, standard deontic logic (SDL)[5] with agent-indexed modal operators and a simple action logic that Chellas [2] called ET (using the operator E in and agent-indexed way with the intended meaning of E_x as x *sees to it that*, containing only the axiom T in order have successful actions).

2.1 Language and Semantics

Let's start with the following language set-up:

Definition 1. *Our modal language is given by*

$$p \in \Phi \mid \phi \wedge \psi \mid \neg\phi \mid \bot \mid E_a\phi \mid \mathbf{O}_{a \to b}\phi$$

for $a, b \in A$ set of agents, where Φ is the set of propositional letters.

Definition 2. *Frames are given as the following: for a set W of possible worlds and set A of agents write*

$$\mathfrak{F} = \langle W, f_a, R^O_{a,b} \rangle_{a,b \in A}$$

where $f_a : \wp(W) \to \wp(W)$ is a function and $R^O_{a,b} \subseteq W^2$ is a binary relation. Models are structures

$$\mathfrak{M} = \langle W, f_a, R^O_{a,b}, V \rangle_{a,b \in A}$$

where V is a valuation function for atomic propositions: $V : \Phi \to \wp(W)$

Definition 3. *For $\mathfrak{M} = \langle W, f_a, R^O_{a,b}, V \rangle_{a,b \in A}$ and $w \in W$ we let*

- $V(\bot) = \emptyset$
- $w \models p \Leftrightarrow w \in V(p)$ *for propositional letters $p \in \Phi$.*
- $w \models \varphi \wedge \psi \Leftrightarrow w \models \varphi$ *AND* $w \models \psi$.

---

[5]The originally used deontic logic in the theory of normative positions is slightly differ what we usually call SDL as does not contain the derivation rule of modal generalization, but this difference has no effect on the current formalization

- $w \models \neg\varphi \Leftrightarrow w \not\models \varphi$.

It is convenient to extend the valuation V to arbitrary formulas:

$$V(\varphi) := \{w : \mathfrak{M}, w \models \varphi\}$$

and we add the following:

- $w \models E_a\varphi \Leftrightarrow w \in f_a(V(\varphi))$

- $w \models \mathbf{O}_{a \to b}\varphi \Leftrightarrow \forall w'(wR^O_{a,b}w' \Rightarrow w' \models \varphi)$

Constraints

- *constraint on* f : $f_a(X) \subseteq X$ *for all* $X \subseteq W$, *in particular* $f_a(V(\varphi)) \subseteq V(\varphi)$

- *constraint on* $R^O_{a,b}$: $\forall w \exists w'\ wR^O_{a,b}w'$

In order to have the equivalence between the correlative pair of claim-right and duty (as the Hohfeldian correlativity's logical interpretation in the reception is unanimously equivalence[6]) we need to add the direction of the duty indicating the other agent who has the claim-right:

(1) $\mathbf{CR}_x E_y F \Leftrightarrow \mathbf{O}_{y \to x} E_y F$

(where F is a propositional letter with the intended meaning of 'given state of affairs').[7] As *claim-right* is a so-called passive right as the right owner and the actor (the agent whose action is the subject of the normative position) are different, the formalization presents both agent, while its correlative pair, *duty* is an active positions having the same agent in the indices of the deontic and action operator needs to made up with the direction to formally exhibit the counterparties making the equivalence hold.

While the notion of *duty* as directed obligation is generally accepted in the literature, the relationality of the other Hohfeldian conceptions as positions is practically overlooked. But Hohfeld was very consequential in this directedness: all legal relations he considers are two agents'—counterparties'—relation: the Hohfeldian privilege does not mean being free from any claim-right, but only the freedom from the specific

---

[6] As it has been indicated above already, the Hohfeldian notion of 'opposition' is less precise in the logical sense: while it means one negation considering *claim-right* and *no-claim*, we need two negations for moving from *duty* to *privilege*: I have a privilege to see to it that A if and only if I *don't* have a duty *not* to see to it that A. It will be visible in the later formulae.

[7] Thanks to the equivalences, we don't need to add $\mathbf{CR}_x$, neither the later modalities, to the language.

other party's claim-right. This difference is formally visible if we want to formalize the correlativity of *privilege* and *claim-right*: while $\mathbf{PR}_x E_x F \Leftrightarrow \mathbf{NC}_y \neg E_x F$ does not hold, the following does:

$$(2) \qquad \mathbf{PR}_{x \succ y} E_x F \Leftrightarrow \mathbf{NC}_y \neg E_x F^8$$

In this paper we won't work with the second square of Hohfeld (containing the so-called higher order modalities), but it needs to be told that the same is true for the Hohfeldian *power-liability* and *immunity-disability* pairs, too.[9] *Power* and *disability* are active positions: the right owner (and who lacks it in case of *disability*) is the same as the acting agent, therefore the relationality needs to be indicated formally, too:

$$(3) \qquad \mathbf{P}_{x \to y} E_x F \Leftrightarrow \mathbf{L}_y E_x F$$
$$(4) \qquad \mathbf{I}_x E_y F \Leftrightarrow \mathbf{D}_{y \to x} E_x F$$

2.2 Undirected rights and duties?

Does this mean that the Hohfeldian theory is designed to directed rights and duties and, *therefore*, is indeed unable to grasp undirected rights and duties? Just the opposite: this resolutely relational structure enables us to clearly refer to those right positions which seems to be—and often simply considered as—*absolute*. What does it mean in terms of the formulae above? If we want to "generate" absolute or general rights and duties we only need to take the conjunction of the given relations with each other agent (given a finite set of agents). This formally looks like the following in the case of active rights:

---

[8]The choice of the symbol $\succ$ is intended to reflect the "similarity" between its form and the notion being free *from* something; has no relation to the usual use of it.

[9]It may seem strange at first glance that *power* is a relational thing. This feeling might come from that, in everyday life, we usually refer to *power* as something which is not relational: we usually say the someone has *the* power to do something, that is, the only thing we mention is the right-owner. But—as it has been emphasized above—the Hohfeldian system is consequently relational: with power we can change *someone's* rights or duties. If I go to the registrar in order to get married, the question is not that whether she has the power to marry two people in general, the real question is whether she has the power to marry *us*—which is not obvious since it can be the question of jurisdiction restricted to a given district or state. Another source of considering *power* general is that the point in the *Searlian* power in [17] is exactly the generally accepted feature, but the Hohfeldian notion, while strongly connected, is different. But this discussion would lead us out from the scope of this paper, for further arguments and details see Markovich [16]).

(5) $\qquad \bigwedge_{y \in A} \mathbf{O}_{x \to y} E_x F \Leftrightarrow \mathbf{O}_x E_x F$

(6) $\qquad \bigwedge_{y \in A} \mathbf{PR}_{x \succ y} E_x F \Leftrightarrow \mathbf{PR}_x E_x F$

(7) $\qquad \bigwedge_{y \in A} \mathbf{P}_{x \to y} E_x F \Leftrightarrow \mathbf{P}_x E_x F$

(8) $\qquad \bigwedge_{y \in A} \mathbf{D}_{x \to y} E_x F \Leftrightarrow \mathbf{D}_x E_x F$

Also in the case of passive rights, here, though, we need to interpret the action operator slightly differently than we did above as we remove the agent-index, so EF needs to be interpreted as '*it is seen to it that F*':

(9) $\qquad \mathbf{CR}_x \bigwedge_{y \in A} E_y F \Leftrightarrow \mathbf{CR}_x EF$

(10) $\qquad \mathbf{NC}_x \bigwedge_{y \in A} E_y F \Leftrightarrow \mathbf{NC}_x EF$

(11) $\qquad \mathbf{L}_x \bigwedge_{y \in A} E_y F \Leftrightarrow \mathbf{L}_x EF$

(12) $\qquad \mathbf{I}_x \bigwedge_{y \in A} E_y F \Leftrightarrow \mathbf{I}_x EF$

And this not just does not go against the Hohfeldian intentions, but perfectly fits. Why we can be so sure?

There is a second part of the often cited famous essay *Fundamental Legal Conception as Applied in Judicial Reasoning* Hohfeld wrote three years later (both parts can be found in Hohfeld [7]). This second part is not so well-known and much less often cited (or even mentioned), while it is crucial in understanding Hohfeld's intentions and theory. In this second part he differentiates between the so-called *paucital* and *multital* rights. This differentiation fits the classical one between 'relations in personam' and 'relations in rem': renaming it happens exactly to emphasize that the so-called 'in rem' legal relations are also between people so the classical name is misleading. Nigel Simmonds picks two picturing examples in [19] of each category: "Suppose that I have a contract with you whereby you are obliged to manufacture a quantity of widgets. I have a claim-right against you and you have a correlative duty to manufacture the widgets. I might have a similar contract with another widget manufacturer, with similar consequences in terms of our claim-rights and duties. However many such contracts I have, however, my claim-rights are essentially limited to a definite number of persons. These are what Hohfeld calls 'paucital' claim-rights. (...) Suppose on the other hand, that I am the owner of Blackacre. I have a claim-right that you should not enter the land without my consent. I have the identical claim-right against your mother, my employer, the Bishop of Ely, and anyone else that you care to mention. Each of these claim-rights is a consequence of my ownership of Blackacre. These are 'multital rights'."

Hohfeld gives a short summary description of each type's features: "A paucital right, or claim, is either a unique right residing in a person (or group of persons) and availing against a single person (or single group of persons); or else it is a one of a *few* fundamentally similar, yet separate, rights availing respectively against a few definite persons. A multital right, or claim, is always *one* of a large *class* of *fundamentally similar* yet separate rights, actual and potential, residing in *a single* person (or single group of persons) but availing *respectively* against persons constituting a very large and indefinite class of people." There is another good example of Simmonds in [19]: "my claim-right that you should not assault me is a multital right, since it is only one member of a large class of similar rights holding against an indefinite number of people (i.e. I have a right that your mother should not assault me, a right that the Bishop of Ely should not assault me, and so on)", for which Hohfeld would add the example of a patentee's right that any other person shall not manufacture articles covered by the patent.

That is, using the series (the conjunction) of a—directed—right, gives us a (seemingly) undirected one. This is exactly what we did with the formulae above.

3   Meaning of Legal Rights

To understand what a right means in criminal law we first need to provide some general description of what a legal right in general means. As it is suggested in Markovich [15, 16], we can start from the intuitive sounding informal definition provided by Makinson in [14]:[10]

x bears an obligation to y that F under the system N of norms
iff
in the case that F is not true then y has the power under the code N to initiate legal action against x for non-fulfillment of F

---

[10]Makinson aimed at defining the notion of the counterparty, but as we can see, this definition is a definition of what a directed obligation (duty) is, that is—having the equivalence in (1)—also a definition of what a claim-right is.

Whatever intuitive it is, the right-to-left direction of the biconditional does not work: the fact that I have the power to initiate a legal action against someone does not imply that I had a claim-right against him (that is, he had a duty towards me). If this was the case, the court would not need to carry out the proceeding: the fact of initiating the legal action would mean winning it. But sometimes people lose their cases and the reason is exactly that they did not have the claim-right originally. Sergot in [18] suggests to add "with some expectation of success" to the definition. Even if this approaches reality well, this amended definition still would not tell us anything about what a claim-right is. The following biconditional, though, serves well in terms of providing the sufficient and necessary conditions of talking about a—legal—claim-right.[11]

$$(13) \qquad \mathbf{O}_{x \to y} E_x F \leftrightarrow \Box(\neg E_x F \to \mathbf{CR}_y E_j E_x F)$$

that is, a directed duty to see to it that F (and because of the equivalence, its correlative claim-right) means that if the duty bearer does not fulfill it then its counterparty has a claim-right against the judiciary (indicated with the agent constant j) to see to it that the original duty bearer see to it that F is the case. We need to introduce and use a necessity operator to make the conditional strict instead of material, it comes with the modal logic S5[12] and with the taste of talking about legal metaphysics: this is how things are in law, this is what a right *is* in law. This description in (13) is still based on the state enforcement but does not use the notion of *power*. The reason is to keep two crucial notions distinct: these are the ability/position to have rights and the ability/position of changing them. These two

---

[11]The formula in (13) is not a proper definition in terms of not having the definiendum at the side of the definiens, but this practically follows from the Hohfeldian intentions: he considered these conceptions *sui generis*, that is, he refused to reduce them to something else.

[12]This changes our language and models in the following way:

$$p \in \Phi \mid \varphi \wedge \psi \mid \neg\varphi \mid \bot \mid E_a\varphi \mid O_{a \to b}\varphi \mid \Box\varphi$$

for $a, b \in A$, where Φ is the set of propositional letters.
Frames are now defined as it follows: for a set W of possible worlds and the set A of agents we write

$$\mathfrak{F} = \langle W, f_a, R^O_{a,b}, R^\Box \rangle_{a,b \in A}$$

where $f_a : \wp(W) \to \wp(W)$ is a function and $R^O_{a,b}, R^\Box \subseteq W^2$ are binary relations.

Models now are structures: $\mathfrak{M} = \langle W, f_a, R^O_{a,b}, R^\Box, V \rangle_{a,b \in A}$ where v is a valuation function for atomic propositions: $V : \Phi \to \wp(W)$
The truth conditions for $\Box$: $w \models \Box\varphi \Leftrightarrow \forall w'(wR^\Box w' \Rightarrow w' \models \varphi)$

notions have clear terminology in languages of countries having continental legal systems ('Rechtsfähigkeit' and 'Handlungsfähigkeit' in German, or 'zdolność prawna' and 'zdolność do czynności prawnych' in Polish, 'capcité juridique' and 'capacité d'agir' in French, 'capacitá giuridica' and 'capacitá di agire' in Italian, all respectively), while the English terminology seems to be a bit loose in distinguishing them (maybe 'legal capacity' is the best version to the first and 'legal competence' or 'capacity to act' to the second). Keeping them distinct is essential: every human has the first one, but not the second (infants and people lacking mental soundness partly or completely lack the capacity to act.) If we use the notion of *power* in defining what a claim-right is (as it happened in the Makinsonian definition), these notions collapse. It absolutely does not mean that the notion of *power* would be eliminable or less important. It has a crucial role in legal systems, and Hohfeld was right to take it as one of the fundamentals. The point here is only that we should not involve it into defining claim-rights. Neither can be reduced to the other (just like Hohfeld said: these are sui generis notions). *Power* is about the ability to change someone's rights (for instance, put a duty on him), and while we won't analyze its notion in this paper (neither conceptually, nor formally), we will refer to and rely on the role it plays in legal systems.

3.1 Legal Rights in Private Law

The formula (13) is not perfect yet, it needs some refinement—which refinement depends on the area of law in which we would like to use it. In private law (paradigmatically, law of contracts) actually it is not the *original* duty whose fulfillment is enforced by the judiciary: the original duty had a deadline which is per definitionem over when the whole enforcment comes to the picture, so the description of the state of affairs that needs to be seen to must differ at least in a date from the original one. That is, the enforced state of affairs is a *compensation* (C) of (not fulfilling) the original duty. Therefore, we need to refine the formula above:

$$(14) \qquad \mathbf{O}_{x \to y} E_x F \leftrightarrow \Box(\neg E_x F \to \mathbf{CR}_y E_j E_x C F)$$

But at this point we stop and do not go into the details of what a compensation is, how it behaves formally, as our topic here is not private law, but criminal law.

3.2 Legal Rights in Criminal Law

Considering rights that are handled, protected by the means of criminal law sheds some light on further difficulties of a right-definition building on *power* to initiate a legal action. Consider, for instance, the right to physical integrity. There is a crime called assault which is obviously about the violation of one's right to physical integrity. What happens after a serious assault (grievous body harm)? The police starts investigation ex officio, the public prosecutor brings charges ex officio and represents the prosecution. The person whose right to physical integrity has been violated actually gives testimony but has no power to be considered. But, as we saw above, a claim-right "definition" without involving power avoids the problem raised by the ex officio steps of authorities, so for us, this won't be a problem. It might raise the question, though: why is it not the person with the violated right who stands in front of the court opposing the perpetrator's defense? This is something we need to be able to answer if we want our model to work in criminal law.

Another right protected in criminal law makes the situation even more difficult: the right to life. The obvious crime violating one's right to life is murder after which the person whose right has been violated is dead so it is problematic to speak about his right against the judiciary as in the most legal systems we cannot even consider a dead person's rights. But the structure of rights that are handled and protected by the means criminal law is different. In case of a criminal action it is not a specific person who opposes the perpetrator: it is everyone in the given society. Just consider what the court clerk says when trial starts: after naming the case with the name of the accused person, he says 'vs. the people of the given state' (for instance, as it happens in the movie *Goodfellas*: 'Henry Hill. People of the State of New York vs. Henry Hill'). This is because everyone's right in that society (concerned by a given legal system) has been violated: their right that no felony (murder, assault, etc.) be committed. The right to life is a clear value whose *legal form* in criminal law is a right of everyone that no one commit a murder. If we think about directed graphs where the nods are agents and the edges are legal relations, for instance an agent a has a claim-right against another agent b iff there is an edge going from a to b (of course, the edge going the other way around would represent the correlative duty). Figure 1 below shows a claim-right resulting from e.g. a contract between the people it connects (on the left), and how a claim-right looks like in criminal law, e.g. the right to life which actually means that everyone has a claim-right against everyone else to that no one see to it that a murder is committed (on the right). That is, rights in criminal law are complete directed graphs.

Putting this into the "definition" we provided above, we get the description of how rights in criminal law work: all of us have a claim-right against everyone else that no one commit a felony (murder) iff it is (necessarily) the case that if anyone does commit a felony (murder) then all of use have a claim-right against the judiciary that it punish (sanction, S) them for committing the felony. That is:

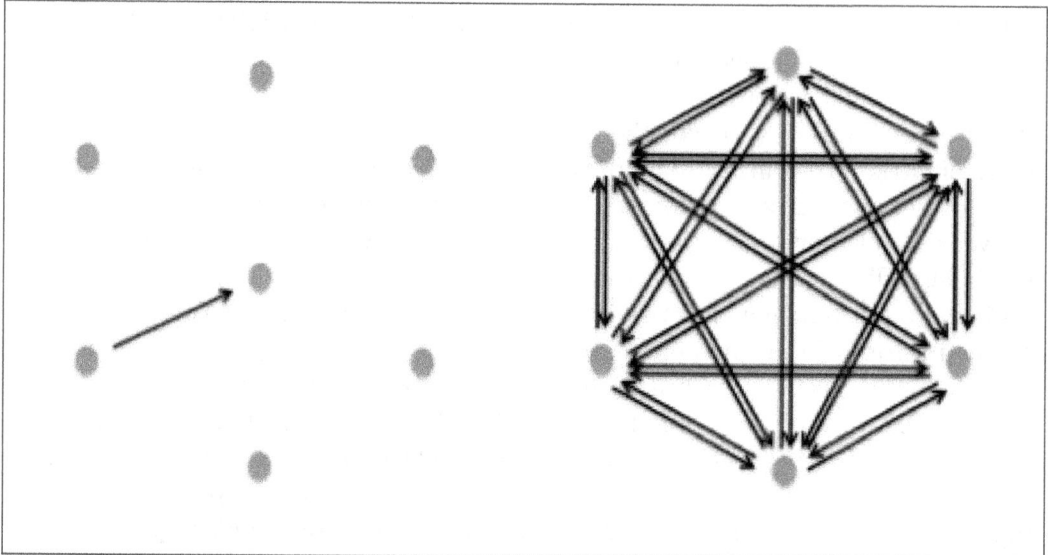

Figure 1: The left graph shows the paradigmatic case of a Hohfeldian right relation between agents (that he calls paucital right). The graph on the right hand side shows what rights handled in criminal law look like (a case of the one Hohfeld called multital right).

$$(15) \qquad \bigwedge_{x \in A} \mathbf{CR}_x \neg \bigvee_{y \in A} E_y F \leftrightarrow \Box \bigwedge_{y \in A} (E_y F \rightarrow \bigwedge_{x \in A} \mathbf{CR}_x E_j S_y F)$$

Please note that the sanction is considered not just as a punishment but also as a tool of enforcing that the convicted person refrain from committing the felony again—so we still build on enforcement. But what does a sanction mean?[13] This is what we pursue to answer in the second half of this paper.

4 Sanction in Terms of Rights and Duties

In order to get some insight what a sanction is it seems to be practical to check what criminal codes declare to be a sanction. The Chapter 3 of German Criminal Code says:

Title 1: Punishments:

a) imprisonment;

---

[13]The word 'sanction' is sometimes used to refer generally to the third part of a legal norm (hypothesis, disposition, sanction), but in this paper it is to denote a narrower sense: the negative legal consequence, the punishment (or penalty). These are used as synonyms here.

b) fine

c) property fine;

d) driving ban;

e) loss of the capacity to hold, or be elected to public office and the right to vote.

Section 9 in the Dutch Criminal Code looks like:

The Punishments are:

a) Principal punishments

1. imprisonment;

2. detention;

3. community service;

4. fine.

Section 33 in the Hungarian Criminal Code lists the followings:

(1) Penalties are:

a) imprisonment;

b) custodial arrest;

c) community service work;

d) fine;

e) prohibition to exercise professional activity;

f) driving ban;

g) prohibition from residing in a particular area;

h) ban from visiting sport events;

i) expulsion.

The lists of these different countries are pretty similar, but what are these things listed?

4.1 Punishment as Duties

What is sure that punishment comes with imposing a sentence: the judge (the judiciary) has the *power* to change our rights and duties, and this is the *tool* of enforcement. But what changes are these? In the case of driving ban it is clear that the punishment is a prohibition: the convicted must not drive (which is an obligation to refrain from driving). Same in the case of prohibition to exercise (a given) professional activity. But there are noun phrases in the lists, too, where there is no direct reference to the deontic nature. It seems obvious, though, that in the case of a fine and community work it is also an obligation: an obligation to pay and an obligation to conduct community work. It is general to consider the sanction as an obligation, this is how it is carachterized in LegalRuleML, too, see [1].

This seems to be the most obvious interpretation of imprisonment, too: it is an obligation of the convicted person to go to prison (and stay there for a while). In terms of duty, that is, a directed obligation, it seems to be reasonable to talk about an "undirected", multital duty: a duty towards everyone else in the given society. We, all of us, have a claim-right against convicted people that they stay in prison while. It fits the image of prison break being a *felony*: all of us have a claim-right against everyone else that no one escape from the prison, which also means according to the description we gave above that it is necessary that if anyone escapes then all of us will have a claim-right against the judiciary to sanction him or her. Checking the Hungarian Criminal Code we find indeed that escaping from prison is a felony: *Section 283: Any person who escapes from the custody of the authority in the course of criminal proceedings or from imprisonment or custodial arrest is guilty of a felony punishable by imprisonment not exceeding three years.*

4.2   Right to Freedom vs. Right to Escape

We might think at this point that all that have been said here is straightforward, no wonder the presupposition that the imprisonment is a duty to go and stay in prison works so well. But while—in accordance with this presupposition—it is a felony to escape from the jail in most of the countries, there are some—Germany, Belgium and the Netherlands, among others—in which *prison break is not a crime.* These countries' criminal codes do not declare prison break a felony. In some of these countries the prisoners are punishable for causing any damage or committing another felony while escaping—without which it is pretty difficult to manage an escape of course, still: *the escape itself is not penalized.* The reason is: it is basic human instinct to want to be free, so it should not be punished. Gold says in [22] that this kind of regulation (she mentions the German one) reflects to the philosophy of Hobbes according to which the law should not impose impossible obligations and refraining from pursuing freedom is impossible (see [6]), being in accordance with the Kantian principle of 'ought implies can'.

Until a year ago, this was the case in Mexico, too, where the penal code not just missed to list the prison break among the crimes, but directly declared in Section 154 that the person who escapes won't be punished because of this action, and the explanation referred to freedom as an ideal brought with the French Revolution.[14] (In 2017, after 86 years, Mexico amended this section making the prison break punishable. While the official argument[15] raises the question whether the earlier regulation was justifiable at all—as an inmate who evades compliance with the punishment im-

---

[14]https://bit.ly/2R7vZxu
[15]https://bit.ly/2DSO27E

posed on him by the State attacks the rule of law and goes against the sovereign will—they refer to the direct reason leading to the amendment: the third—as the document refers to it: shameful—escape of C. Joaquín Guzmán Loera.[16]) Some online articles[17] refer to this regulation as those in which one has the right to escape from prison. Taking the Hohfeldian model, this is true: if we don't have a duty to refrain from something, we have a privilege to do that, which a type of right at Hohfeld.

This means, though, that the presupposition that imprisonment means the duty of the convicted person to go to and stay in prison cannot be upheld—given, of course, that we would not like to consider these legal systems as contradictory ones. What else the imprisonment as a sanction can then be? The answer is also in the Hohfeldian system. All of us have a right to freedom. This right is a *claim-right* normally: all of us have a claim-right toward everyone else that no one see to it that our freedom is restricted (that is, that everyone refrain from restricting our freedom)—this is why freedom restricting actions like illegal detention and kidnapping are crimes: everyone has a duty toward everyone else (which we usually consider it as a general obligation) to refrain from kidnapping. In case of imprisonment, this claim-right to freedom *turns—weakens—into a privilege*: the convict still does not have a duty to go and stay in prison but loses his claim-right against others that they refrain from detaining him. At the same time (as this change in his right would not ensure that the imprisonment will be, or at least pursued to be, realized), one—practically the penal institution as such—has a change in their duties: concerning the convict's detention, their *duty not to* turns into a *duty to*. That is, while they, like everyone else, had a duty to refrain from detaining the person, when this person becomes a convict (getting imprisonment), they are going to have a duty to detain him.[18] And this duty remains until the end of the term of imprisonment, this is why, basic instinct or not, once the convict is caught, he is put back to prison.

---

[16] Joaquín "El Chapo" Guzmán is a very powerful Mexican drug trafficker who first spent 20 years in prison, then escaped. He was captured and imprisoned in 2014, but in 2015 he escaped again through a 1.5 km tunnel equipped with artificial light leading to a construction site. After his third arrest, Mexico extradited him to the US—and changed the penal code.

[17] This phrasing appears on the Washington Post (https://wapo.st/2PMZzMf), and gives the title on the Hungarian Index.hu (https://bit.ly/2R6je6i), what is more, in the text 'human right' is mentioned.

[18] This duty, in practice, is of course a bundle of different duties and duty-bearers, e.g. it is the duty of the police to caught and get the convict into the prison.

5 Conclusion

Areas of law where we consider general or absolute rights and obligations do not fall out of the scope of the Hohfeldian theory. The basis of this latter is the relationality, the directedness, but handling this feature systematically enables us to talk about and handle clearly the so-called absolute normative positions. An amended formalism based on the one used by the classical formalizations can be easily used to show this as interpreting the absolute position as a series, that is, a conjunction of each directed one. In criminal law, the rights relations go from everyone to everyone else, that is, they create a complete directed graph on the given society.

The Hohfeldian theory is also useful in understanding the notion of sanction or punishment in terms of rights and duties: using the theory enables us to explain how it can be the case that there are countries where it is not forbidden to escape from the prison, that is, convicts have such a right. This can be so because this kind of right is a weakened one: not a claim-right anymore protecting (that is, a claim-right against everyone else not to interfere) my freedom, but only a privilege in the Hohfeldian sense: there is no duty of the convicts to stay in the prison in these countries. What still makes it realizing a punishment is a joint change in the penal institutions' duties: while they had a duty to not detain me, after my conviction, they have a duty to do so.

References

[1] Tara Athan, Harold Boley, Guido Governatori, Monica Palmirani, Adrian Paschke, and Adam Wyner. Oasis LegalRuleML. In *Proceedings of the Fourteenth International Conference on Artificial Intelligence and Law*, ICAIL '13, pages 3–12, New York, NY, USA, 2013. ACM.

[2] Brian F. Chellas. *Modal Logic. An Introduction.* Cambridge University Press, 1980.

[3] Bengt Hansson. Deontic logic and different levels of generality. *Theoria*, 36:241–248, 1970.

[4] H. L. A. Hart. Legal rights. In *Essays on Bentham: Studies in Jurisprudence and Political Theory*, pages 162–193. Clarendon Press, Oxford, 1982.

[5] Henning Herrestad and Christen Krogh. Obligations directed from bearers to counter-parties. In *Proceedings of the International Conference on Artificial Intelligence and Law*, pages 210–218. ACM, 1995.

[6] Thomas Hobbes. *Leviathan, parts I and II.* Bobbs-Merrill, 1958.

[7] Wesley Newcomb Hohfeld. Fundamental legal conceptions applied in judicial reasoning. In Walter Wheeler Cook, editor, *Fundamental Legal Conceptions Applied in Judicial Reasoning and Other Legal Essays*, pages 23–64. New Haven : Yale University Press, 1923.

[8] Stig Kanger. New foundations of ethical theory. In Risto Hilpinen, editor, *Deontic Logic: Introductory and Systematic Readings*, pages 36–58. D. Reidel, Dordrecht, 1971.

[9] Stig Kanger. Law and logic. *Theoria*, 38:105–132, 1972.

[10] Stig Kanger. On realization of human rights. *Acta Philosophica Fennica*, 38, 1985.

[11] Albert Kocourek. Hohfeld system of fundamental legal concepts, the. *Illinois Law Review*, 15, January 1920.

[12] Lars Lindahl. Stig Kanger's Theory of Rights. In D. Werserstahl D. Prawitz, B. Skyrms, editor, *Logic, Methodology and Philosophy of Science IX*, pages 889–911. Elsevier Science Publsiher, New York, 1994.

[13] David Lyons, Marcus Singer, and David Braybrooke. The correlativity of rights and duties. *Noûs*, 4(1), February 1970.

[14] David Makinson. On the formal representation of rights relations: Remarks on the work of Stig Kanger and Lars Lindahl. *Journal of Philosophical Logic*, 15(4):403–425, November 1986.

[15] Réka Markovich. No match-making but biconditionals: Agents and the role of the state in legal relations. In *Legal Knowledge and Information Systems - JURIX 2015: The Twenty-Eighth Annual Conference, Braga, Portual, December 10-11, 2015*, pages 161–164, 2015.

[16] Réka Markovich. Understanding Hohfeld and formalizing legal rights: the Hohfeldian conceptions and their conditional consequences. *Studia Logica*, 2019, to appear.

[17] John R Searle. *The construction of social reality*. Penguin, London, 1996.

[18] Marek Sergot. Normative Positions. In Xavier Parent Ron van der Meyden Dov Gabbay, John Horty and Leendert van der Torre, editors, *Handbook of Deontic Logic and Normative Systems*, pages 353–406. College Publications, 2013.

[19] Nigel Simmonds. Introduction. In *Hohfeld: Fundamental legal conceptions as applied in judicial reasoning*, Classical Jurisprudence series. Ashgate, Aldershot, new ed. / edited by David Campbell and Philip Thomas. edition, 2001.

[20] Gopal Sreenivasan. Duties and their direction. *Ethics*, 120(3):465–494, 2010.

[21] Miklós Szabó. A jogosultság logikai státusáról. In Györfi Tamás Ficsor Krisztina and Szabó Miklós, editors, *Jogosultságok: Elmélet és gyakorlat*, pages 35–46. Miskolc: Bíbor Kiadó, 2009.

[22] Judith Zubrin Gold. Prison escape and defenses based on conditions: A theory of social preference. *California Law Review*, 67(5), 1979.

 Received 17 January 2019

PROBABILISTIC LEGAL DECISION STANDARDS STILL FAIL

(AT LEAST WHEN IT COMES TO THE DIFFICULTY ABOUT CONJUNCTION AND THE GATECRASHER PARADOX)

RAFAL URBANIAK

University of Gdansk (Poland) & Ghent University (Belgium)

Abstract

Various probabilistic explications of the phrase *the court's decision regarding a fact, given the evidence, is justified* have been proposed. In this paper I evaluate them against two conceptual challenges: the difficulty about conjunction and the gatecrasher paradox. I argue that despite arguments to the opposite, all proposed models fail to solve these two problems.

1 Introduction

Imagine you are a trier of fact in a legal proceeding in which the defendant's guilt is identified as equivalent to a certain factual statement G and that somehow you succeeded in properly evaluating $P(G|E)$ – the probability of G given the total evidence presented to you, E (and perhaps some other relevant probabilities). For various reasons, some of which will be mentioned soon, this is an idealized situation. One question that arises in such a situation is: *when should you decide against the defendant? when is the evidence good enough?*

What we are after here is a condition Ψ, formulated in (primarily) probabilistic terms, such that the trier of fact, at least ideally, should accept any relevant claim (including G) just in case $\Psi(A, E)$. The requirement that the condition should apply to any relevant claim whatsoever (and not just a selected claim, such as G) will be called the **equal treatment requirement.**[1]

This research has been funded by Narodowe Centrum Nauki (grant no. 2016/22/E/HS1/00304) and Fonds Wetenschappelijk Onderzoek. The author would like to express his gratitude to Marcello Di Bello, Pavel Janda, Alicja Kowalewska, and two anonymous referees for detailed comments on earlier drafts of this paper.

[1]The requirement is not explicitly mentioned in the discussion, but it is tacitly assumed, so it is useful to have a name for it. Moreover, it will turn out crucial when it comes to finding a resolution of the difficulties, but further details need to wait till the last section of this paper.

For instance, one straightforward attempt might be to say: convict if $P(G|E)$ is above a certain threshold, otherwise acquit. From this perspective, whether assessment of facts leading to conviction is justified is a matter of whether the factual statement corresponding to guilt is sufficiently probable given the evidence.

As it turns out, the idea that such a probabilistic explication Ψ can be given does not play nicely with some other desiderata that we might want to put forward for what a rational trier of fact should think about facts and evidence.

A large-scale attack on probabilistic approach to legal decisions has been launched quite some time ago by [6], and some of the developments in probabilistic evidence scholarship are to some extent a reaction to some of Cohen's objections. My goal here is to focus on two of them – the **difficulty about conjunction** and the **gatecrasher paradox**. They correspond to two requirements. One, that Ψ should be such that for any relevant A and B there should be no difference between the trier's acceptance A and B separately, and her acceptance of their conjunction, $A \wedge B$, that is, that $\Psi(A, E)$ and $\Psi(B, E)$ just in case $\Psi(A \wedge B, E)$. Two, that any such explication should help us make sense of cases in which the probability of guilt given the evidence is high, and yet, conviction is not justified.

I will argue that even most recent proposals of what such a Ψ should be have failed to address these difficulties. This, however, does not mean that I side with Cohen and claim that thinking of evidence in legal context in terms of probabilities is doomed. Quite the contrary: probabilistic tools are highly useful, and their utility can be increased (and defended) by addressing Cohen's concerns properly. In this paper, however, I leave this positive task for a later occasion, restricting myself to a negative task of showing that legal probabilism so far has not reached this stage.

The paper is somewhat uneven: parts of it contain mostly philosophical discussion, while some other parts get involved with rather detailed and lengthy mathematical arguments whose philosophical points can be summed up in one or two sentences. To avoid discouraging the philosophically minded reader and disappointing the more mathematically oriented one, let me suggest three different reading strategies.

A quick philosophical look You do not want to get bogged down in mathematical details, but you want to find out what the philosophical gist of the paper is. In this case, after reading this section, read *Section* 3 to find out what the basic variant of the view under criticism is, *Section* 4 for a description of the difficulty about conjunction, *Section* 7 - which introduces the gatecrasher paradox, and *Section* 14 for an informal survey of the results and a philosophical discussion thereof. This is the shortest strategy.

The middle way You want to understand the key mathematical aspects, but you do not care too much about detailed arguments and digressions that can be skipped. Read the whole paper, except for the parts marked in the following manner.

OPTIONAL CONTENT STARTS

This is how inessential technicalities and longer side remarks are marked.

OPTIONAL CONTENT ENDS

The meticulous reader strategy Read everything. Or, start with a quick philosophical look as described above, and then read everything.

The structure of this paper is as follows. I start with a more detailed discussion of legal probabilism in *Section* 2. Then, I go through various ways legal probabilism so far has tried to navigate around the difficulties that I am concerned with, each time explaining why it failed. To illustrate the idea of a probabilistic model of a decision standard, and to get considerations of conceptual difficulties started, in *Section 3* I move to the first and most straightforward candidate: *Threshold-based LP*. In *Section* 4 I explain how this view falls prey to the difficulty about conjunction. Next, in *Section* 5 I look at an important attempt to save threshold-based legal probabilism from the difficulty about conjunction, and in *Section* 6 I argue that it fails. In *Section* 7 I introduce the other conceptual difficulty that will be used as a measuring stick in the assessment of probabilistic models: the gatecrasher paradox. In *Section 8* I outline Cheng's Relative LP. In *Section 9* I explain how it is supposed to handle DAC, and in *section* 10 I look at Cheng's attempt to avoid the gatecrasher paradox. In *Section 11* I argue that the approach fails on both counts. In *Section 12* I introduce the last candidate to be discussed, Kaplow's Decision-Theoretic LP, and in *Section* 13 I argue that it also cannot handle the difficulties. *Section* 14 contains an informal overview and discussion.

2   Legal probabilism and its motivations

In multiple domains of applications, probabilistic methods have been exceedingly successful. Some of the successful applications involve forensic and judiciary contexts.[2] What they seem to have in common is that they pertain to the interpretation or weighing of particular pieces of evidence, or an evaluation of a particular argument involving probabilities or statistics.

What is somehow more contentious is whether a more general and all-encompassing probabilistic model of evidence evaluation and decision about conviction can

---

[2] See for instance [11, 1, 26, 19, 23].

be successfully built: one which would explicate the phrase *given the evidence, the conviction is justified* in probabilistic terms (where the claim under consideration is a factual claim considered as equivalent to guilt according to the law). On one hand, *prima facie*, what the judge or the jury seem to evaluate is the probability of guilt given total available evidence and common sense knowledge, and so, probabilistic tools seem fit to model, at least in abstraction, such phenomena. On the other hand, particular pieces of forensic evidence aside, precise probabilities are hard to come by, and conceptual and practical difficulties with developing such a model abound.

To avoid setting the bar too high, let me get clear on what, on the present approach, a successful probabilistic model is *not* required to achieve. Namely, I am putting aside most of the issues that have to do with practicality.[3] I will not be concerned with the lack of real data to support certain probability assessment, I will not be concerned with people being bad at reasoning about probabilities, etc. Basically, I will not be concerned with those practical issues that would arise if one would like to deploy a probabilistic model directly, by writing down numerical values for all the probabilities relevant in a given case and simply calculating the probability of guilt. I will simply grant that at least for now, successful deployments of this type are not viable.

This, however, does not mean that developing a general probabilistic model is pointless. There are multiple ways in which such a model, even if unfit for direct deployment, could be useful. Once we have a probabilistic model, a vast array of mathematical results pertaining to probability can be used to deepen our understanding of the rationality of legal decisions. If at least in abstraction adequate, the model could be useful for diagnosing various types of biases that humans are susceptible to in such contexts; it could be useful as a measuring stick against which various qualitative inference patterns are assessed, and it could be useful as a source of insights about various aspects of legal decisions and evidence presentation methods. Just as understanding physics might be useful for deepening our understanding of how things work, and for building things or moving them around without performing direct exact calculations, a general probabilistic model – again, if adequate – could help us get better at understanding and making legal decisions without its direct deployment in practice.

Just because I put strong practicality requirements aside, it does not mean that I put no constraints on the probabilistic model to be developed. While *sufficient* conditions of adequacy of such a model are somewhat hard to explicate and I will not get into a deeper discussion thereof, there is at least a fairly clear *necessary* condition.

---

[3]My impression is that with a few exceptions, most of the arguments about legal probabilism in the early stage of the debate were concerned mostly with practicality. See [2, 14, 8, 24, 30, 28, 18, 16, 27].

A successful probabilistic model should either avoid or explain away what seem to be important conceptual difficulties that it runs into. And this is what I will be focusing on in this paper: investigating whether available probabilistic models of legal decision standards avoid or explain away the conceptual difficulties that – it seems – they should be able to handle. In particular, I will focus on two pieces of paradoxical flavor, the **difficulty about conjunction (DAC)** and the **paradox of the gatecrasher**.

One reason to choose these two is that they are easy to explain: and it would be nice if we could handle basic conceptual difficulties before we move to more complex issues. Another reason is that in some variant or another, these have been widely discussed in literature. DAC has a very close cousin named the lottery paradox, which occupied the minds of many, and the gatecrasher paradox and related thought experiments and real cases have been extensively discussed by philosophers trying to identify the factor that makes naked statistical evidence actionable.

Let us start with a very general assumption that all the approaches that will be discussed in what follows share; we will call it *Legal Probabilism* (LP). It is the view that the legal notion of probability is to be governed by the mathematical principles of probability theory, and that the decision process in juridical fact-finding is to be explicated by means of probabilistic tools. LP is fairly general: it does not tell us *how exactly* the decision standards are to be explicated in probabilistic terms, it only tells us that somehow they should.

3   Threshold-based legal probabilism (TLP)

LP comes in various shapes. It is one thing to say that the standards of juridical proof are to be explicated in probabilistic terms, it is another to provide such an explication. The threshold-based legal probabilism has it that once the probability of guilt (or, to be more precise, the factual statement that according to law is equivalent to guilt) given the total evidence available is assessed, conviction is justified just in case this probability is above a certain threshold.[4]

Classical Legal Probabilism (CLP), stemming from [4], keeps the threshold constant:[5]

---

[4]In the Anglo-Saxon tradition there is a distinction between decision standards in civil and in criminal cases. In the former, decision is to be made on the preponderance of probability, and in criminal cases, the guilt statement is supposed to be beyond reasonable doubt. Assuming these are to be modeled by probability thresholds different from 1, there is no essential difference here as far as the conceptual difficulties to be discussed in this paper are involved.

[5]Again, we are going to ignore the difference between civil and criminal litigation here. If one wants to keep this distinction in mind, CLP can be easily revised by positing one threshold for

(CLP)   There is a certain probability of guilt threshold t, such that in any particular case, if the probability of guilt conditional on all the evidence is above t, convict; otherwise acquit.

A slightly weaker (and perhaps more common among evidence scholars) variant of this view, let us call it the *Sensitive Legal Probabilism* (SLP), also embraces the idea that what is to be evaluated is the probability of guilt given the evidence, but abandons the requirement that there should be a single threshold for all cases; rather, SLP suggests that the context of each particular case will determine which threshold is appropriate for it.

(SLP)   For any particular case, there is a contextually determined probability threshold t such that if the probability of guilt conditional on all the evidence is above t, convict; otherwise acquit.

Before we move to the discussion of the two key difficulties that we will be interested in, let me briefly mention one issue about TLP that I will not be concerned with. A careful reader might already have the following complaint: *if you are saying that there is a conviction probability threshold, what exactly is it and why?* And indeed, it seems quite difficult to point to any particular choice of value and argue that the choice is not to a large extent arbitrary.

One reason why I will not be concerned with this problem is that this is an issue that seems to pertain to TLP only, while I would like to focus on problems that seem to be more general.

Another reason is that once decision-theoretic tools are allowed, there might be reasons to think that the choice of threshold is not that arbitrary [17]. Say the probability of guilt (or responsibility) is p, the disutility of acquitting a guilty person is d_g, and the disutility of convicting an innocent person is d_i. From the perspective of minimalization of expected disutility, we would like to convict, or find for the plaintiff, just in case the expected disutility of a mistaken acquittal is greater than the expected disutility of an incorrect conviction:

$$pd_g > (1-p)d_i$$

---

criminal cases, and one for civil ones.

Now, solving for p gives us:

$$pd_g > d_i - pd_i$$
$$pd_g + pd_i > d_i$$
$$p(d_g + d_i) > d_i$$
$$p > \frac{d_i}{d_g + d_i}$$

So, as long as you can quantify these disutilities, the probability threshold can be determined. But since I want to focus on probabilistic considerations, I will not pursue this discussion.

Finally, as practice (such as conviction decisions based on DNA identification) indicates, there are probabilities of guilt clearly considered high enough for conviction, and there are ones which clearly are not high enough for conviction. Perhaps, there are some borderline cases, but these are not too many. From this perspective, the phrase *probability of guilt sufficient for conviction* can be argued to be vague, but the vagueness does not seem too damaging in practice (at least, not more than the vagueness that is already there, even without probabilistic tools). Moreover, it is a rather common practice to theorize about notions which in practice are somewhat vague using idealized mathematical tools which do not tolerate vagueness. As long as the results obtained hold independently of any particular choice of the precisification of a given vague notion, the initial vagueness is not a deep obstacle to the utility of these theoretical considerations.

Now that we know what the first explication is, let us move to the first of the two conceptual difficulties that we actually will be concerned with – the difficulty about conjunction.

4   The difficulty about conjunction

The *Difficulty About Conjunction* (DAC) proceeds as follows. Say we focus on a civil suit where a plaintiff is required to prove their case on the balance of probability, which for the sake of argument we construe as passing the 0.5 probability threshold.[6] Suppose the plaintiff's claim to be proven based on total evidence E is composed of two elements, A and B, independent conditionally on E.[7] The question is, what exactly is the plaintiff supposed to establish? It seems we have two possible readings:

---

[6]This is a natural choice given that the plaintiff is supposed to show that their claim is more probable than the defendant's. The assumption is not essential. DAC can be deployed against any $\neq 1$ guilt probability threshold.

[7]These assumptions, again, are not too essential. In fact, the difficulties become more severe as the number of elements grows, and, extreme cases aside, do not tend to disappear if the elements

Requirement 1	$P(A \wedge B|E) > 0.5$
Requirement 2	$P(A|E) > 0.5$ and $P(B|E) > 0.5$

Requirement 1 says that the plaintiff should show that their *whole* claim is more likely than its negation. There are strong intuitions that this is what they should do. But the problem is, this requirement is not equivalent to **Requirement 2**. In fact, if we need $P(A \wedge B|E) = P(A|E) \times P(B|E) > 0.5$ (the identity being justified by the independence assumption), satisfying **Requirement 2** is not sufficient for this purpose. For instance, if $P(A|E) = P(B|E) = 0.51$, $P(A|E) \times P(B|E) \approx 0.26$, and so the plaintiff's claim as a whole still fails to be established. This means that requiring the proof of $A \wedge B$ on the balance of probability puts an importantly higher requirement on the separate probabilities of the conjuncts.

Moreover, what is required exactly for one of them depends on what has been achieved for the other. If I already established that $P(A|E) = 0.8$, I need $P(B|E) \geq 0.635$ to end up with $P(A \wedge B|E) \geq 0.51$. If, however, $P(A|E) = 0.6$, I need $P(B|E) \geq 0.85$ to reach the same threshold. This would mean that standards of proof for a given claim could vary depending on how well a different claim has been argued for and on whether it is a part of a more complex claim that one is defending, and this does not seem very intuitive. At least, this goes strongly against the equal treatment requirement mentioned already in the introduction.

Should we then abandon **Requirement 1** and remain content with **Requirement 2**? Cohen [6, 66] convincingly argues that we should not. Not evaluating a complex civil case as a whole is the opposite of what the courts themselves normally do. There are good reasons to think that every common law system subscribes to a sort of conjunction principle, which states that if A and B are established on the balance of probabilities, then so is $A \wedge B$.

So, on one hand, if we take our decision standard from **Requirement 2**, our acceptance standard will not involve closure under conjunction, and might lead to conviction in cases where $P(G|E)$ is quite low, just because G is a conjunction of elements which separately satisfy the standard of proof – and this seems unintuitive. On the other hand, following Cohen, if we take our decision standard from **Requirement 1**, we will put seemingly unnecessarily high requirements sensitive to fairly contingent and irrelevant facts on the prosecution, and treat various elements to be proven unevenly. Neither seems desirable.

---

are dependent.

5 Dawid's likelihood strategy

One well-known attempt to handle DAC from the probabilistic perspective without any drastic changes to the probabilistic model is due to Dawid [9]. Here is how it proceeds (the considerations that follow apply to other sorts of uncertain evidence; we'll focus on witnesses for the sake of simplicity). Imagine the plaintiff produces two independent witnesses: W_A attesting to A, and W_B attesting to B. Say the witnesses are regarded as 70% reliable and A and B are probabilistically independent, so we infer $P(A) = P(B) = 0.7$ and $P(A \wedge B) = 0.7^2 = 0.49$.

But, Dawid argues, this is misleading, because to reach this result we misrepresented the reliability of the witnesses: 70% reliability of a witness, he continues, doesn't mean that if the witness testifies that A we should believe that $P(A) = 0.7$. To see his point, consider two potential testimonies:

A_1	The sun rose today.
A_2	The sun moved backwards through the sky today.

Intuitively, after hearing them, we would still take $P(A_1)$ to be close to 1 and $P(A_2)$ to be close to 0, because we already have fairly strong convictions about the issues at hand. In general, how we should revise our beliefs in light of a testimony depends not only on the reliability of the witness, but also on our prior convictions.[8] And this is as it should be: as indicated by Bayes' Theorem, one and the same testimony with different priors might lead to different posterior probabilities.

So far so good. But how should we represent evidence (or testimony) strength then? Well, one pretty standard way to go is to focus on how much it contributes to the change in our beliefs in a way independent of any particular choice of prior beliefs. Let a be the event that the witness testified that A. It is useful to think about the problem in terms of *odds, conditional odds (O)* and *likelihood ratios (LR)*:

$$O(A) = \frac{P(A)}{P(\neg A)}$$

$$O(A|a) = \frac{P(A|a)}{P(\neg A|a)}$$

$$LR(a|A) = \frac{P(a|A)}{P(a|\neg A)}.$$

---

[8]An issue that Dawid does not bring up is the interplay between our priors and our assessment of the reliability of the witnesses. Clearly, our posterior assessment of the credibility of the witness who testified A_2 will be lower than that of the other witness. But a deeper discussion goes beyond the scope of this paper.

Suppose our prior beliefs and background knowledge, before hearing a testimony, are captured by the prior probability measure $P_{prior}(\cdot)$, and the only thing that we learn is a. We're interested in what our *posterior* probability measure, $P_{posterior}(\cdot)$, and posterior odds should then be. If we're to proceed with Bayesian updating, we should have:

$$\frac{P_{posterior}(A)}{P_{posterior}(\neg A)} = \frac{P_{prior}(A|a)}{P_{prior}(\neg A|a)} = \frac{P_{prior}(a|A)}{P_{prior}(a|\neg A)} \times \frac{P_{prior}(A)}{P_{prior}(\neg A)}$$

that is,

$$O_{posterior}(A) = O_{prior}(A|a) = \underbrace{LR_{prior}(a|A)}_{\text{conditional likelihood ratio}} \times O_{prior}(A) \qquad (1)$$

The conditional likelihood ratio seems to be a much more direct measure of the value of a, independent of our priors regarding A itself. In general, the posterior probability of an event will equal to the witness's reliability in the sense introduced above only if the prior is $1/2$.

OPTIONAL CONTENT STARTS

Dawid gives no general argument, but it is not too hard to give one. Let $rel(a) = P(a|A) = P(\neg a|\neg A)$. We have in the background $P(a|\neg A) = 1 - P(\neg a|\neg A) = 1 - rel(a)$.

We want to find the condition under which $P(A|a) = P(a|A)$. Set $P(A) = p$ and start with Bayes' Theorem and the law of total probability, and go from there:

$$P(A|a) = P(a|A)$$

$$\frac{P(a|A)p}{P(a|A)p + P(a|\neg A)(1-p)} = P(a|A)$$

$$P(a|A)p = P(a|A)[P(a|A)p + P(a|\neg A)(1-p)]$$

$$p = P(a|A)p + P(a|\neg A) - P(a|\neg A)p$$

$$p = rel(a)p + 1 - rel(a) - (1 - rel(a))p$$

$$p = rel(a)p + 1 - rel(a) - p + rel(a)p$$

$$2p = 2rel(a)p + 1 - rel(a)$$

$$2p - 2rel(a)p = 1 - rel(a)$$

$$2p(1 - rel(a)) = 1 - rel(a)$$

$$2p = 1$$

First we multiplied both sides by the denominator. Then we divided both sides by $P(a|A)$ and multiplied on the right side. Then we used our background notation and information. Next, we manipulated the right-hand side algebraically and moved $-p$ to the left-hand side. Move $2rel(a)p$ to the left and manipulate the result algebraically to get to the last line last line (naturally, we assume $rel(a) \neq 1$).

OPTIONAL CONTENT ENDS

But how does our preference for the likelihood ratio as a measure of evidence strength relate to DAC? Let's go through Dawid's reasoning.

A sensible way to probabilistically interpret the 70% reliability of a witness who testifies that A is to take it to consist in the fact that the probability of a positive testimony if A is the case, just as the probability of a negative testimony (that is, testimony that A is false) if A isn't the case, is 0.7:[9]

$$P_{prior}(a|A) = P_{prior}(\neg a|\neg A) = 0.7.$$

$P_{prior}(a|\neg A) = 1 - P_{prior}(\neg a|\neg A) = 0.3$, and so the same information is encoded in the appropriate likelihood ratio:

$$LR_{prior}(a|A) = \frac{P_{prior}(a|A)}{P_{prior}(a|\neg A)} = \frac{0.7}{0.3}$$

Let's say that a *provides (positive) support* for A in case

$$O_{posterior}(A) = O_{prior}(A|a) > O_{prior}(A)$$

that is, a testimony a supports A just in case the posterior odds of A given a are greater than the prior odds of A (this happens just in case $P_{posterior}(A) > P_{prior}(A)$). By (1), this will be the case if and only if $LR_{prior}(a|A) > 1$.

One question that Dawid addresses is this: assuming reliability of witnesses 0.7, and assuming that a and b, taken separately, provide positive support for their respective claims, does it follow that $a \wedge b$ provides positive support for $A \wedge B$?

Assuming the independence of the witnesses, this will hold in non-degenerate cases that do not involve extreme probabilities, on the assumption of independence of a and b conditional on all combinations: $A \wedge B$, $A \wedge \neg B$, $\neg A \wedge B$ and $\neg A \wedge \neg B$.[10, 11]

---

[9]In general setting, these are called the *sensitivity* and *specificity* of a test (respectively), and they don't have to be equal. For instance, a degenerate test for an illness which always responds positively, diagnoses everyone as ill, and so has sensitivity 1, but specificity 0.

[10]Dawid only talks about the independence of witnesses without reference to conditional independence. Conditional independence does not follow from independence, and it is the former that is needed here (also, four non-equivalent different versions of it).

[11]In terms of notation and derivation in the optional content that will follow, the claim holds if and only if $28 > 28p_{11} - 12p_{00}$. This inequality is not true for all admissible values of p_{11} and p_{00}. If $p_{11} = 1$ and $p_{00} = 0$, the sides are equal. However, this is a rather degenerate example. Normally, we are interested in cases where $p_{11} < 1$. And indeed, on this assumption, the inequality holds.

Let us see why the above claim holds. The calculations are my reconstruction and are not due to Dawid. The reader might be annoyed with me working out the mundane details of Dawid's claims, but it turns out that in the case of Dawid's strategy, the devil is in the details. The independence of witnesses gives us:

$$P(a \wedge b | A \wedge B) = 0.7^2 = 0.49$$
$$P(a \wedge b | A \wedge \neg B) = 0.7 \times 0.3 = 0.21$$
$$P(a \wedge b | \neg A \wedge B) = 0.3 \times 0.7 = 0.21$$
$$P(a \wedge b | \neg A \wedge \neg B) = 0.3 \times 0.3 = 0.09$$

Without assuming A and B to be independent, let the probabilities of $A \wedge B$, $\neg A \wedge B$, $A \wedge \neg B$, $\neg A \wedge \neg B$ be $p_{11}, p_{01}, p_{10}, p_{00}$. First, let's see what $P(a \wedge b)$ boils down to.

By the law of total probability we have:

$$P(a \wedge b) = P(a \wedge b | A \wedge B)P(A \wedge B)+ \qquad (2)$$
$$+ P(a \wedge b | A \wedge \neg B)P(A \wedge \neg B)$$
$$+ P(a \wedge b | \neg A \wedge B)P(\neg A \wedge B)+$$
$$+ P(a \wedge b | \neg A \wedge \neg B)P(\neg A \wedge \neg B)$$

which, when we substitute our values and constants, results in:

$$= 0.49p_{11} + 0.21(p_{10} + p_{01}) + 0.09p_{00}$$

Now, note that because p_{ii}s add up to one, we have $p_{10} + p_{01} = 1 - p_{00} - p_{11}$. Let us continue.

$$= 0.49p_{11} + 0.21(1 - p_{00} - p_{11}) + 0.09p_{00}$$
$$= 0.21 + 0.28p_{11} - 0.12p_{00}$$

Next, we ask what the posterior of $A \wedge B$ given $a \wedge b$ is (in the last line, we also multiply the numerator and the denominator by 100).

$$P(A \wedge B | a \wedge b) = \frac{P(a \wedge b | A \wedge B)P(A \wedge B)}{P(a \wedge b)}$$
$$= \frac{49p_{11}}{21 + 28p_{11} - 12p_{00}}$$

In this particular case, then, our question whether $P(A \wedge B | a \wedge b) > P(A \wedge B)$ boils down to asking whether

$$\frac{49p_{11}}{21 + 28p_{11} - 12p_{00}} > p_{11}$$

that is, whether $28 > 28p_{11} - 12p_{00}$ (just divide both sides by p_{11}, multiply by the denominator, and manipulate algebraically).

OPTIONAL CONTENT ENDS

Dawid continues working with particular choices of values and provides neither a general statement of the fact that the above considerations instantiate nor a proof of it. In the middle of the paper he says:

> Even under prior dependence, the combined support is always positive, in the sense that the posterior probability of the case always exceeds its prior probability... When the problem is analysed carefully, the 'paradox' evaporates [pp. 95-7]

where he still means the case with the particular values that he has given, but he seems to suggest that the claim generalizes to a large array of cases.

The paper does not contain a precise statement making the conditions required explicit and, *a fortriori*, does not contain a proof of it. Given the example above and Dawid's informal reading, let us develop a more precise statement of the claim and a proof thereof.

Fact 1. *Suppose that $rel(a), rel(b) > 0.5$ and witnesses are independent conditional on all Boolean combinations of A and B (in a sense to be specified), and that none of the Boolean combinations of A and B has an extreme probability (of 0 or 1). It follows that $P(A \wedge B | a \wedge b) > P(A \wedge B)$. (Independence of A and B is not required.)*

Roughly, the theorem says that if independent and reliable witnesses provide positive support of their separate claims, their joint testimony provides positive support of the conjunction of their claims.

OPTIONAL CONTENT STARTS

Let us see why the claim holds. First, we introduce an abbreviation for witness reliability:

$$\mathbf{a} = rel(a) = P(a|A) = P(\neg a|\neg A) > 0.5$$
$$\mathbf{b} = rel(b) = P(b|B) = P(\neg b|\neg A) > 0.5$$

Our independence assumption means:

$$P(a \wedge b | A \wedge B) = \mathbf{ab}$$
$$P(a \wedge b | A \wedge \neg B) = \mathbf{a(1 - b)}$$
$$P(a \wedge b | \neg A \wedge B) = \mathbf{(1 - a)b}$$
$$P(a \wedge b | \neg A \wedge \neg B) = \mathbf{(1 - a)(1 - b)}$$

Abbreviate the probabilities the way we already did:

$$P(A \wedge B) = p_{11} \qquad P(A \wedge \neg B) = p_{10}$$
$$P(\neg A \wedge B) = p_{01} \qquad P(\neg A \wedge \neg B) = p_{00}$$

Our assumptions entail $0 \neq p_{ij} \neq 1$ for $i, j \in \{0, 1\}$ and:

$$p_{11} + p_{10} + p_{01} + p_{00} = 1 \tag{3}$$

So, we can use this with (2) to get:

$$P(a \wedge b) = \mathbf{ab}p_{11} + \mathbf{a(1-b)}p_{10} + \mathbf{(1-a)b}p_{01} + \mathbf{(1-a)(1-b)}p_{00} \tag{4}$$
$$= p_{11}\mathbf{ab} + p_{10}(\mathbf{a - ab}) + p_{01}(\mathbf{b - ab}) + p_{00}(\mathbf{1 - b - a + ab})$$

Let's now work out what the posterior of $A \wedge B$ will be, starting with an application of the Bayes' Theorem:

$$P(A \wedge B | a \wedge b) = \frac{P(a \wedge b | A \wedge B)P(A \wedge B)}{P(a \wedge b)}$$

$$= \frac{\mathbf{ab}p_{11}}{p_{11}\mathbf{ab} + p_{10}(\mathbf{a - ab}) + p_{01}(\mathbf{b - ab}) + p_{00}(\mathbf{1 - b - a + ab})} \tag{5}$$

To answer our question we therefore have to compare the content of (5) to p_{11} and our claim holds just in case:

$$\frac{\mathbf{ab}p_{11}}{p_{11}\mathbf{ab} + p_{10}(\mathbf{a - ab}) + p_{01}(\mathbf{b - ab}) + p_{00}(\mathbf{1 - b - a + ab})} > p_{11}$$

$$\frac{\mathbf{ab}}{p_{11}\mathbf{ab} + p_{10}(\mathbf{a - ab}) + p_{01}(\mathbf{b - ab}) + p_{00}(\mathbf{1 - b - a + ab})} > 1$$

$$p_{11}\mathbf{ab} + p_{10}(\mathbf{a - ab}) + p_{01}(\mathbf{b - ab}) + p_{00}(\mathbf{1 - b - a + ab}) < \mathbf{ab} \tag{6}$$

Proving (6) is therefore our goal for now. This is achieved by the following reasoning:[12]

1.	$b > 0.5$, $a > 0.5$	assumption
2.	$2b > 1$, $2a > 1$	from 1.
3.	$2ab > a$, $2ab > b$	multiplying by a and b respectively
4.	$p_{10}2ab > p_{10}a$, $p_{01}2ab > p_{01}b$	multiplying by p_{10} and p_{01} respectively
5.	$p_{10}2ab + p_{01}2ab > p_{10}a + p_{01}b$	adding by sides, 3., 4.
6.	$1 - b - a < 0$	from 1.
7.	$p_{00}(1 - b - a) < 0$	From 6., because $p_{00} > 0$
8.	$p_{10}2ab + p_{01}2ab > p_{10}a + p_{01}b + p_{00}(1 - b - a)$	from 5. and 7.
9.	$p_{10}ab + p_{10}ab + p_{01}ab + p_{01}ab + p_{00}ab - p_{00}ab > p_{10}a + p_{01}b + p_{00}(1 - b - a)$	8., rewriting left-hand side
10.	$p_{10}ab + p_{01}ab + p_{00}ab > -p_{10}ab - p_{01}ab + p_{00}ab + p_{10}a + p_{01}b + p_{00}(1 - b - a)$	9., moving from left to right
11.	$ab(p_{10} + p_{01} + p_{00}) > p_{10}(a - ab) + p_{01}(b - ab) + p_{00}(1 - b - a + ab)$	10., algebraic manipulation
12.	$ab(1 - p_{11}) > p_{10}(a - ab) + p_{01}(b - ab) + p_{00}(1 - b - a + ab)$	11. and equation (3)
13.	$ab - abp_{11} > p_{10}(a - ab) + p_{01}(b - ab) + p_{00}(1 - b - a + ab)$	12., algebraic manipulation
14.	$ab > abp_{11} + p_{10}(a - ab) + p_{01}(b - ab) + p_{00}(1 - b - a + ab)$	13., moving from left to right

---

[12]Thanks to Pawel Pawlowski for working on this proof with me.

The last line is what we have been after.

Now that we have as a theorem an explication of what Dawid informally suggested, let's see whether it helps the probabilist handling of DAC.

6 Troubles with the likelihood strategy

Recall that DAC was a problem posed for the decision standard proposed by TLP, and the real question is how the information resulting from Fact 1 can help to avoid that problem. Dawid does not mention any decision standard, and so addresses quite a different question, and so it is not clear that "the 'paradox' evaporates", as Dawid suggests.

What Dawid correctly suggests (and we establish in general as Fact 1) is that the support of the conjunction by two witnesses will be positive as soon as their separate support for the conjuncts is positive. That is, that the posterior of the conjunction will be higher that its prior. But the critic of probabilism never denied that the conjunction of testimonies might raise the probability of the conjunction if the testimonies taken separately support the conjuncts taken separately. Such a critic can still insist that Fact 1 does nothing to alleviate her concern. After all, at least *prima facie* it still might be the case that:

- the posterior probabilities of the conjuncts are above a given threshold,

- the posterior probability of the conjunction is higher than the prior probability of the conjunction,

- the posterior probability of the conjunction is still below the threshold.

That is, Fact 1 does not entail that once the conjuncts satisfy a decision standard, so does the conjunction.

At some point, Dawid makes a general claim that is somewhat stronger than the one already cited:

> When the problem is analysed carefully, the 'paradox' evaporates: suitably measured, the support supplied by the conjunction of several independent testimonies exceeds that supplied by any of its constituents.
>
> [p. 97]

This is quite a different claim from the content of Fact 1, because previously the joint probability was claimed only to increase as compared to the prior, and here it is claimed to increase above the level of the separate increases provided by separate testimonies. Regarding this issue Dawid elaborates (we still use the p_{ij}-notation that we've already introduced):

"More generally, let $P(a|A)/P(a|\neg A) = \lambda$, $P(b|B)/P(b|\neg B) = \mu$, with $\lambda, \mu > 0.7$, as might arise, for example, when there are several available testimonies. If the witnesses are independent, then

$$P(A \wedge B|a \wedge b) = \lambda\mu p_{11}/(\lambda\mu p_{11} + \lambda p_{10} + \mu p_{01} + p_{00})$$

which increases with each of λ and μ, and is never less than the larger of $\lambda p_{11}/(1 - p_{11} + \lambda p_{11})$, $\mu p_{11}/(1 - p_{11}1 + \mu p_{11})$, the posterior probabilities appropriate to the individual testimonies." [p. 95]

This claim, however, is false.

_____
OPTIONAL CONTENT STARTS
_____

Let us see why. The quoted passage is a bit dense. It contains four claims for which no arguments are given in the paper. The first three are listed below as (7), the fourth is that if the conditions in (7) hold, $P(A \wedge B|a \wedge b) > max(P(A|a), P(B|b))$. Notice that $\lambda = LR(a|A)$ and $\mu = LR(b|B)$. Suppose the first three claims hold, that is:

$$P(A \wedge B|a \wedge b) = \lambda\mu p_{11}/(\lambda\mu p_{11} + \lambda p_{10} + \mu p_{01} + p_{00}) \tag{7}$$

$$P(A|a) = \frac{\lambda p_{11}}{1 - p_{11} + \lambda p_{11}}$$

$$P(B|b) = \frac{\mu p_{11}}{1 - p_{11} + \mu p_{11}}$$

Is it really the case that $P(A \wedge B|a \wedge b) > P(A|a), P(B|b)$? It does not seem so. Let $\mathbf{a} = \mathbf{b} = 0.6$, $pr = \langle p_{11}, p_{10}, p_{01}, p_{00} \rangle = \langle 0.1, 0.7, 0.1, 0.1 \rangle$. Then, $\lambda = \mu = 1.5 > 0.7$ so the assumption is satisfied. Then we have $P(A) = p_{11} + p_{10} = 0.8$, $P(B) = p_{11} + p_{01} = 0.2$. We can also easily compute $P(a) = \mathbf{a}P(A) + (1 - \mathbf{a})P(\neg A) = 0.56$ and $P(b) = \mathbf{b}P(B) + (1 - \mathbf{b})P(\neg B) = 0.44$. Yet:

$$P(A|a) = \frac{P(a|A)P(A)}{P(a)} = \frac{0.6 \times 0.8}{0.6 \times 0.8 + 0.4 \times 0.2} \approx 0.8571$$

$$P(B|b) = \frac{P(b|B)P(B)}{P(b)} = \frac{0.6 \times 0.2}{0.6 \times 0.2 + 0.4 \times 0.8} \approx 0.272$$

$$P(A \wedge B|a \wedge b) = \frac{P(a \wedge b|A \wedge B)P(A \wedge B)}{\begin{array}{c} P(a \wedge b|A \wedge B)P(A \wedge B) + P(a \wedge b|A \wedge \neg B)P(A \wedge \neg B) + \\ + P(a \wedge b|\neg A \wedge B)P(\neg A \wedge B) + P(a \wedge b|\neg A \wedge \neg B)P(\neg A \wedge \neg B) \end{array}}$$

$$= \frac{\mathbf{ab}p_{11}}{\mathbf{ab}p_{11} + \mathbf{a}(1-\mathbf{b})p_{10} + (1-\mathbf{a})\mathbf{b}p_{01} + (1-\mathbf{a})(1-\mathbf{b})p_{00}} \approx 0.147$$

The posterior probability of $A \wedge B$ is not only lower than the larger of the individual posteriors, but also lower than any of them!

So what went wrong in Dawid's calculations in (7)? Well, the first formula is correct. However, let us take a look at what the second one says (the problem with the third one is pretty much the same):

$$P(A|a) = \frac{\frac{P(a|A)}{P(\neg a|A)} \times P(A \wedge B)}{P(\neg(A \wedge B)) + \frac{P(a|A)}{P(\neg a|A)} \times P(A \wedge B)}$$

Quite surprisingly, in Dawid's formula for $P(A|a)$, the probability of $A \wedge B$ plays a role. To see that it should not take any B that excludes A and the formula will lead to the conclusion that *always* $P(A|a)$ is undefined. The problem with Dawid's formula is that instead of $p_{11} = P(A \wedge B)$ he should have used $P(A) = p_{11} + p_{10}$, in which case the formula would rather say this:

$$P(A|a) = \frac{\frac{P(a|A)}{P(\neg a|A)} \times P(A)}{P(\neg A) + \frac{P(a|A)}{P(\neg a|A)} \times P(A)}$$

$$= \frac{\frac{P(a|A)P(A)}{P(\neg a|A)}}{\frac{P(\neg a|A)P(\neg A)}{P(\neg a|A)} + \frac{P(a|A)P(A)}{P(\neg a|A)}}$$

$$= \frac{P(a|A)P(A)}{P(\neg a|A)P(\neg A) + P(a|A)P(A)}$$

Now, on the assumption that witness' sensitivity is equal to their specificity, we have

$P(a|\neg A) = P(\neg a|A)$ and can substitute this in the denominator:

$$= \frac{P(a|A)P(A)}{P(a|\neg A)P(\neg A) + P(a|A)P(A)}$$

and this would be a formulation of Bayes' theorem. And indeed with $P(A) = p_{11}+p_{10}$ the formula works (albeit its adequacy rests on the identity of $P(a|\neg A)$ and $P(\neg a|A)$), and yields the result that we already obtained:

$$P(A|a) = \frac{\lambda(p_{11} + p_{10})}{1 - (p_{11} + p_{10}) + \lambda(p_{11} + p_{10})}$$

$$= \frac{1.5 \times 0.8}{1 - 0.8 + 1.5 \times 0.8} \approx 0.8571$$

The situation cannot be much improved by taking **a** and **b** to be high. For instance, if they're both 0.9 and $pr = \langle 0.1, 0.7, 0.1, 0.1 \rangle$, the posterior of A is ≈ 0.972, the posterior of B is ≈ 0.692, and yet the joint posterior of $A \wedge B$ is 0.525.

The situation cannot also be improved by saying that at least if the threshold is 0.5, then as soon as **a** and **b** are above 0.7 (and, *a fortriori*, so are λ and μ), the individual posteriors being above 0.5 entails the joint posterior being above 0.5 as well. For instance, for **a** = 0.7 and **b** = 0.9 with $pr = \langle 0.1, 0.3, 0.5, 0.1 \rangle$, the individual posteriors of A and B are ≈ 0.608 and ≈ 0.931 respectively, while the joint posterior of $A \wedge B$ is ≈ 0.283.

OPTIONAL CONTENT ENDS

The situation cannot be improved by saying that what was meant was rather that the joint likelihood is going to be at least as high as the maximum of the individual likelihoods, because quite the opposite is the case: the joint likelihood is going to be lower than any of the individual ones.

OPTIONAL CONTENT STARTS

Let us make sure this is the case. We have:

$$LR(a|A) = \frac{P(a|A)}{P(a|\neg A)}$$

$$= \frac{P(a|A)}{P(\neg a|A)}$$

$$= \frac{\mathbf{a}}{1 - \mathbf{a}}.$$

where the substitution in the denominator is legitimate only because witness' sensitivity is identical to their specificity.

With the joint likelihood, the reasoning is just a bit more tricky. We will need to know what $P(a \wedge b|\neg(A \wedge B))$ is. There are three disjoint possible conditions in which the condition holds: $A \wedge \neg B$, $\neg A \wedge B$, and $\neg A \wedge \neg B$. The probabilities of $a \wedge b$ in these three scenarios are respectively $\mathbf{a}(1-\mathbf{b}), (1-\mathbf{a})\mathbf{b}, (1-\mathbf{a})(1-\mathbf{b})$ (again, the assumption of independence is important), and so on the assumption $\neg(A \wedge B)$ the probability of $a \wedge b$ is:

$$
\begin{aligned}
P(a \wedge b | \neg(A \wedge B)) &= \mathbf{a}(1-\mathbf{b}) + (1-\mathbf{a})\mathbf{b} + (1-\mathbf{a})(1-\mathbf{b}) \\
&= \mathbf{a}(1-\mathbf{b}) + (1-\mathbf{a})(\mathbf{b}+1-\mathbf{b}) \\
&= \mathbf{a}(1-\mathbf{b}) + (1-\mathbf{a}) \\
&= \mathbf{a} - \mathbf{ab} + 1 - \mathbf{a} = 1 - \mathbf{ab}
\end{aligned}
$$

So, on the assumption of witness independence, we have:

$$
\begin{aligned}
LR(a \wedge b | A \wedge B) &= \frac{P(a \wedge b | A \wedge B)}{P(a \wedge b | \neg(A \wedge B))} \\
&= \frac{\mathbf{ab}}{1 - \mathbf{ab}}
\end{aligned}
$$

With $0 < \mathbf{a}, \mathbf{b} < 1$ we have $\mathbf{ab} < \mathbf{a}$, $1 - \mathbf{ab} > 1 - \mathbf{a}$, and consequently:

$$
\frac{\mathbf{ab}}{1 - \mathbf{ab}} < \frac{\mathbf{a}}{1 - \mathbf{a}}
$$

which means that the joint likelihood is going to be lower than any of the individual ones.

OPTIONAL CONTENT ENDS

Fact 1 is so far the most optimistic reading of the claim that if witnesses are independent and fairly reliable, their testimonies are going to provide positive support for the conjunction,[13] and any stronger reading of Dawid's suggestions fails. But Fact 1 is not too exciting when it comes to answering the original DAC. The original question focused on the adjudication model according to which the deciding agents are to evaluate the posterior probability of the whole case conditional on all evidence, and to convict if it is above a certain threshold. The problem, generally, is that it might be the case that the pieces of evidence for particular elements of the claim can have high likelihood and posterior probabilities of particular elements can be above the threshold while the posterior joint probability will still fail to meet the threshold. The fact that the joint posterior will be higher than the joint prior does not help much. For instance, if $\mathbf{a} = \mathbf{b} = 0.7$, $pr = \langle 0.1, 0.5, 0.3, 0.1 \rangle$, the posterior of

---

[13] And this is the reading that Dawid in passing suggests: "the combined support is always positive, in the sense that the posterior probability of the case always exceeds its prior probability." [9, 95]

A is ≈ 0.777, the posterior of B is ≈ 0.608 and the joint posterior is ≈ 0.216 (yes, it is higher than the joint prior $= 0.1$, but this does not help the conjunction to satisfy the decision standard).

To see the extent to which Dawid's strategy is helpful here, perhaps the following analogy might be useful. Imagine it is winter, the heating does not work in my office and I am quite cold. I pick up the phone and call maintenance. A rather cheerful fellow picks up the phone. I tell him what my problem is, and he reacts:

— Oh, don't worry.
— What do you mean? It's cold in here!
— No no, everything is fine, don't worry.
— It's not fine! I'm cold here!
— Look, sir, my notion of it being warm in your office is that the building provides some improvement to what the situation would be if it wasn't there. And you agree that you're definitely warmer than you'd be if your desk was standing outside, don't you? Your, so to speak, posterior warmth is higher than your prior warmth, right?

Dawid's discussion is in the vein of the above conversation. In response to a problem with the adjudication model under consideration Dawid simply invites us to abandon thinking in terms of it and to abandon requirements crucial for the model. Instead, he puts forward a fairly weak notion of support (analogous to a fairly weak sense of the building providing improvement), according to which, assuming witnesses are fairly reliable, if separate fairly reliable witnesses provide positive support to the conjuncts, then their joint testimony provides positive support for the conjunction.

As far as our assessment of the original adjudication model and dealing with DAC, this leaves us hanging. Yes, if we abandon the model, DAC does not worry us anymore. But should we? And if we do, what should we change it to, if we do not want to be banished from the paradise of probabilistic methods?

Having said this, let me emphasize that Dawid's paper is important in the development of the debate, since it shifts focus on the likelihood ratios, which for various reasons are much better measures of evidential support provided by particular pieces of evidence than mere posterior probabilities.

Before we move to another attempt at a probabilistic formulation of the decision standard, let us introduce the other hero of our story: the gatecrasher paradox. It is against DAC and this paradox that the next model will be judged.

OPTIONAL CONTENT STARTS

In fact, Cohen replied to Dawid's paper [7]. His reply, however, does not have much to do with the workings of Dawid's strategy, and is rather unusual. Cohen's first point is that the calculations of posteriors require odds about unique events, whose meaning is usually given in terms of potential wagers – and the key criticism here is that in practice such wagers cannot be decided. This is not a convincing criticism, because the betting-odds interpretations of subjective probability do not require that on each occasion the bet should really be practically decidable. It rather invites one to imagine a possible situation in which the truth could be found out and asks: how much would we bet on a certain claim in such a situation? In some cases, this assumption is false, but there is nothing in principle wrong with thinking about the consequences of false assumptions.

Second, Cohen says that Dawid's argument works only for testimonial evidence, not for other types thereof. But this claim is simply false – just because Dawid used testimonial evidence as an example that he worked through it by no means follows that the approach cannot be extended. After all, as long as we can talk about sensitivity and specificity of a given piece of evidence, everything that Dawid said about testimonies can be repeated *mutatis mutandis*.

Third, Cohen complaints that Dawid in his example worked with rather high priors, which according to Cohen would be too high to correspond to the presumption of innocence. This also is not a very successful rejoinder. Cohen picked his priors in the example for the ease of calculations, and the reasoning can be run with lower priors. Moreover, instead of discussing the conjunction problem, Cohen brings in quite a different problem: how to probabilistically model the presumption of innocence, and what priors of guilt should be appropriate? This, indeed, is an important problem; but it does not have much to do with DAC, and should be discussed separately.

OPTIONAL CONTENT ENDS

7 The gatecrasher paradox

Here's another problem with TLP, the *paradox of the gatecrasher* [6, 21]. A variant of the paradox goes as follows:

> Suppose our guilt threshold is high, say at 0.99. Consider the situation in which 1000 fans enter a football stadium, and 991 of them avoid paying for their tickets. A random spectator is tried for not paying. The probability that the spectator under trial did not pay exceeds 0.99. Yet, intuitively, a spectator cannot be considered guilty on the sole basis of the number of people who did and did not pay.[14]

---

[14]The thought experiment that in the absence of any other evidence, the only source of proba-

The thought experiment can be adapted to match any particular threshold that a proponent of TLP might suggest, as long as it is < 1. For any such a choice of a threshold, it seems, we can think of a situation where all available evidence increases the probability of guilt above it, and yet, conviction seems unjustified.

The problem is not only that TLP leads to a conviction that intuitively seems unjustified and might be wrong. Once we notice that our evidence about each spectator is exactly the same, TLP seems to commit us to the conclusion that all of them should be punished, including the nine that actually paid, as long as we can't tell them apart. And arguably, there is something disturbing in the idea of a system of justice which pretty much explicitly admits that some innocent people should be punished.

The gatecrasher paradox can be considered (or at least has been considered by some scholars) illustrative of a wider phenomenon. According to at least some approaches, there is an important distinction between *naked statistical evidence*, such as the evidence involved in the Gatecrasher Paradox, and *individualized evidence* (such as, say, eyewitness testimony) [13]. Seemingly, judges and human subjects are less willing to convict based on naked statistical evidence than when individualized evidence is available, despite the subjective probability of guilt being the same [31].

Philosophers accepting this distinction have proposed many different explications of what this supposed difference consists in exactly, without much agreement being reached.[15] However, the underdevelopment of philosophical theories aside, as the gatecrasher paradox and some real cases based solely on DNA cold hits that got thrown away indicate, there are at least some cases in which the probability of guilt given the evidence might be high, and the conviction still is not justified. Arguably, a probabilistic explication of judiciary decision standard should at least allow for this possibility and specify the conditions under which this might happen.

8 Cheng's Relative Legal Probabilism (RLP)

Let us think about juridical decisions in analogy to statistical hypothesis testing. We have two hypotheses under consideration: defendant's H_Δ and plaintiff's H_Π, and we are to pick one: D_Δ stands for the decision for H_Δ and D_Π is the decision that H_Π. If we are right, no costs result, but incorrect decisions have their price. Let us say that if the defendant is right and we find against them, the cost is c_1,

---

bilistic information is the statistics, and so that the probability of guilt corresponds to the frequency of unpaid admissions. If the reader does not agree, I ask her to play along, and to notice that in such a case a principled story of what the probability of guilt is and why is needed.

[15]See [22] for a critical survey and [10] and [25] for more recent proposals.

and if the plaintiff is right and we find against them, the cost is c_2:

		Decision	
		D_Δ	D_Π
Truth	H_Δ	0	c_1
	H_Π	c_2	0

Arguably, we need a decision rule which minimizes the expected cost. Say that given our total evidence E we have the corresponding probabilities:

$$p_\Delta = \mathrm{P}(H_\Delta|E)$$
$$p_\Pi = \mathrm{P}(H_\Pi|E)$$

where P stands for the prior probability (this will be the case throughout our discussion of Cheng). The expected costs for deciding that H_Δ and H_Π, respectively, are:

$$E(D_\Delta) = p_\Delta 0 + p_\Pi c_2 = c_2 p_\Pi$$
$$E(D_\Pi) = p_\Delta c_1 + p_\Pi 0 = c_1 p_\Delta$$

so, assuming that we are minimizing expected cost, we would like to choose H_Π just in case $E(D_\Pi) < E(D_\Delta)$. This condition is equivalent to:

$$c_1 p_\Delta < c_2 p_\Pi$$
$$c_1 < \frac{c_2 p_\Pi}{p_\Delta}$$
$$\frac{c_1}{c_2} < \frac{p_\Pi}{p_\Delta} \tag{8}$$

[5, 1261] insists:

> At the same time, in a civil trial, the legal system expresses no preference between finding erroneously for the plaintiff (false positives) and finding erroneously for the defendant (false negatives). The costs c_1 and c_2 are thus equal...

If we grant this assumption, (8) reduces to:

$$1 < \frac{p_\Pi}{p_\Delta}$$
$$p_\Pi > p_\Delta \tag{9}$$

That is, in standard civil litigation we are to find for the plaintiff just in case H_Π is more probable given the evidence than H_Δ, which doesn't seem like an insane conclusion.[16]

So on this approach, rather than directly evaluating the probability of H_Π given the evidence and comparing it to a threshold, we compare the support that the evidence provides for alternative hypotheses H_Π and H_Δ (where, let's emphasize again, the latter doesn't have to be the negation of the former), and decide for the better supported one. Let's call this decision standard **Relative Legal Probabilism (RLP)**.[17]

9 RLP vs. DAC

How is RLP supposed to handle DAC? Consider an imaginary case, used by Cheng to discuss this issue. In it, the plaintiff claims that the defendant was speeding (S) and that the crash caused her neck injury (C). Thus, H_Π is $S \wedge C$. Suppose that given total evidence E, the conjuncts, taken separately, meet the decision standard of RLP:

$$\frac{P(S|E)}{P(\neg S|E)} > 1 \qquad\qquad \frac{P(C|E)}{P(\neg C|E)} > 1$$

The question, clearly, is whether $\frac{P(S \wedge C|E)}{H_\Delta|E} > 1$. But to answer it, we have to decide what H_Δ is. This is the point where Cheng's remark that H_Δ isn't normally simply $\neg H_\Pi$. Instead, he insists, there are three alternative defense scenarios: $H_{\Delta_1} = S \wedge \neg C$, $H_{\Delta_2} = \neg S \wedge C$, and $H_{\Delta_3} = \neg S \wedge \neg C$. How does H_Π compare to each of them? Cheng (assuming independence) argues:

$$\frac{P(S \wedge C|E)}{P(S \wedge \neg C|E)} = \frac{P(S|E)P(C|E)}{P(S|E)P(\neg C|E)} = \frac{P(C|E)}{P(\neg C|E)} > 1 \tag{10}$$

$$\frac{P(S \wedge C|E)}{P(\neg S \wedge C|E)} = \frac{P(S|E)P(C|E)}{P(\neg S|E)P(C|E)} = \frac{P(S|E)}{P(\neg S|E)} > 1$$

$$\frac{P(S \wedge C|E)}{P(\neg S \wedge \neg C|E)} = \frac{P(S|E)P(C|E)}{P(\neg S|E)P(\neg C|E)} > 1$$

---

[16]Notice that this instruction is somewhat more general than the usual suggestion of the preponderance standard in civil litigation, according to which the court should find for the plaintiff just in case $P(H_\Pi|E) > 0.5$. This threshold, however, results from (9) if it so happens that H_Δ is $\neg H_\Pi$, that is, if the defendant's claim is simply the negation of the plaintiff's thesis. By no means, Cheng argues, this is always the case: often the defendant offers a story which is much more than simply the denial of what the opposite side said.

[17]I was not aware of any particular name for Cheng's model so we came up with this one. We're not particularly attached to it, and it is not standard terminology.

It seems that whatever the defense story is, it is less plausible than the plaintiff's claim. So, at least in this case, whenever elements of a plaintiff's claim satisfy the decision standard proposed by RLP, then so does their conjunction.

10 RLP vs. the Gatecrasher Paradox

Similarly, RLP is claimed to handle the gatecrasher paradox. It is useful to think about the problem in terms of odds and likelikoods, where the *prior odds* (before evidence E) of H_Π as compared to H_Δ, are $\frac{P(H_\Pi)}{P(H_\Delta)}$, the posterior odds of H_Δ given E are $\frac{P(H_\Pi|E)}{P(H_\Delta|E)}$, and the corresponding likelihood ratio is $\frac{P(E|H_\Pi)}{P(E|H_\Delta)}$.

Now, with this notation the *odds form of Bayes' Theorem* tells us that the posterior odds equal the likelihood ratio multiplied by prior odds:

$$\frac{P(H_\Pi|E)}{P(H_\Delta|E)} = \frac{P(E|H_\Pi)}{P(E|H_\Delta)} \times \frac{P(H_\Pi)}{P(H_\Delta)}$$

[5, 1267] insists that in civil trials the prior probabilities should be equal. Granted this assumption, prior odds are 1, and we have:

$$\frac{P(H_\Pi|E)}{P(H_\Delta|E)} = \frac{P(E|H_\Pi)}{P(E|H_\Delta)} \tag{11}$$

This means that our original task of establishing that the left-hand side is greater than 1 now reduces to establishing that so is the right-hand side, which means that RLP tells us to convict just in case:

$$P(E|H_\Pi) > P(E|H_\Delta) \tag{12}$$

Thus, (12) tells us to convict just in case $LR(E) > 1$.

In the case of the gatecrasher paradox, our evidence is statistical. In our variant E="991 out of 1000 spectators gatecrashed". Now pick a random spectator, call him Tom, and let H_Π="Tom gatecrashed." [5, 1270] insists:

> But whether the audience member is a lawful patron or a gatecrasher
> does not change the probability of observing the evidence presented.

So, on his view, in such a case, $P(E|H_\Pi) = P(E|H_\Delta)$, the posterior odds are, by (11), equal to 1, and conviction is unjustified.

11   Troubles with RLP

There are various issues with how RLP has been deployed to resolve the difficulties that CLP and TLP run into. First of all, to move from (8) to (9), Cheng assumes that the costs of wrongful decision is the same, be it conviction or acquittal. This is by no means obvious. If a poor elderly lady sues a large company for serious health damage that it supposedly caused, leaving her penniless if the company is liable is definitely not on a par with mistakenly making the company lose a small percent of their funds. Even in cases where such costs are equal, careful consideration and separate argument is needed. If, for instance, $c_1 = 5c_2$, we are to convict just in case $5 < \frac{p_\Pi}{p_\Delta}$. This limits the applicability of Cheng's reasoning about DAC, because his reasoning, if correct (and I will argue that it is not correct later on), yields only the result that the relevant posterior odds are greater than 1, not that they are greater than 5. The difficulty, however, will not have much impact on Cheng's solution of the gatecrasher paradox, as long as $c_1 \leq c_2$. This is because his reasoning, if correct, establishes that the relevant posterior odds are below 1, and so below any higher threshold as well.

Secondly, Cheng's resolution of DAC uses another suspicious assumption. For (10) to be acceptable we need to assume that the following pairs of events are independent conditionally on E: $\langle S, C \rangle$, $\langle S, \neg C \rangle$, $\langle \neg S, C \rangle$, $\langle \neg S, \neg C \rangle$. Otherwise, Cheng would not be able to replace conditional probabilities of corresponding conjunctions with the result of multiplication of conditional probabilities of the conjuncts. But it is far from obvious that speeding and neck injury are independent. If, for instance, the evidence makes it certain that if the car was not speeding, the neck injury was not caused by the accident, $P(\neg S \wedge C|E) = 0$, despite the fact that $P(\neg S|E)P(C|E)$ does not have to be 0!

Without independence, the best that we can get, say for the first line of (10), is:

$$P(S \wedge C|E) = P(C|E)P(S|C \wedge E)$$
$$P(S \wedge \neg C|E) = P(\neg C|E)P(S|\neg C \wedge E)$$

and even if we know that $P(C|E) > P(\neg C|E)$, this tells us nothing about the comparison of $P(S \wedge C|E)$ and $P(S \wedge \neg C|E)$, because the remaining factors can make up for the former inequality.

Perhaps even more importantly, much of the heavy lifting here is done by the strategic splitting of the defense line into multiple scenarios. The result is rather paradoxical. For suppose $P(H_\Pi|E) = 0.37$ and the probability of each of the defense lines given E is 0.21. This means that H_Π wins with each of the scenarios, so, according to RLP, we should find for the plaintiff. On the other hand, how eager are we to convict once we notice that given the evidence, the accusation is rather false, because $P(\neg H_\Pi | E) = 0.63$?

The problem generalizes. If, as here, we individualize scenarios by boolean combinations of elements of a case, the more elements there are, into more scenarios $\neg H_\Pi$ needs to be divided. This normally would lead to the probability of each of them being even lower (because now $P(\neg H_\Pi)$ needs to be "split" between more different scenarios). So, if we take this approach seriously, the more elements a case has, the more at disadvantage the defense is. This is clearly undesirable.

In the process of solving the gatecrasher paradox, to reach (11), Cheng makes another controversial assumption: that the prior odds should be one, that is, that before any evidence specific to the case is obtained, $P(H_\Pi) = P(H_\Delta)$. One problem with this assumption is that it is not clear how to square this with how Cheng handles DAC. For there, he insisted we need to consider *three different* defense scenarios, which we marked as $H_{\Delta_1}, H_{\Delta_2}$ and H_{Δ_3}. Now, do we take Cheng's suggestion to be that we should have

$$P(H_\Pi) = P(H_{\Delta_1}) = P(H_{\Delta_2}) = P(H_{\Delta_3})?$$

Given that the scenarios are jointly exhaustive and pairwise exclusive this would mean that each of them should have prior probability 0.25 and, in principle that the prior probability of guilt can be made lower simply by the addition of elements under consideration. This conclusion seems suboptimal.

If, on the other hand, we read Cheng as saying that we should have $P(H_\Pi) = P(\neg H_\Pi)$, the side-effect is that even a slightest evidence in support of H_Π will make the posterior probability of H_Π larger than that of $\neg H_\Pi$, and so the plaintiff can win their case way too easily. Worse still, if $P(\neg H_\Pi)$ is to be divided between multiple defense scenarios against which H_Π is to be compared, then as soon as this division proceeds in a non-extreme fashion, the prior of each defense scenario will be lower than the prior of H_Π, and so from the perspective of RLP, the plaintiff does not have to do anything to win (as long as the defense does not provide absolving evidence), because his case is won without any evidence already!

Finally, let us play along and assume that in the gatecrasher scenario the conviction is justified just in case (12) holds. Cheng insists that it does not, because $P(E|H_\Pi) = P(E|H_\Delta)$. This supposedly captures the intuition that whether Tom paid has no impact on the statistics that we have.

But this is not obvious. Here is one way to think about this. Tom either paid the entrance fee or did not. Consider these two options, assuming nothing else about the case changes. If he did pay, then he is among the 9 innocent spectators. But this means that if he had not paid, there would have been 992 gatecrashers, and so

E would be false (because it says there was 991 of them). If, on the other hand, Tom in reality did not pay (and so is among the 991 gatecrashers), then had he paid, there would have been only 990 gatecrashers and E would have been false, again!

So whether conviction is justified and what the relevant ratios are depends on whether Tom really paid. Cheng's criterion (12) results in the conclusion that Tom should be penalized if and only if he did not pay. But this does not help us much when it comes to handling the paradox, because the reason why we needed to rely on E was exactly that we did not know whether Tom paid.

If you are not buying into the above argument, here is another way to state the problem. Say your priors are $P(E) = e$, $P(H_\Pi) = \pi$. By Bayes' Theorem we have:

$$P(E|H_\Pi) = \frac{P(H_\Pi|E)e}{\pi}$$

$$P(E|H_\Delta) = \frac{P(H_\Delta|E)e}{1 - \pi}$$

Assuming our posteriors are taken from the statistical evidence, we have $P(H_\Pi|E) = 0.991$ and $P(H_\Delta|E) = 0.009$. So we have:

$$
\begin{aligned}
LR(E) &= \frac{P(H_\Pi|E)e}{\pi} \times \frac{1 - \pi}{P(H_\Delta|E)e} \\
&= \frac{P(H_\Pi|E) - P(H_\Pi|E)\pi}{P(H_\Delta|E)\pi} \\
&= \frac{0.991 - 0.991\pi}{0.009\pi}
\end{aligned}
\tag{13}
$$

and $LR(E)$ will be > 1 as soon as $\pi < 0.991$. This means that contrary to what Cheng suggested, in any situation in which the prior probability of guilt is less than the posterior probability of guilt, RLP tells us to convict. This, however, does not seem desirable.

12 Kaplow's Decision-Theoretic LP (DTLP)

On RLP, at least in certain cases, the decision rule leads us to (12), which tells us to decide the case based on whether the likelihood ratio is greater than 1. Quite independently, Kaplow [15] suggested another approach to juridical decisions which focuses on likelihood ratios, of which Cheng's suggestion is only a particular case.[18]

---

[18] Again, the name of the view is by no means standard, it is just a term I coined to refer to various types of legal probabilism in a fairly uniform manner.

While Kaplow did not discuss DAC or the gatecrasher paradox, it is only fair to evaluate Kaplow's proposal from the perspective of these difficulties.

Let $LR(E) = P(E|H_\Pi)/P(E|H_\Delta)$. In whole generality, DTLP invites us to convict just in case $LR(E) > LR^\star$, where $LR^\star$ is some critical value of the likelihood ratio.

Say we want to formulate the usual preponderance rule: convict iff $P(H_\Pi|E) > 0.5$, that is, iff $\dfrac{P(H_\Pi|E)}{P(H_\Delta|E)} > 1$. By Bayes' Theorem we have:

$$\frac{P(H_\Pi|E)}{P(H_\Delta|E)} = \frac{P(H_\Pi)}{P(H_\Delta)} \times \frac{P(E|H_\Pi)}{P(E|H_\Delta)} > 1 \Leftrightarrow$$

$$\Leftrightarrow \frac{P(E|H_\Pi)}{P(E|H_\Delta)} > \frac{P(H_\Delta)}{P(H_\Pi)}$$

So, as expected, $LR^\star$ is not unique and depends on priors. Analogous reformulations are available for thresholds other than 0.5.

However, Kaplow's point is not that we can reformulate threshold decision rules in terms of priors-sensitive likelihood ratio thresholds. Rather, he insists, when we make a decision, we should factor in its consequences. Let G represent potential gain from correct conviction, and L stand for the potential loss resulting from mistaken conviction. Taking them into account, Kaplow suggests, we should convict if and only if:

$$P(H_\Pi|E) \times G > P(H_\Delta|E) \times L \tag{14}$$

Now, (14) is equivalent to:

$$\frac{P(H_\Pi|E)}{P(H_\Delta|E)} > \frac{L}{G}$$

$$\frac{P(H_\Pi)}{P(H_\Delta)} \times \frac{P(E|H_\Pi)}{P(E|H_\Delta)} > \frac{L}{G}$$

$$\frac{P(E|H_\Pi)}{P(E|H_\Delta)} > \frac{P(H_\Delta)}{P(H_\Pi)} \times \frac{L}{G}$$

$$LR(E) > \frac{P(H_\Delta)}{P(H_\Pi)} \times \frac{L}{G} \tag{15}$$

This is the general format of Kaplow's decision standard. Now, let us see how it fares when it comes to DAC and the gatecrasher paradox.

13   Troubles with DTLP

Kaplow does not discuss the conceptual difficulties that we are concerned with, but this will not stop us from asking whether DTLP can handle them (and answering to the negative). Let us start with DAC.

Say we consider two claims, A and B. Is it generally the case that if they separately satisfy the decision rule, then so does $A \land B$? That is, do the assumptions:

$$\frac{P(E|A)}{P(E|\neg A)} > \frac{P(\neg A)}{P(A)} \times \frac{L}{G}$$

$$\frac{P(E|B)}{P(E|\neg B)} > \frac{P(\neg B)}{P(B)} \times \frac{L}{G}$$

entail

$$\frac{P(E|A \land B)}{P(E|\neg(A \land B))} > \frac{P(\neg(A \land B))}{P(A \land B)} \times \frac{L}{G}?$$

Alas, the answer is negative.

OPTIONAL CONTENT STARTS

This can be seen from the following example. Suppose a random digit from 0-9 is drawn; we do not know the result; we are told that the result is < 7 ($E =$ 'the result is < 7'), and we are to decide whether to accept the following claims:

A	the result is < 5.
B	the result is an even number.
$A \land B$	the result is an even number < 5.

Suppose that $L = G$ (this is for simplicity only — nothing hinges on this, counterexamples for when this condition fails are analogous). First, notice that A and B taken separately satisfy (15). $P(A) = P(\neg A) = 0.5$, $P(\neg A)/P(A) = 1$ $P(E|A) = 1$, $P(E|\neg A) = 0.4$. (15) tells us to check:

$$\frac{P(E|A)}{P(E|\neg A)} > \frac{L}{G} \times \frac{P(\neg A)}{P(A)}$$

$$\frac{1}{0.4} > 1$$

so, following DTLP, we should accept A. For analogous reasons, we should also accept B. $P(B) = P(\neg B) = 0.5$, $P(\neg B)/P(B) = 1$ $P(E|B) = 0.8$, $P(E|\neg B) = 0.6$, so we need to check that indeed:

$$\frac{P(E|B)}{P(E|\neg B)} > \frac{L}{G} \times \frac{P(\neg B)}{P(B)}$$

$$\frac{0.8}{0.6} > 1$$

But now, $P(A \wedge B) = 0.3$, $P(\neg(A \wedge B)) = 0.7$, $P(\neg(A \wedge B))/P(A \wedge B) = 2\frac{1}{3}$, $P(E|A \wedge B) = 1$, $P(E|\neg(A \wedge B)) = 4/7$ and it is false that:

$$\frac{P(E|A \wedge B)}{P(E|\neg(A \wedge B))} > \frac{L}{G} \times \frac{P(\neg(A \wedge B))}{P(A \wedge B)}$$

$$\frac{7}{4} > \frac{7}{3}$$

The example was easy, but the conjuncts are probabilistically dependent. One might ask: are there counterexamples that involve claims which are probabilistically independent?[19]

Consider an experiment in which someone tosses a six-sided die twice. Let the result of the first toss be X and the result of the second one Y. Your evidence is that the results of both tosses are greater than one ($E =: X > 1 \wedge Y > 1$). Now, let A say that $X < 5$ and B say that $Y < 5$.

The prior probability of A is $2/3$ and the prior probability of $\neg A$ is $1/3$ and so $\frac{P(\neg A)}{P(A)} = 0.5$. Further, $P(E|A) = 0.625$, $P(E|\neg A) = 5/6$ and so $\frac{P(E|A)}{P(E|\neg A)} = 0.75$. Clearly, $0.75 > 0.5$, so A satisfies the decision standard. Since the situation with B is symmetric, so does B.

Now, $P(A \wedge B) = (2/3)^2 = 4/9$ and $P(\neg(A \wedge B)) = 5/9$. So $\frac{P(\neg(A \wedge B))}{P(A \wedge B)} = 5/4$. Out of 16 outcomes for which $A \wedge B$ holds, E holds in 9, so $P(E|A \wedge B) = 9/16$. Out of 20 remaining outcomes for which $A \wedge B$ fails, E holds in 16, so $P(E|\neg(A \wedge B)) = 4/5$. Thus, $\frac{P(E|A \wedge B)}{P(E|\neg(A \wedge B))} = 45/64 < 5/4$, so the conjunction does not satisfy the decision standard.

OPTIONAL CONTENT ENDS

Let us turn to the gatecrasher paradox.

Suppose $L = G$ and recall our abbreviations: $P(E) = e$, $P(H_\Pi) = \pi$. DTLP tells us to convict just in case:

$$LR(E) > \frac{1 - \pi}{\pi}$$

---

[19]Thanks to Alicja Kowalewska for pressing me on this.

From (13) we already now that

$$LR(E) = \frac{0.991 - 0.991\pi}{0.009\pi}$$

so we need to see whether there are any $0 < \pi < 1$ for which

$$\frac{0.991 - 0.991\pi}{0.009\pi} > \frac{1 - \pi}{\pi}$$

Multiply both sides first by 009π and then by π:

$$0.991\pi - 0.991\pi^2 > 0.09\pi - 0.009\pi^2$$

Simplify and call the resulting function f:

$$f(\pi) = -0.982\pi^2 + 0.982\pi > 0$$

The above condition is satisfied for any $0 < \pi < 1$ (f has two zeros: $\pi = 0$ and $\pi = 1$). Here is a plot of f:

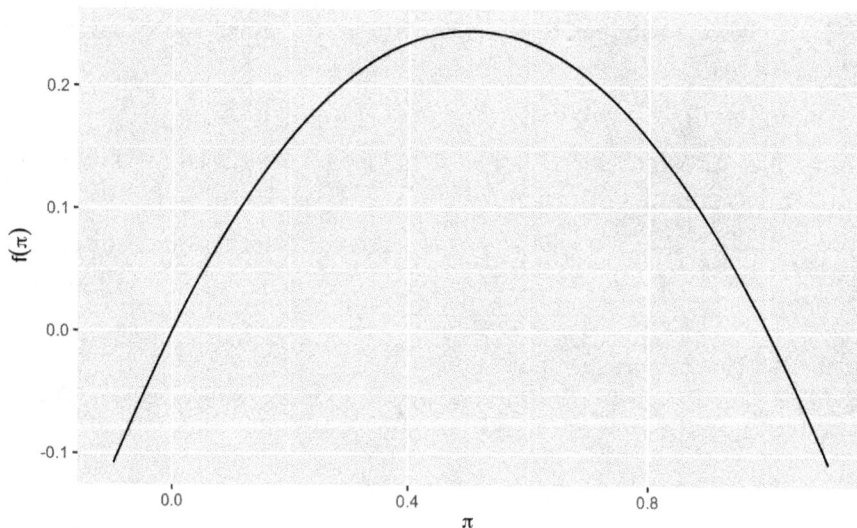

Similarly, $LR(E) > 1$ for any $0 < \pi < 1$. Here is a plot of $LR(E)$ against π:

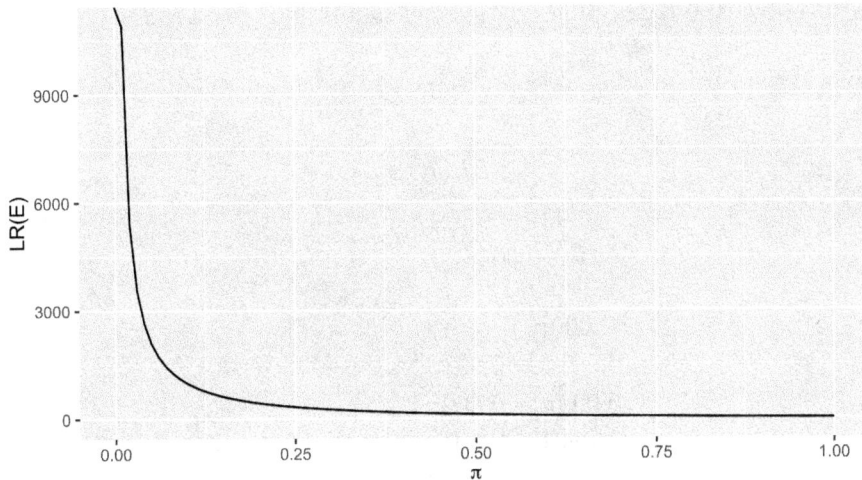

Notice that $LR(E)$ does not go below 1. This means that for $L = G$ in the gate-crasher scenario DTLP wold tell us to convict for any prior probability of guilt $\pi \neq 0, 1$.

One might ask: is the conclusion very sensitive to the choice of L and G? The answer is, not too much.

OPTIONAL CONTENT STARTS

How sensitive is our analysis to the choice of L/G? Well, $LR(E)$ does not change at all, only the threshold moves. For instance, if $L/G = 4$, instead of f we end up with

$$f'(\pi) = -0.955\pi^2 + 0.955\pi > 0$$

and the function still takes positive values on the interval $(0, 1)$. In fact, the decision won't change until L/G increases to ≈ 111. Denote L/G as ρ, and let us start with the general decision standard, plugging in our calculations for $LR(E)$:

$$LR(E) > \frac{P(H_\Delta)}{P(H_\Pi)}\rho$$

$$LR(E) > \frac{1-\pi}{\pi}\rho$$

$$\frac{0.991 - 0.991\pi}{0.009\pi} > \frac{1-\pi}{\pi}\rho$$

$$\frac{0.991 - 0.991\pi}{0.009\pi} \cdot \frac{\pi}{1-\pi} > \rho$$

$$\frac{0.991\pi - 0.991\pi^2}{0.009\pi - 0.009\pi^2} > \rho$$

$$\frac{\pi(0.991 - 0.991\pi)}{\pi(0.009 - 0.009\pi)} > \rho$$

$$\frac{0.991 - 0.991\pi}{0.009 - 0.009\pi} > \rho$$

$$\frac{0.991(1 - \pi)}{0.009(1 - \pi)} > \rho$$

$$\frac{0.991}{0.009} > \rho$$

$$110.1111 > \rho$$

OPTIONAL CONTENT ENDS

So, we conclude, in usual circumstances, DTLP does not handle the gatecrasher paradox.

OPTIONAL CONTENT STARTS

There is another, recent approach due to Miller [20].[20] Instead of using $P(H|E)$, he introduces a new function, Q, which he calls contrapositive probability, and defines it as:

$$Q(H|E) = P(\neg E|\neg H)$$

According to a theorem that Miller stated without a proof, if these assumptions hold

$$Q(H_1|E) > Q(\neg H_1|E)$$
$$Q(H_2|E) > Q(\neg H_2|E)$$

---

[20]The idea is not developed in any of his papers. What follows is an account based on his lecture at the UNILOG '18 conference.

then it follows that:

$$Q(H_1 \wedge H_2 | E) > Q(\neg(H_1 \wedge H_2) | E).$$

Full assessment of this approach will have to wait for a more complete development of the strategy. Note however, that it is far from clear that the above theorem solves the issue. It only applies to cases in which the threshold is 0.5 and says that if conjuncts are above it, then so is the conjunction. It still might be the case that the conjunction has lower "score" than any of the conjuncts, and if so, shifting the threshold might not preserve the value of the theorem.

Moreover, the measure has very unintuitive properties. Consider a single toss of a die and its result. $P(< 2| < 3)$ (that is, the probability that the result is 1 given it is less than 3) is $1/2$, and so $Q(\geq 3| \geq 2) = 1/2$. So, on this approach, the evidence that the result is one of $2, 3, 4, 5, 6$ supports the claim that it is one of $3, 4, 5, 6$ only at the level of $1/2$, despite the corresponding probability being $4/5$. Similarly, $P(< 3| < 5) = 1/2$ and so $Q(\geq 5| \geq 3) = 1/2$ (the same level as before), despite the corresponding probability this time being $1/4$.

Optional Content Ends

14   Informal overview

Where are we, how did we get here, and where can we go from here? We were looking for a probabilistically explicated condition Ψ such that the trier of fact, at least ideally, should accept any relevant claim (including G) just in case $\Psi(A, E)$.

From the discussion that transpired it should be clear that we were looking for a Ψ satisfying the following desiderata:

conjunction closure If $\Psi(A, E)$ and $\Psi(B, E)$, then $\Psi(A \wedge B, E)$.

naked statistics The account should at least make it possible for convictions based on strong, but naked statistical evidence to be unjustified.

equal treatment the condition should apply to any relevant claim whatsoever (and not just a selected claim, such as G).

Throughout the paper we focused on the first two conditions (formulated in terms of the difficulty about conjunction (DAC), and the gatecrasher paradox), going over various proposals of what Ψ should be like and evaluating how they fare. The results can be summed up in the following table:

View	Convict iff	DAC	Gatecrasher
Threshold-based LP (TLP)	Probability of guilt given the evidence is above a certain threshold	fails	fails
Dawid's likelihood strategy	No condition given, focus on $\dfrac{\mathrm{P}(H\mid E)}{\mathrm{P}(H\mid\neg E)}$	- If evidence is fairly reliable, the posterior of $A\wedge B$ will be greater than the prior. - The posterior of $A\wedge B$ can still be lower than the posterior of any of A and B. - Joint likelihood, contrary do Dawid's claim, can also be lower than any of the individual likelihoods.	fails
Cheng's relative LP (RLP)	Posterior of guilt higher than the posterior of any of the defending narrations	The solution assumes equal costs of errors and independence of A and B conditional on E. It also relies on there being multiple defending scenarios individualized in terms of combinations of literals involving A and B.	Assumes that the prior odds of guilt are 1, and that the statistics is not sensitive to guilt (which is dubious). If the latter fails, tells to convict as long as the prior of guilt < 0.991.
Kaplow's decision-theoretic LP (DTLP)	The likelihood of the evidence is higher than the odds of innocence multiplied by the cost of error ratio	fails	convict if cost ratio $<$ 110.1111

Thus, each account either simply fails to satisfy the desiderata, or succeeds on rather unrealistic assumptions. Does this mean that a probabilistic approach to legal evidence evaluation should be abandoned? No. This only means that if we are to develop a general probabilistic model of legal decision standards, we have to do better. One promising direction is to go back to Cohen's pressure against **Requirement 1** and push against it. A brief paper suggesting this direction is [3], where the idea is that the probabilistic standard (be it a threshold or a comparative standard wrt. defending narrations) should be applied to the whole claim put forward by the plaintiff, and not to its elements. In such a context, DAC does not arise, but **equal treatment** is violated. Perhaps, there are independent reasons to abandon it, but the issue deserves further discussion. Another strategy might be to go in the direction of employing probabilistic methods to explicate the narration theory of legal decision standards [29], but a discussion of how this approach relates to DAC and the gatecrasher paradox lies beyond the scope of this paper.

References

[1] Aitken, C. and Taroni, F. (2004). *Statistics and the evaluation of evidence for forensic scientists*, volume 16. Wiley Online Library.

[2] Ball, V. C. (1960). The moment of truth: probability theory and standards of proof. *Vanderbilt Law Review*, 14:807–830.

[3] Bello, M. D. (2019). Plausibility and probability in juridical proof. *The International Journal of Evidence & Proof*, doi 10.1177/1365712718815355.

[4] Bernoulli, J. (1713). *Ars conjectandi*.

[5] Cheng, E. (2012). Reconceptualizing the burden of proof. *Yale LJ*, 122:1254.

[6] Cohen, J. (1977). *The probable and the provable*. Oxford University Press.

[7] Cohen, J. (1988). The difficulty about conjunction in forensic proof. *Journal of the Royal Statistical Society: Series D (The Statistician)*, 37(4-5):415–416

[8] Cullison, A. D. (1969). Probability analysis of judicial fact-finding: A preliminary outline of the subjective approach. *Toledo Law Review*, 1:538–598.

[9] Dawid, A. P. (1987). The difficulty about conjunction. *The Statistician*, pages 91–97.

[10] Enoch, D. and Fisher, T. (2015). Sense and sensitivity: Epistemic and instrumental approaches to statistical evidence. *Stan. L. Rev.*, 67:557.

[11] Finkelstein, M. O. and Levin, B. (2001). *Statistics for lawyers*. Springer.

[12] Haack, S. (2014a). *Evidence Matters: Science, Proof, and Truth in the Law*. Cambridge University Press.

[13] Haack, S. (2014b). Legal probabilism: an epistemological dissent. In *[12]*, pages 47–77.

[14] Kaplan, J. (1968). Decision theory and the factfinding process. *Stanford Law Review*, 20:1065–1092.

[15] Kaplow, L. (2014). Likelihood ratio tests and legal decision rules. *American Law and Economics Review*, 16(1):1–39.

[16] Kaye, D. (1979). The paradox of the gatecrasher and other stories. *Arizona State Law Journal*, pages 101–110.

[17] Kaye, D. H. (1986). Do we need a calculus of weight to understand proof beyond a reasonable doubt? *Boston University Law Review*, 66(3-4).

[18] Lempert, R. O. (1977). Modeling relevance. *Michigan Law Review*, 75:1021–1057.

[19] Lucy, D. (2013). *Introduction to statistics for forensic scientists*. John Wiley & Sons.

[20] Miller, D. (2018). Cohen's criticisms of the use of probability in the law. Lecture at the Sixth World Congress on Universal Logic, Vichy.

[21] Nesson, C. R. (1979). Reasonable doubt and permissive inferences: The value of complexity. *Harvard Law Review*, 92(6):1187–1225.

[22] Redmayne, M. (2008). Exploring the proof paradoxes. *Legal Theory*, 14(4):281–309.

[23] Robertson, B., Vignaux, G., and Berger, C. (2016). *Interpreting evidence: evaluating forensic science in the courtroom*. John Wiley & Sons.

[24] Simon, R. J. and Mahan, L. (1970). Quantifying burdens of proof-a view from the bench, the jury, and the classroom. *Law and Society Review*, 5(3):319–330.

[25] Smith, M. (2017). When does evidence suffice for conviction? *Mind*.

[26] Taroni, F., Biedermann, A., Bozza, S., Garbolino, P., and Aitken, C. (2006). *Bayesian networks for probabilistic inference and decision analysis in forensic science*. John Wiley & Sons.

[27] Tillers, P. and Green, E. D., editors (1988). *Probability and Inference in the Law of Evidence. The Uses and Limits of Bayesianism*, volume 109 of *Boston studies in the philosophy of science*. Springer.

[28] Tribe, L. H. (1971a). A further critique of mathematical proof. *Harvard Law Review*, 84:1810–1820.

[29] Urbaniak, R. (2018). Narration in judiciary fact-finding: a probabilistic explication. *Artificial Intelligence and Law*, 26(4):345-376.

[30] Tribe, L. H. (1971b). Trial by mathematics: Precision and ritual in the legal process. *Harvard Law Review*, 84(6):1329–1393.

[31] Wells, G. (1992). Naked statistical evidence of liability: Is subjective probability enough? *Journal of Personality and Social Psychology*, 62(5):739–752.

Received 3 August 2018

A Deontic Argumentation Framework Towards Doctrine Reification

Régis Riveret
CSIRO, Australia
regis.riveret@data61.csiro.au

Antonino Rotolo
University of Bologna, Italy
antonino.rotolo@unibo.it

Giovanni Sartor
European University Institute, Italy
giovanni.sartor@eui.eu

Abstract

A modular rule-based argumentation system is proposed to represent and reason upon conditional norms featuring obligations, prohibitions, and (strong or weak) permissions. The approach is based on common constructs in computational models of argument: rule-based arguments, argumentation graphs, argument labelling semantics and statement labelling semantics. Deontic reasoning patterns are captured with defeasible rule schemata to the greatest extent, towards the reification of doctrinal pieces. We show then that bivalent statement labellings can fall short to address normative completeness, and for this reason, we propose to use trivalent labelling semantics. Given an argumentation graph, deontic statuses can be computed efficiently. The system is illustrated with a scenario featuring a violation and a contrary-to-duty obligation.

Keywords: Knowledge representation and reasoning, computational argumentation, legal reasoning, normative systems.

We would like to thank anonymous reviewers for their remarks, and Guido Governatori for discussions on the work reported here.

Vol. 6 No. 5 2019
Journal of Applied Logics — IfCoLog Journal of Logics and their Applications

1 Introduction

There exist multiple formalisms to capture deontic reasoning [20,35]. As deontic reasoning is often embedded into legal reasoning, and because legal reasoning is naturally formalised in formal models of arguments, deontic argumentation frameworks can be elegant and convenient formalisms to capture arguments in favour or against deontic statements. Yet, argumentation is more generally used to address defeasible claims raised on the basis of partial and conflicting pieces of information, that is, defeasible claims in uncertain contexts. Hence, argumentation may be well suited to reason upon deontic statements in uncertain contexts.

As interests in deontic systems can vary, diverse requirements or considerations can be put forward to build deontic argumentation frameworks. For example, some emulation of human discourses may be a key desideratum, while computational complexity may be a requisite for practical applications. For our purposes, we will pay attention to human interface (explainability, emulation and isomorphism), as well as more technical considerations (parsimony, modularity and computational efficiency), and a legit requirement regarding normative completeness. These design considerations are exposed in more detail later in the paper.

Bearing in mind specific design considerations, we explore a modular argumentation system. The argumentation system amounts to a knowledge-based system where conditional norms and common sense knowledge are represented as object rules into a knowledge base. The argumentation process then runs in three stages. In the first stage, arguments are built from object rules in the knowledge base by using some inference apparatus. In the second stage, arguments are labelled to reflect their acceptance. In the final stage, statements are labelled at their turn and normative completeness ensured.

To build such a knowledge-based system, we can roughly state that two main approaches can be taken: a knowledge-based (KB) approach and an inference-based (IB) approach. In the KB approach, maximum knowledge is expressed in the knowledge base, while the inference apparatus is restricted and not domain-specific. Possibly, inference schemata too are included in the knowledge base. Archetypes of the KB approach are Hilbert-style systems which usually code knowledge in axioms and axiom schemata with a few inference rules. In contrast, the IB approach tends to capture portions of knowledge in the inference apparatus which handles the knowledge base. Examples of the IB approach are Gentzen-style systems where the number of inference rules is typically larger than in Hilbert-style systems. Thus, we can say that in general the KB approach limits the number of inference patterns, while the IB approach favours a diversity of inference patterns.

Both approaches have strengths and weaknesses. If something cannot be achieved with the KB approach (e.g. the pursuit of some natural reasoning patterns, specific inferences or features such as some quantitative assessment, or a certain computational complexity), then one can take the IB approach and attempt to devise some specific inference patterns. Some trade-off may be obtained, for example by devising systems with an 'inference knowledge base' storing a diversity of inference patterns. This approach may be called here the inference-knowledge-based (IKB) approach. The advantage of the IKB approach in automated systems is that designers and users can have an easy and direct access to reasoning patterns which would be otherwise hidden in the implemented inference machinery.

Any of these options can be adopted to build a computational system where norms are represented as rules from which arguments are formed to decide about the status of deontic statements. Within the IB approach, deontic reasoning is captured with specific inference patterns (e.g. deontic detachment). Examples of such an approach in deontic argumentation are [7, 47]. Within the KB approach, deontic reasoning is written as object rules in the knowledge base with no specific inference patterns. For example, a weak permission can be supported by some arguments built out from object rules (instead of hard-coding its support from the absence of prohibitions). These object rules may be seen as a reification of inference rules into the language of the knowledge base, but they are not inference rules because they are part of the object knowledge base. If the inference rules for deontic reasoning are part of an inference knowledge base, then we have a deontic IKB approach. An example of the IKB approach is [48] which uses ASPIC$^+$ framework [37] and where inference rules are tuned for deontic reasoning.

In this paper, we explore a hybrid approach: first, we use defeasible rule schemata to account for deontic reasoning, towards doctrine reification. In particular, we investigate the reification of the principle of prohibition to address normative completeness. In that sense, we adopt a KB approach. However, for aspects of normative completeness we could not cover with defeasible rule schemata, we exploit the inference apparatus through labelling semantics, and thus we complete our KB approach with labelling inferences.

Contribution. A modular rule-based argumentation system is proposed to capture normative knowledge and reason upon it. To do so, we adopt a KB approach to the greatest extent through defeasible rule schemata, towards doctrine reification. In particular, we explore the reification of the principle of prohibition to address normative completeness. Then, for aspects of normative completeness not covered by defeasible rules schemata, we exploit the inference apparatus through labelling semantics. The system is actually constructed on quite common definitions of rule-based arguments, argumentation graphs, argument and statement labelling semantics. By doing so, we aim at showing that deontic argumentation frameworks can be devised on the basis of common argumentation constructs, with little modifications to standard inference apparatus. As a consequence, the system inherits the modularity of labelling argumentation systems, as it can be tuned to capture diverse refinements on argument and statement labellings; each refinement variant leading to a specific deontic argumentation framework. In that regard, we will see that standard bivalent statement labellings are not sufficient to achieve normative completeness, and for this reason, we propose to use a trivalent labelling semantics. Through this journey, we will retrieve three possible interpretations of the principle of prohibition.

Outline. In Section 2, some design considerations are exposed to build the computational system. In Section 3, we present a simple argumentation setting which is our basis of our deontic argumentation system. In Section 4, deontic knowledge is captured in deontic defeasible theories, from which argumentation graphs are produced, and deontic statements are labelled. Section 5 illustrates the overall approach. Section 6 evaluates the system with respect to the elicited design considerations. We discuss related work in Section 7, before concluding.

2   Some Design Considerations

Some design considerations may be useful to guide the construction of deontic argumentation systems. We will develop our system by bearing in mind 'human interface' and 'inference' issues, and a 'legit' requirement concerning normative completeness.

Interface considerations regard human-centric requirements to facilitate user interactions with the system, and we will value explainability, emulation, and isomorphism.

Explainability Computation of normative accounts should be explainable. Yet, the question 'What is a good explanation?' is quite elusive and goes far beyond the present work, see diverse conceptions in philosophy e.g. [5, 26, 31] or psychology [28, 32]. Here we focus on a specific aspect of explainability, namely, the explanation of why a certain argument is accepted or rejected, in the context of all relevant arguments. Thus, explainability means for our purposes that the output acceptance labelling of statements (i.e. conclusions) should be easily explainable to humans (-in-the-loop) by the interplay of arguments.

Emulation Computational models of argument are often inspired by argumentation as practised by humans. Following this line, argumentation for deontic reasoning should somehow emulate the way humans argue about deontic statements. In particular, we would like to account for full-fledged arguments built from doctrinal pieces, for example those arguments supporting permissions which are implicit in normative systems.

Isomorphism A well-established principle to build knowledge-based systens in the legal domain states that there should be an 'isomorphic' correspondence between the knowledge base and the sources [8]. Isomorphism can facilitate the development, verification, validation and maintenance of the knowledge base, and the provision of more intelligible explanations to end-users. Isomorphism implies here that we have constructs to account for conditional norms featuring obligations, prohibitions and permissions.

Inference considerations are more technical, and we will focus on parsimony, modularity and efficiency.

Parsimony As various deontic argumentation formalisms can be proposed to meet our needs, we shall prefer the simplest or most parsimonious formalisms, that is formalisms where unnecessary elements and constructions are excluded. Beyond formal elegance, a parsimonious deontic formalism is important because of its greater falsifiability, and because it can ease integration with other endeavours in argumentation, such as e.g. judgement aggregation, probabilistic, strategic or machine learning undertakings [9, 40, 41, 43]. Parsimony can be analysed in various ways – we will appreciate it at the level of inference machinery and the knowledge base.

Modularity To facilitate semantics variants, verification, validation and maintenance of eventual 'argument-based software systems', the deontic argumentation system may be, so to say, modular by design. This can be achieved, for example, by developing a module for each labelling stage, such that every module can be tuned. In this view, we prefer to investigate a deontic argumentation system from which various deontic argumentation frameworks can be drawn.

Efficiency Concerning computational complexity, one may seek for systems which can operate efficiently for practical ends, and thus argumentation semantics which can be accompanied with efficient algorithms. Following computational complexity theory, an algorithm is efficient if it can be performed in polynomial time.

Finally, beyond consistency according to which incompatible (deontic) statements should not be accepted together, the legit requirement regards the normative completeness of our formal system.

Completeness We are after deontic argumentation frameworks which are complete. Prescriptions may be such that there exist 'normative gaps' (or 'legal gaps' in legal systems), i.e., in some cases, something is neither explicitly obligatory, permitted nor prohibited. Normative completeness refers here to the completeness quality of the deontic argumentation system: a system is complete if, and only if, anything is eventually obligatory, permitted or prohibited, even though, for example, something is not regulated by any primary norms. To address normative completeness, the 'principle of prohibition', according to which everything that is not prohibited is permitted, can be put forward, and we will do so. Yet, such a principle can be interpreted in various ways, see e.g. [1], which should be accounted for.

Many argumentation settings in the literature cater for various human interface and inference aspects in a way or another. We are thus greatly inspired from these systems to build our deontic argumentation system in the next sections. However, normative completeness has attracted less attention. and we shall thus pay particular attention to it. Above-mentioned design considerations will be used to evaluate the system later in Section 6.

3 Argumentation System

This section presents a lightweight ASPIC$^+$-like argumentation system along with a common sequential model consisting of the following stages [4]: definition of the language, defeasible theories and argumentation graph production, argument acceptance/justification and statement acceptance/justification. These stages are developed in the remainder of the section.

3.1 Language

Building blocks of the formalism are so-called 'literal statements' (which are later further specified to cater for deontic modalities).

Definition 3.1. *A **literal statement** is either a plain literal statement or a modal literal statement, where*
- *a **plain literal statement** is either an atomic proposition p or the negation of an atomic proposition, i.e. $\neg p$, and*
- *a **modal literal statement** is a statement of the form $\Box \gamma$ or $\neg \Box \gamma$, such that $\Box$ is a placeholder for any modal operator and γ is a plain literal statement.*

Notation 3.1. *For any plain literal statement γ, its complement is written $\bar{\gamma}$. Hence, if γ is p then $\bar{\gamma}$ is $\neg p$, and if γ is $\neg p$ then $\bar{\gamma}$ is p.*

Literal statements can be put in relation through defeasible rules. Defeasible rules represent conditionals of the form '*if ... then ... unless ...*'. For the sake of simplicity, we deal with defeasible rules only, i.e. rules that can be defeated by other rules.

Definition 3.2. *A **defeasible rule** over a set of literal statements $\mathscr{S}$ is a construct of the form $r : \varphi_1, \ldots, \varphi_n, \sim \varphi'_1, \ldots, \sim \varphi'_m \Rightarrow \varphi$ with $0 \leq n$ and $0 \leq m$, and where*
- *r is the unique identifier of the rule, and*
- *for any $0 \leq i \leq n$ and $0 \leq j \leq m$, $\varphi_i, \varphi'_j, \varphi \in \mathscr{S}$ are all literal statements.*

Given a rule as in Definition 3.2, the set $\{\varphi_1, \ldots, \varphi_n, \sim \varphi'_1, \ldots, \sim \varphi'_m\}$ is the body of the rule. The singleton $\{\varphi\}$ is the head of the rule.

Notation 3.2. *Given a rule r as in Definition 3.2,*
- *the body of r is denoted $\mathrm{Body}(r)$, i.e. $\mathrm{Body}(r) = \{\varphi_1, \ldots, \varphi_n, \sim \varphi'_1, \ldots, \sim \varphi'_m\}$,*
- *the head of r is denoted $\mathrm{Head}(r)$, i.e. $\mathrm{Head}(r) = \{\varphi\}$,*
- *the set of propositions of r is denoted $\mathrm{Prop}(r)$, i.e. $\mathrm{Prop}(r) = \{p \mid p, \neg p, \sim p, \sim \neg p, \Box p, \neg \Box p, \Box \neg p, \neg \Box \neg p, \sim \Box p, \sim \neg \Box p, \sim \Box \neg p, \sim \neg \Box \neg p \in \mathrm{Body}(r) \cup \mathrm{Head}(r)\}$.*

The set of propositions of a set of rules Rules is denoted $\mathrm{Prop}(Rules)$, i.e. $\mathrm{Prop}(Rules) = \bigcup_{r \in Rules} \mathrm{Prop}(r)$.

A defeasible rule $r : \varphi_1, \ldots, \varphi_n, \sim \varphi_1', \ldots, \sim \varphi_m' \Rightarrow \varphi$ can be roughly read as follows: '*if* φ_1 *and* ... *and* φ_n *are supported then* φ *is defeasibly supported, unless* φ_1' *is supported or* ... *or unless* φ_m' *is supported*'. We specify later what 'supported' means here. The symbol $\sim$ can be viewed as a sort of negation as failure, but it may be rather understood as a point of attack (as we will see soon) to avoid any confusion with formal semantics from the literature on the concept of negation as failure.

Rules may head to incompatible statements. Incompatibilities amongst statements are captured in a conflict relation defined as a binary relation over literal statements.

Definition 3.3. *A **conflict relation** 'Conflicts' over a set of literal statements $\mathscr{S}$ is a binary relation over $\mathscr{S}$, i.e. Conflicts $\subseteq \mathscr{S} \times \mathscr{S}$.*

Notation 3.3. *The set propositions of a conflict relation 'Conflicts' is denoted* Prop(*Conflicts*), *i.e.* Prop(*Conflicts*) $= \{p \mid (\varphi, \varphi') \in$ *Conflicts* $: \varphi = p, \neg p, \Box p, \neg \Box p, \Box \neg p, \neg \Box \neg p,$ *or* $\varphi' = p, \neg p, \Box p, \neg \Box p, \Box \neg p, \neg \Box \neg p\}$.*

The relation is meant to be 'well-formed'. For example, we may constrain conflicts such that for any literal statement γ and its complement $\overline{\gamma}$ we have *Conflicts*$(\gamma, \overline{\gamma})$. Well-formedness is left unspecified at this stage, it will be specified in the deontic development of the relation. We may also further refine the relation with asymmetric and symmetric conflicts to deal with contrary or contradictory statements, but such sophistications are not necessary for our ends.

When two rules have conflicting heads, one rule may prevail over another one. To possibly disentangle such cases, we use a superiority relation $\succ$ over rules. Informally, $s \succ r$ states that rule s prevails over rule r.

Definition 3.4. *A **superiority relation** $\succ$ over a set of rules Rules is an antireflexive and antisymmetric binary relation over Rules, i.e. $\succ \subseteq$ Rules $\times$ Rules.*

As the superiority relation is antireflexive and antisymmetric, for any rule r it does not hold that $r \succ r$, and for two distinct rules r and r' we cannot have both $r \succ r'$ and $r' \succ r$.

3.2 Defeasible theories and argumentation graphs

A defeasible theory lists a set of rules, a conflict relation and a superiority relation.

Definition 3.5. *A **defeasible theory** is a tuple $\langle$Rules, Conflicts, $\succ \rangle$ where*
- *Rules is a set of rules, and*
- *Conflicts is a conflict relation, and*
- *$\succ$ is a superiority relation over Rules.*

Notation 3.4. *Given a defeasible theory $T = \langle Rules, Conflicts, \succ \rangle$,*
- *the set of rules Rules, the relation Conflicts, and the relation $\succ$ are denoted $\mathrm{Rules}(T)$, $\mathrm{Conflicts}(T)$, and $\succ (T)$ respectively,*
- *the set propositions of T is denoted $\mathrm{Prop}(T)$, i.e. $\mathrm{Prop}(T) = \mathrm{Prop}(Rules) \cup \mathrm{Prop}(Conflicts)$.*

By chaining rules of a defeasible theory, we can build arguments. Arguments are captured by the following definition, which is much inspired from other rule-based argumentation frameworks as exposed for example in [11, 37].

Definition 3.6. *An **argument** A constructed from a defeasible theory $\langle Rules, Conflicts, \succ \rangle$ is a finite construct of the form: $A : A_1, \ldots A_n, \sim \varphi_1, \ldots, \sim \varphi_m \Rightarrow_r \varphi$ with $0 \leq n$ and $0 \leq m$ and such that*
- *A is the unique identifier of the argument, and*
- *$A_1, \ldots, A_n$ are arguments constructed from the defeasible theory $\langle Rules, Conflicts, \succ \rangle$, and*
- *φ is the conclusion of argument A; the conclusion of an argument A is denoted $\mathrm{conc}(A)$, i.e. $\mathrm{conc}(A) = \varphi$, and*
- *there exists a rule $r \in Rules$ such that $r : \mathrm{conc}(A_1), \ldots, \mathrm{conc}(A_n), \sim \varphi_1, \ldots, \sim \varphi_m \Rightarrow \varphi$.*

Definition 3.7. *Given an argument $A : A_1, \ldots A_n, \sim \varphi_1, \ldots, \sim \varphi_m \Rightarrow_r \varphi$, the set of its subarguments $\mathrm{Sub}(A)$, the set of its direct subarguments $\mathrm{DirectSub}(A)$, the last inference rule $\mathrm{TopRule}(A)$, and the set of all the rules in the argument $\mathrm{Rules}(A)$ are defined as follows:*
- $\mathrm{Sub}(A) = \mathrm{Sub}(A_1) \cup \ldots \cup \mathrm{Sub}(A_n) \cup \{A\}$,
- $\mathrm{DirectSub}(A) = \{A_1, \ldots, A_n\}$,
- $\mathrm{TopRule}(A) = (r : \mathrm{conc}(A_1), \ldots, \mathrm{conc}(A_n), \sim \varphi_1, \ldots, \sim \varphi_m \Rightarrow \varphi)$,
- $\mathrm{Rules}(A) = \mathrm{Rules}(A_1) \cup \ldots \cup \mathrm{Rules}(A_n) \cup \{\mathrm{TopRule}(A)\}$.

According to Definition 3.6, an argument without direct subarguments has thus the form $A : \sim \varphi_1, \ldots, \sim \varphi_m \Rightarrow_r \varphi$ with $0 \leq m$. Arguments may be infinite, and we may have an infinite set of arguments constructed from a defeasible theory, however, we will work with finite sets of finite arguments.

Last item in Definition 3.6 asserts that arguments are built using one single implicit inference rule, namely (defeasible) modus ponens of the form

$$r : \varphi_1, \ldots, \varphi_n, \sim \varphi'_1, \ldots, \sim \varphi'_m \Rightarrow \varphi$$
$$\frac{\varphi_1, \ldots, \varphi_n}{\varphi}$$

By using this simple and single inference rule pertaining to common sense reasoning, our intention is to build arguments through reasoning steps which come 'as close as possible to actual reasoning'. Yet, by contrast to Gentzen-style calculi (sequent calculus and natural deduction) which have a minimal number of axioms and multiple inference rules, we propose to have one single inference rule, namely a defeasible modus ponens to meet our parsimony requirement. If one wants to bring the framework closer to actual reasoning, then the framework can be certainly developed in that direction by including sundry inference rules to build arguments.

Arguments may conflict and thus attacks between arguments may appear. We reckon two types of attacks: rebuttals (clash of incompatible conclusions) and undercuttings[1] (attacks on negation as failure premises). In regard to rebuttals, we assume a preference relation over arguments determining whether two rebutting arguments mutually attack each other or only one of them (being preferred) attacks the other. The preference relation over arguments can be defined in various ways on the basis of the preference over rules. We adopt a simple last-link ordering (back to e.g. [38]), according to which an argument A is preferred over another argument B, denoted as $A \succ B$, if, and only if, the rule TopRule(A) is superior to the rule TopRule(B), i.e. TopRule(A) $\succ$ TopRule(B). This leads us to adopt the following definition of attack relation, cf. other formulations e.g. in [37].

Definition 3.8. *An **attack relation** $\rightsquigarrow$ over a set of arguments $\mathscr{A}$ is a binary relation over $\mathscr{A}$, i.e. $\rightsquigarrow \subseteq \mathscr{A} \times \mathscr{A}$. An argument B attacks an argument A, i.e. $B\rightsquigarrow A$, iff B rebuts or undercuts A, where*
- *B rebuts A (on A') iff $\exists A' \in$ Sub(A) such that conc(B) and conc(A') are in conflict, i.e. Conflicts(conc(B), conc(A')), and $A' \not\succ B$;*
- *B undercuts A (on A') iff $\exists A' \in$ Sub(A) such that $\sim$ conc(B) belongs to the body of TopRule(A'), i.e. ($\sim$ conc(B)) $\in$ Body(TopRule(A')).*

Arguments and attack relations can be then captured in Dung's abstract argumentation graphs, originally called abstract argumentation frameworks in [16].

Definition 3.9. *An **argumentation graph** is a pair $\langle \mathscr{A}, \rightsquigarrow \rangle$ where $\mathscr{A}$ is a set of arguments, and $\rightsquigarrow \subseteq \mathscr{A} \times \mathscr{A}$ is a binary relation of attack.*

Notation 3.5. *Given an argumentation graph $G = \langle \mathscr{A}, \rightsquigarrow \rangle$, we may denote $\mathscr{A}$ as $\mathscr{A}_G$ and $\rightsquigarrow$ as $\rightsquigarrow_G$.*

Definition 3.10. *An argumentation graph $\langle \mathscr{A}, \rightsquigarrow \rangle$ is an **argumentation graph constructed from a defeasible theory** iff $\mathscr{A}$ is the set of all arguments constructed from the defeasible theory.*

---

[1]The term undercutting is overloaded in argumentation literature and is used with different meanings in different contexts, cf. [37].

Clearly, the number of arguments in an argumentation graph constructed from a defeasible theory may not be polynomial in the number of rules of the theory. As complexity matters, we may focus on those theories from which argumentation graphs can be constructed efficiently. However, we may also be given an argumentation graph, and check its 'well-formedness', presumably against some theory. Anyhow, given an argumentation graph built (possibly efficiently) from a defeasible theory, we can then label arguments following argument labelling semantics.

3.3 Labelling semantics

Given an argumentation graph, the sets of arguments that are accepted or rejected, that is, those arguments that will survive or not to possible attacks, are computed using some semantics. For our purposes, we resort to labelling semantics as reviewed in [2,3]. Accordingly, we endorse $\{$IN, OUT, UND$\}$-labellings where each argument is associated with one label which is either IN, OUT, or UND, respectively meaning that the argument is accepted, rejected, or undecided.

Definition 3.11. *A $\{$IN, OUT, UND$\}$-**labelling** of an argumentation graph G is a total function $L : \mathscr{A}_G \rightarrow \{IN, OUT, UND\}$.*

Notation 3.6. *A $\{$IN, OUT, UND$\}$-labelling L may be represented as a tuple $\langle$IN(L), OUT(L), UND$(L)\rangle$ where IN(L) stands for $\{A \mid L(A) =$ IN$\}$, OUT(L) for $\{A \mid L(A) =$ OUT$\}$, and UND(L) for $\{A \mid L(A) =$ UND$\}$.*

Most argument labellings studied in the literature are *complete labellings* [2]. An argumentation graph may have several complete $\{$IN, OUT, UND$\}$-labellings, we will focus on the unique complete labelling with the smallest set of labels IN, namely the grounded $\{$IN, OUT, UND$\}$-labelling.

Definition 3.12. *Let G denote an argumentation graph. A **complete** $\{$IN, OUT, UND$\}$-**labelling** of G is a $\{$IN, OUT, UND$\}$-labelling such that for every argument A in $\mathscr{A}_G$:*
- *A is labelled IN iff all attackers of A are labelled OUT,*
- *A is labelled OUT iff A has an attacker labelled IN.*

Definition 3.13. *A complete $\{$IN, OUT, UND$\}$-labelling L is a **grounded** $\{$IN, OUT, UND$\}$-**labelling** of an argumentation graph G if IN(L) is minimal (w.r.t. set inclusion) amongst all complete $\{$IN, OUT, UND$\}$-labellings of G.*

Since complete or grounded $\{$IN, OUT, UND$\}$-labellings are total functions, if an argument is not labelled IN or OUT, then it is labelled UND.

The reason to focus on the grounded $\{$IN, OUT, UND$\}$-labelling is that it is unique and it can be computed in a polynomial time, using e.g. Algorithm 1 [36]. The algorithm begins by labelling IN all arguments not being attacked or whose attackers are OUT (line 4), and then it iteratively labels OUT any argument attacked by an argument labelled IN (line 5). The iteration continues until no more arguments can be labelled IN or OUT (line 6); and if the argumentation graph is finite, then it terminates by labelling UND unlabelled arguments (line 7).

Algorithm 1 Computation of a grounded {IN, OUT, UND}-labelling.

1: **input** An argumentation graph G,

2: $L_0 = (\emptyset, \emptyset, \emptyset)$,

3: **repeat**

4:     $\text{IN}(L_{i+1}) \leftarrow \text{IN}(L_i) \cup \{A \mid A \in \mathscr{A}_G$ is not labelled in L_i, and $\forall B \in \mathscr{A}_G$: if B attacks A then $B \in \text{OUT}(L_i)\}$

5:     $\text{OUT}(L_{i+1}) \leftarrow \text{OUT}(L_i) \cup \{A \mid A \in \mathscr{A}_G$ is not labelled in L_i, and $\exists B \in \mathscr{A}_G$: B attacks A and $B \in \text{IN}(L_{i+1})\}$

6: **until** $L_i = L_{i+1}$

7: **return** $(\text{IN}(L_i), \text{OUT}(L_i), \mathscr{A}_G \backslash (\text{IN}(L_i) \cup \text{OUT}(L_i)))$

Algorithm 1 is given here to show that the computation of the grounded labelling of an argumentation graph can be performed in polynomial time. It does not pre-empt the use of more efficient algortihms, see e.g. [14, 15].

Given a set of statements, a labelling of this set is a (preferably total) function associating any statement with a label. Different specifications for statement labellings are possible, see e.g. [4] where statement *acceptance* labellings are distinguished from statement *justification* labellings. For our purposes, we will work with acceptance labellings, and we first turn to the acceptance labelling semantics which is perhaps the simplest in a meaningful way, namely bivalent labelling semantics, according to which a statement is either accepted or not, without further sophistication. If a statement is accepted then it is labelled 'in', otherwise it is labelled 'ni'. As statements are labelled relatively to so-called argument acceptance labelling semantics, we have acceptance bivalent {in, ni}-labellings, but we will simply call them bivalent {in, ni}-labellings.

Definition 3.14. *Let $\mathfrak{L}$ be a set of* {IN, OUT, UND}*-labellings, $\mathscr{S}$ a set of literal statements. A **bivalent** {in, ni}**-labelling** of $\mathscr{S}$ and from $\mathfrak{L}$ is a total function $K : \mathfrak{L}, \mathscr{S} \rightarrow$ {in, ni} such that for any argument labelling $L \in \mathfrak{L}$ and $\varphi \in \mathscr{S}$:*

- *$K(L, \varphi) = $ in iff $\exists A \in \text{IN}(L) : \text{conc}(A) = \varphi$,*
- *$K(L, \varphi) = $ ni otherwise.*

We can also take statement labellings which better exploit statuses of {IN, OUT, UND}-labellings [4]. For our very investigation into deontic argumentation, we consider trivalent {in, und, niund}-labellings which reckon undecided statements.

Definition 3.15. *Let $\mathfrak{L}$ be a set of $\{$IN, OUT, UND$\}$-labellings, $\mathscr{S}$ a set of literal statements. A **trivalent** $\{$in, und, niund$\}$-labelling of $\mathscr{S}$ and from $\mathfrak{L}$ is a total function $K : \mathfrak{L}, \mathscr{S} \rightarrow \{$in, und, niund$\}$ such that for any argument labelling $L \in \mathfrak{L}$ and $\varphi \in \mathscr{S}$:*
- *$K(L, \varphi) = $ in iff $\exists A \in $ IN$(L) : $conc$(A) = \varphi$, and*
- *$K(L, \varphi) = $ und iff $\exists A \in $ UND$(L) : $conc$(A) = \varphi$ and $\nexists A \in $ IN$(L) : $conc$(A) = \varphi$, and*
- *$K(L, \varphi) = $ niund otherwise.*

Notation 3.7. *A bivalent $\{$in, ni$\}$-labelling or a trivalent $\{$in, und, niund$\}$-labelling K may be represented as a tuple $\langle$in(K), ni$(K)\rangle$ and $\langle$in(K), und(K), niund$(K)\rangle$ respectively, with the obvious meaning.*

For the sake of simplicity, since we will work with grounded $\{$IN, OUT, UND$\}$-labelling semantics and because every argumentation graph has a unique grounded $\{$IN, OUT, UND$\}$-labelling, we assume that any statement justification labelling simply corresponds to a statement acceptance labelling which can be a bivalent $\{$in, ni$\}$-labelling or a trivalent $\{$in, und, niund$\}$-labelling. The distinction between bivalent and trivalent labellings is later exploited in our deontic setting, in particular to address normative completeness.

4 Deontic Argumentation System

Having laid out a simple rule-based argumentation system, we can now specify a deontic version of it. To do so, we first specify our deontic statements, and then adopted labelling semantics are discussed.

4.1 Deontic language

Legal and deontic reasoning expose varied concepts – ranging from basic obligations and permissions to liberties and rights. For our purposes, we focus on basic concepts in deontic reasoning, namely obligations, prohibitions, and permissions.

Obligations are at the core of our deontic system, and prohibitions are viewed as a by-product of obligations: 'something is prohibited' is equivalently expressed by stating that its opposite is obligatory. Permissions can be understood in terms of obligations too: a permission for something expresses that the opposite is not obligatory.

Accordingly, and for the sake of simplicity, the attention is restricted to a propositional language which is supplemented with a single deontic operator O which indicates an obligation. Hence, we assume a language $\mathscr{L}_D$ whose literal statements are defined as follows, cf. Definition 3.1.

Definition 4.1. *A literal statement of a language $\mathcal{L}_D$ is either a plain literal statement or a deontic literal statement, where:*

- *a **plain literal statement** is either an atomic proposition p or the negation of an atomic proposition, i.e. ¬p, and*
- *a **deontic literal statement** is a statement of the form $O\gamma$ or $\neg O\gamma$ such that γ is a plain literal statement.*

Prohibitions and permissions are captured by assuming that a prohibition $F\gamma$ is equivalently expressed by the obligation $O\bar{\gamma}$, and a permission $P\gamma$ is syntactically equivalent to $\neg O\bar{\gamma}$.

Notation 4.1. *As syntactic sugar, we may write $O\bar{\gamma}$ as $F\gamma$, and $\neg O\bar{\gamma}$ as $P\gamma$ (and vice versa). Accordingly, Op stands for $F\neg p$, $O\neg p$ for Fp, Pp for $\neg O\neg p$, and $P\neg p$ for $\neg Op$.*

Anything is meant to be either obligatory, prohibited or permitted. However, in practice, and as remarked by legal scholars, there can be gaps (see e.g. [1]), i.e. in some cases, something is neither obligatory, permitted nor prohibited. Normative completeness refers here to the completeness quality of the deontic argumentation system: a system is complete if, and only if, anything is eventually obligatory, permitted nor prohibited, even though, for example, something is not regulated by any primary norms.

To address normative completeness, we will exploit the well-established principle of prohibition, which can be formulated as follows: 'everything that is not prohibited is permitted'. The principle is shared by various legal systems (for instance it corresponds to the 'norma generale exclusiva' or 'principio di libertá' in Italian system); however, we have to note that the principle does not apply in all legal systems.

In criminal law for example (in civil law systems), if being permitted to do an action means not being subject to sanction for it, then there is the idea that every criminal sanction must be explicitly stated by positive law (the sanction cannot be obtained by analogy). This principle, according to European Court of Human Rights (case Ozturk v. Germany, 1984), goes beyond criminal law strictly understood, applying to any sanction having punitive character, namely, going beyond compensation of damage (e.g. administative fines). Therefore if an action is not punished by a norm (it is not prohibited), one can conclude that the legal system is committed not to punish (the judge should not punish it), so that the action is definitely permitted, as long as the explicit rules remain the same (the law says implicitly that the action is subject to no sanction) according to criminal law. The idea can be retrieved in the principle 'nullum crimen sine lege' (no crime without law), i.e. everything that is not explicitly prohibited should be considered as permitted.

In private law, on the contrary, it is possible for the judge to establish a sanction (compensation for damages) using analogy or other legal constructions even for actions that are not explicitly prohibited by the law. In civil law, the fact that no norm explicitly establishes a sanction for an action does not ensure that the action will not be sanctioned, there is just a gap: we do not know what will happen. For instance, in many legal systems, before consumer protection laws were enacted, judges started to condemn producers to compensate damages caused by defective products. This required overcoming, in the absence of an explicit rule, the idea that producers owned no duty of care to consumers.

Hence, in general, the principle of prohibition and our upcoming account of it apply to normative systems where it makes sense to use it.

Besides, the principle can be diversely interpreted. In a first interpretation, the principle can be read to stress a mere tautology: no prohibition ($\neg O \bar{\gamma}$) is equivalent to a permission ($P \gamma$). Such a tautology can be, for example, syntactically captured by writing $\neg O \bar{\gamma}$ as $P \gamma$ and vice versa (as we do in Notation 4.1), but this will not appear to successfully fill any gaps in our formal framework. In a second interpretation, we can adopt the reading according to which a thing is permitted unless it is prohibited. Following this interpretation the principle of prohibition is not a tautology anymore, but rather a normative principle included in the normative system being considered, to effectively fill gaps by producing permissions.

In this context, we may distinguish strong and weak permissions (similarly as notably retained by G. H. von Wright [49]), where a strong permission is a permission derived from permissive norms, and a weak permission for φ is a permission which is accepted if the prohibition of or on φ is not accepted. This conception makes reference, for example, to C.E. Alchourrón and E. Bulygin who state: 'Weak permission differs from strong permission in an important way: strong permission expresses a positive fact (the existence of a permissive norm), whereas weak permission refers to a negative fact: the non-existence of a prohibitive norm' [1], see also [33, 45, 46]. In this view, a strong permission does not fully correspond to what we may call an 'explicit permission' (i.e. a permission which is explicitly formulated in a permissive norm): for example, if one derives in the system the acceptance of Pp and Pq, one might infer $P(p \text{ and } q)$, which is not explicit in the sense that it is derived, but is nevertheless strong. The strengths of permissions are not exclusive: we can have a strong and weak permission for the same thing, the two permissions would not be incompatible. In our framework, the strengths of permissions are not directly represented in the language, essentially because laypersons or jurists do not usually specify the strength of permissions in their discourses. In this view, a weak or a strong permission is seen as a permission tout court.

A defeasible rule can specify varied relationships amongst (deontic) literal statements of a given language $\mathcal{L}_D$. Such rules are called normative defeasible rules.

Definition 4.2. *Given a language $\mathcal{L}_D$, a **normative defeasible rule** is a defeasible rule over a set of literal statements in $\mathcal{L}_D$.*

Normative rules are partitioned into *foreground rules* and *background rules*. Foreground rules provide substantive legal regulations, for particular domains of the law, while background rules express deontic assumptions underlying the normative system being dealt with.

Foreground rules are domain-dependent rules. They are meant here to represent primary norms, and thus they can be classified as either *constitutive rules* or *regulative rules*. The effect of a constitutive rule is to define a term as understood in a given situation or to 'create' an institutional entity from a set of brute or institutional facts. A regulative rule, on the other hand, triggers a 'deontic' effect (obligation, prohibition, permission) when certain conditions are established. While constitutive and regulative norms have been formally approached in various (and sometimes sophisticated) ways in the literature [20], the distinction is simply addressed in the present system: the consequent of the rule is a plain literal for constitutive rules, and a deontic literal for regulative rules. In that regard, we can note that a regulative rule heading to a (strong) permission would typically specify an exception to an obligation, as notably discussed by A. Ross [44], but such a rule can also be used to stress a permission and clarify its conditions.

Background rules are domain-independent. They underpin core deontic reasoning. These background rules can be viewed as defeasible rule schemata which are isomorphic to some pieces of (possibly very basic) legal doctrines. Instead of giving a formal definition of background defeasible rules, we give here some examples of such schemata:

$$d\_\gamma : O\gamma \;\Rightarrow P\gamma \quad \text{An obligation } O\gamma \text{ implies a permission } P\gamma$$
$$\text{(cf. Axiom 'D' in deontic logics).}$$

$$p\_\gamma : \quad\;\; \Rightarrow P\gamma \quad \text{Anything is permitted prima facie.}$$

$$k\_\gamma : \sim O\overline{\gamma} \Rightarrow P\gamma \quad \text{Anything that is not prohibited is permitted.}$$

These background rules can be essentially employed to build arguments supporting permissions. In particular, the second and third rules can be used to derive permissions even if there are no applicable foreground permissive rules heading to such permissions. As to the terminology, although such permissions are the consequent of some rules, we may say for now that these background rules head to weak permissions because these rules are meant to ensure that a permission can be accepted if any contrary prohibition is non-existent or rejected, and even though there is no foreground permissive rule heading to such a permission (as we will see later).

Given some background rules, sets of background rules can be formed, possibly to account for various doctrinal systems. In our case, and for our purposes, we will work with the sets $\{d\_\gamma, p\_\gamma\}$ and $\{d\_\gamma, k\_\gamma\}$.

Definition 4.3. *A set of background defeasible rule schemata B is*
• *a **permissive by default set of background defeasible rule schemata** iff B = $\{d\_\gamma, p\_\gamma\}$;*

- a **Kelsenian permissive set of background defeasible rule schemata** iff $B = \{d\_\gamma, k\_\gamma\}$;
- a **permissive set of background defeasible rule schemata** iff $B = \{d\_\gamma, p\_\gamma\}$ or $B = \{d\_\gamma, k\_\gamma\}$.

A permissive by default set or a Kelsenian permissive set indicates that anything is defeasibly permitted. They are distinct in that a Kelsenian permissive set may better reflect the principle of prohibition as exposed by Kelsen (thus its name)[2]. Moreover, this set may appear weaker than the permissive by default set because the rule k_γ features a point of attack $\sim O\overline{\gamma}$ which rule p_γ does not present. We will use in this paper this permissive set of background defeasible rules for our illustrations. For both sets, although one may presuppose that their rules can be used to fill normative gaps, we will see that such background rules are actually not enough to obtain normative completeness when using bivalent statement labelling semantics. Our solution to this issue will turn out to yield the same results for both sets when determining acceptance statuses of statements.

Whatever the set of background defeasible rules, we will ground the rules over a set of propositions. For our purposes, we do so over propositions of an input (domain-dependent) defeasible theory.

Definition 4.4. *A set of rules is a **set of background rules** with respect to a defeasible theory T and a set of background defeasible rule schemata B, denoted* $\mathrm{BackRules}(T, B)$, *iff*
- $\mathrm{BackRules}(T, B) = \{d\_\gamma, d\_\overline{\gamma}, p\_\gamma, p\_\overline{\gamma} \mid \gamma \in \mathrm{Prop}(T)\}$ *if B is a permissive by default set of background defeasible rule schemata;*
- $\mathrm{BackRules}(T, B) = \{d\_\gamma, d\_\overline{\gamma}, k\_\gamma, k\_\overline{\gamma} \mid \gamma \in \mathrm{Prop}(T)\}$ *if B is a Kelsenian permissive set of background defeasible rule schemata.*

For the sake of simplicity, norms potentially captured by 'modalised rules' (e.g, rules of the form $O(r : \varphi_1, \ldots, \varphi_n, \sim \varphi'_1, \ldots, \sim \varphi'_m \Rightarrow \varphi))$ are not accounted for in this paper. Such constructs and their meanings are left to possible developments of the system, towards for example secondary norms as 'meta-rules'.

Background rules are at the domain-independent core of deontic reasoning, and they are employed to background sets of rules.

Definition 4.5. *A set of rules Rules is a **backgrounded set of rules** with respect to a defeasible theory T and a set of background defeasible rule schemata B iff Rules =* $\mathrm{Rules}(T) \cup \mathrm{BackRules}(T, B)$.

---

[2]'As a sanction-prescribing social order, the law regulates human behavior in two ways: in a positive sense, commanding such behavior and thereby prohibiting the opposite behavior; and, negatively, by not attaching a coercive act to a certain behavior, therefore not prohibiting this behavior and not commanding the opposite behavior. Behavior that legally is not prohibited is legally permitted in this negative sense.' [29].

Example 1. *Let us adapt H.L.A Hart's hypothetical [25] for our purposes: assume a policy stating that it is forbidden to enter in a park with a vehicle, unless there is an emergency. This policy may be formalised by the (foreground) defeasible theory $\langle\{r\},\emptyset,\emptyset\rangle$ where*

$\quad$ r : $\quad$ vehi, $\sim$ emer $\Rightarrow$ Fenter

The Kelsenian permissive set of background rules with respect to the theory $\langle\{r\},\emptyset,\emptyset\rangle$ includes all the following rules.

d_vehi :	Ovehi	$\Rightarrow$ Pvehi		d_¬vehi :	O¬vehi	$\Rightarrow$ P¬vehi
k_vehi :	$\sim$ Fvehi	$\Rightarrow$ Pvehi		k_¬vehi :	$\sim$ F¬vehi	$\Rightarrow$ P¬vehi
d_emer :	Oemer	$\Rightarrow$ Pemer		d_¬emer :	O¬emer	$\Rightarrow$ P¬emer
k_emer :	$\sim$ Femer	$\Rightarrow$ Pemer		k_¬emer :	$\sim$ F¬emer	$\Rightarrow$ P¬emer
d_enter :	Oenter	$\Rightarrow$ Penter		d_¬enter :	O¬enter	$\Rightarrow$ P¬enter
k_enter :	$\sim$ Fenter	$\Rightarrow$ Penter		k_¬enter :	$\sim$ F¬enter	$\Rightarrow$ P¬enter

$\square$

As a set of background rules is defined with respect to a defeasible theory T and a set of background defeasible rule schemata B, the cardinality of the set of background rules is $2 \cdot |B| \cdot |\mathrm{Prop}(T)|$ (as illustrated in Example 1). Yet, for practical matters and especially implementation matters, we may employ only background defeasible rule schemata which are instantiated where necessary for the considered computation.

Concerning conflicts, we distinguish foreground conflicts and background conflicts. Foreground conflicts can be any kind of conflicts of the form $(\gamma,\overline{\gamma})$ or $(O\gamma,O\overline{\gamma})$ or $(\neg O\gamma,O\gamma)$ or $(O\gamma,\neg O\gamma)$; deontic conflicts can be visualised in the deontic square drawn in Figure 1.

Definition 4.6. *A conflict is a **foreground conflict** iff it is of the form $(\gamma,\overline{\gamma})$ or $(O\gamma,O\overline{\gamma})$ or $(\neg O\gamma,O\gamma)$ or $(O\gamma,\neg O\gamma)$.*

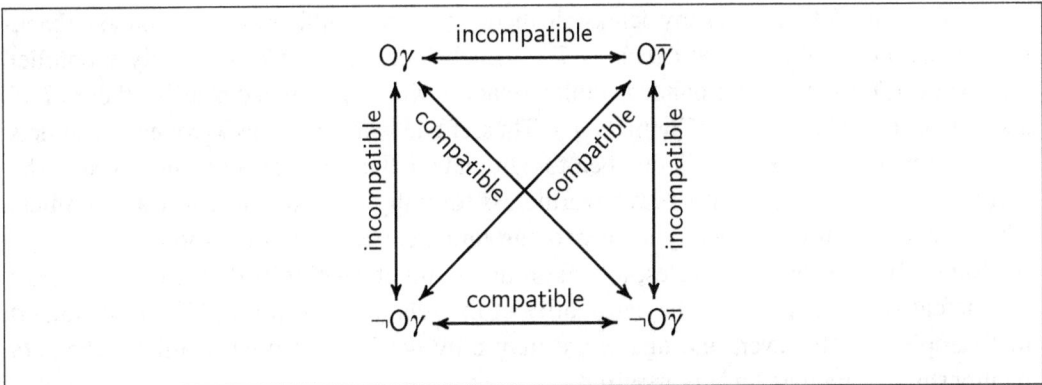

Figure 1: Deontic square of compatibility relation.

Foreground conflicts allow a knowledge engineer to specify particular conflicts between literal statements. However, specified foreground conflicts may appear incomplete in that inevitable conflicts may not be included in the foreground set. To ensure completeness of conflicts, we assume thus background conflicts.

Definition 4.7. *A set of conflicts is a **set of background conflicts** with respect to a defeasible theory T, denoted* $\mathrm{BackConflicts}(T)$, *iff* $\mathrm{BackConflicts}(T) = \{(\gamma, \bar{\gamma}), (\bar{\gamma}, \gamma), (O\gamma, O\bar{\gamma}), (O\bar{\gamma}, O\gamma), (O\gamma, \neg O\gamma), (\neg O\gamma, O\gamma), (O\bar{\gamma}, \neg O\bar{\gamma}), (\neg O\bar{\gamma}, O\bar{\gamma}) \mid \gamma \in \mathrm{Prop}(T)\}$.

Background conflicts are, so to say, domain-independent in deontic reasoning, and a conflict relation is backgrounded by such conflicts if, and only if, they are included in the relation.

Definition 4.8. *A conflict relation $Conflicts$ is a **backgrounded conflict relation** with respect to a defeasible theory T iff* $Conflicts = Conflicts(T) \cup \mathrm{BackConflicts}(T)$.

Example 1 (continued). *The background conflict pairs are as follows.*

(vehi, ¬vehi)	(emer, ¬emer)	(enter, ¬enter)
(¬vehi, vehi)	(¬emer, emer)	(¬enter, enter)
(Ovehi, O¬vehi)	(Oemer, O¬emer)	(Oenter, O¬enter)
(O¬vehi, Ovehi)	(O¬emer, Oemer)	(O¬enter, Oenter)
(Ovehi, ¬Ovehi)	(Oemer, ¬Oemer)	(Oenter, ¬Oenter)
(¬Ovehi, Ovehi)	(¬Oemer, Oemer)	(¬Oenter, Oenter)
(O¬vehi, ¬O¬vehi)	(O¬emer, ¬O¬emer)	(O¬enter, ¬O¬enter)
(¬O¬vehi, O¬vehi)	(¬O¬emer, O¬emer)	(¬O¬enter, O¬enter)

□

We can remark that, given any defeasible theory T where conflicts are foreground or background conflicts, we have that $Conflicts(T) \subseteq \mathrm{BackConflicts}(T)$. Consequently, a conflict relation $Conflicts$ is a backgrounded conflict relation with respect to a defeasible theory T if, and only if, $Conflicts = \mathrm{BackConflicts}(T)$. Thus, if one works with backgrounded conflicts of a defeasible theory, as we will do, then foreground conflicts may appear unnecessary. The definition of foreground conflicts is nevertheless formally necessary to constrain conflicts which can be given when specifying any foreground theories (as defined soon).

Similarly as background rules, a background conflict relation is defined with respect to a defeasible theory T, and the cardinality of the relation is $8 \cdot |\mathrm{Prop}(T)|$ (as illustrated in Example 1). However, and again, we may only need background conflict schemata relationships instantiated where required.

Foreground and background deontic rules may have conflicting heads, and to ensure correct reasoning patterns, background superiorities can be proposed. At first sight, background rule d_γ could be viewed as a very strong rule, that is, in our context a defeasible rule which is superior to any other rule and such that there exist no superior foreground rules. However, such a superiority setting does not fit well with the adopted last-link preference over arguments, as some anomalies may appear. For example, if d_γ is superior to any foreground rules, then any argument C whose top rule is d_γ could defeat any arguments defeating the direct subargument of C.

Example 2. *For instance, suppose the following arguments:*

$O1:$ $\quad\quad\quad\quad \Rightarrow_r Oa$ $\quad\quad$ $O2:$ $\quad\quad\quad\quad \Rightarrow_{r'} O\neg a$
$P1:$ $\quad O1 \quad \Rightarrow_{d\_a} Pa$ $\quad\quad$ $P2:$ $\quad O2 \quad \Rightarrow_{d\_\neg a} P\neg a$
$W1:$ $\quad \sim Fa \Rightarrow_{k\_a} Pa$ $\quad\quad$ $W2:$ $\quad \sim F\neg a \Rightarrow_{k\_\neg a} P\neg a$

Let us assume that rule r' is superior to r, i.e. $r' \succ r$, so that argument O2 attacks O1. Consequently, argument O2 rebuts P1 (on O1). Moreover, if rule d_γ has superior or equal strength to any foreground rules, then argument P1 rebuts O2 and P2 (on O2), and P2 rebuts O1 and P1 (on O1). As a result, all arguments would be labelled UND as illustrated in Figure 2 on the left, but this labelling is not satisfactory. Instead, if d_γ has inferior strength to any foreground rules, then we have the argumentation graph as illustrated in Figure 2 on the right, whose grounded {IN, OUT, UND}-labelling is more satisfactory.

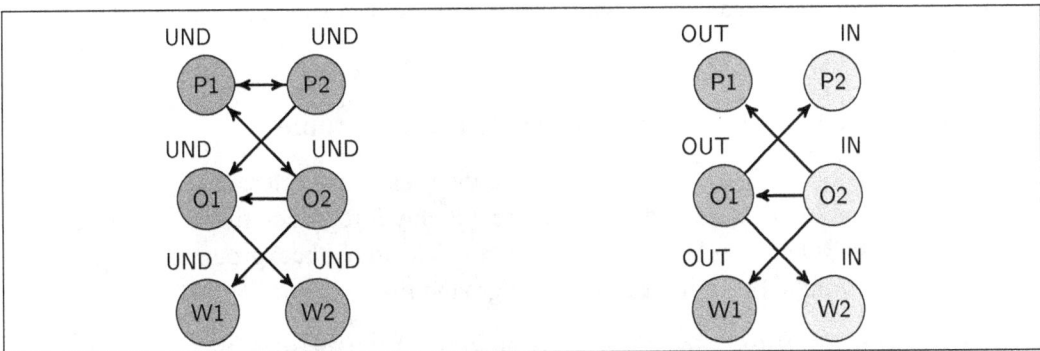

Figure 2: Grounded {IN, OUT, UND}-labellings.

□

Example 2 shows that background rule d_γ should not be viewed as a strong rule. On the contrary, it should be conceived as a very weak rule, so that arguments with such a top rule attack no foreground arguments. In general, we assume in this paper that background rules are inferior to any foreground rules.

Definition 4.9. *A background superiority relation is a **background superiority relation** with respect to a defeasible theory T and a set of background defeasible rule schemata B, denoted* $\mathrm{BackSup}(T, B)$, *iff* $\mathrm{BackSup}(T, B) = \{(s, r) \mid s \in \mathrm{Rules}(T), r \in \mathrm{BackRules}(T, B)\}$.

Definition 4.10. *A superiority relation $\succ$ is a **backgrounded superiority relation** with respect to a defeasible theory T and a set of background defeasible rule schemata B iff* $\succ = \succ (T) \cup \mathrm{BackSup}(T, B)$.

The definition of a backgrounded superiority relation might be simplified if strict rules were employed. However, the use of strict rules can be quite elusive [11,17], and would requisite a larger inference apparatus which would not be congruent with our design consideration on parsimony.

Example 1 (continued). *The background pairs in the superiority relation are as follows.*

(r, d_vehi)	(r, d_emer)	(r, d_enter)
(r, k_vehi)	(r, k_emer)	(r, k_enter)
$(r, d\_\neg vehi)$	$(r, d\_\neg emer)$	$(r, d\_\neg enter)$
$(r, k\_\neg vehi)$	$(r, k\_\neg emer)$	$(r, k\_\neg enter)$

□

A background superiority relation is defined with respect to a defeasible theory T and a set of background defeasible rule schemata B, thus the cardinality of the relation is greater than $|B| \cdot |\mathrm{Prop}(T)| \cdot |\mathrm{Rules}(T)|$. However, similarly as background rules and background conflicts, we may only need background superiority schemata which are instantiated where necessary.

4.2 Deontic defeasible theory and argumentation graphs

We now can propose to 'background' defeasible theories where rules, conflicts and superiority relationships are backgrounded with respect to any foreground theory. A foreground defeasible theory is here a defasible theory where rules are not background rules, i.e. rules whose identifiers are not identifiers of any background rules.

Definition 4.11. *A defeasible theory $\langle Rules, Conflicts, \succ \rangle$ is a **foreground defeasible theory** iff*
- *every defeasible rule in Rules is a (foreground) normative defeasible rule which is not a background defeasible rule, and*
- *every conflict in Conflicts is a foreground conflict.*

Definition 4.12. *A defeasible theory $\langle Rules, Conflicts, \succ \rangle$ is a **backgrounded defeasible theory** of a foreground defeasible theory T with a set of background defeasible rule schemata B iff*

- *Rules is a backgrounded set of rules with respect to T and B, and*
- *Conflicts is a backgrounded conflict relation with respect to T, and*
- $\succ$ *is a backgrounded superiority relation with respect to T and B.*

In practice, we will first write a foreground defeasible theory to then hold a backgrounded defeasible theory. In the remainder, we assume that any defeasible theory is backgrounded with a permissive set of background defeasible rule schemata, to obtain a permissive defeasible theory.

Definition 4.13. *A defeasible theory is a **permissive defeasible theory** iff it is a backgrounded defeasible theory with a permissive set of background defeasible rule schemata.*

From a backgrounded defeasible theory, we can build arguments. When building arguments, chaining rules implicitly uses the detachment of the consequent of rules. In that regard, we can note that deontic studies usually distinguish factual detachments and deontic detachments. For our purposes, we consider factual detachments only, leaving (defeasible) deontic detachments (if accepted) to future developments.

Once arguments are built, we can form an argumentation graph, and then label arguments and (deontic) statements to determine their statuses, as discussed next.

4.3 Deontic labelling semantics

On the basis of an argumentation graph built from a backgrounded defeasible theory, we can now look at semantics for (deontic) literal statements. By semantics, we mean labelling semantics as put forward in the previous section so that acceptance labellings of (deontic) literal statements are defined with respect to acceptance labellings of arguments in terms of 'if, and only if'.

First, we have to note that, sometimes, imperatives are deemed to bear no truth values, and thus no semantics in terms of truth values can be devised [27]. For instance, there is no truth value in an imperative such as 'Do not enter!'. Alternatively, however, we may evaluate the acceptance of normative statements with respect to the normative system. For example, we can evaluate whether an obligation holds in particular situation. We adopt such an epistemic view in the rest of the paper by labelling arguments and statements.

To label deontic arguments and statements, we first resort to labelling semantics as previously exposed. Hence, given the argumentation graph built from a backgrounded defeasible theory, arguments are labelled according to the grounded $\{\mathsf{IN}, \mathsf{OUT}, \mathsf{UND}\}$-labelling semantics. Then, literal statements are labelled with statement acceptance labelling semantics. Such labellings are thus a straightforward application of standard labelling semantics corresponding to some common sense reasoning.

As such, the labelling framework satisfies some intuitive properties pertaining to deontic consistency. Let us first observe that if two arguments A and B have conflicting conclusions and A is labelled IN then argument B is labelled OUT. Thus, for instance, if an IN-labelled argument has an obligation $O\gamma$ or a permission $P\gamma$ for conclusion, then any argument whose conclusion is a prohibition $F\gamma$ is labelled OUT.

Lemma 4.1. *Let L be a grounded $\{$IN, OUT, UND$\}$-labelling of an argumentation graph G constructed from a permissive defeasible theory $\langle Rules, Conflicts, \succ \rangle$, and A, B any arguments in $\mathscr{A}_G$ such that $Conflicts(\mathrm{conc}(A), \mathrm{conc}(B))$. If $L(A) = $ IN then $L(B) = $ OUT.*

It follows that the bivalent $\{$in, und, niund$\}$-labelling and trivalent $\{$in, und, niund$\}$-labelling semantics trivially imply that two conflicting deontic statements cannot be labelled in: if a deontic statement is labelled in then any conflicting statement is labelled ni or niund depending on the selected statement labelling semantics.

Proposition 4.1. *Let L be a grounded $\{$IN, OUT, UND$\}$-labelling of an argumentation graph constructed from a permissive defeasible theory T, and $\mathscr{S}$ a set of literal statements such that $\mathscr{S} = \{p, \neg p \mid p \in \mathrm{Prop}(T)\}$.*
- *Let K be a bivalent $\{$in, ni$\}$-labelling of $\mathscr{S}$ and from $\{L\}$. For any $\gamma \in \mathscr{S}$:*
  - *if $K(L, O\gamma) = $ in then $K(L, O\bar{\gamma}) = $ ni;*
  - *if $K(L, \neg O\gamma) = $ in then $K(L, O\gamma) = $ ni;*
  - *if $K(L, O\gamma) = $ in then $K(L, \neg O\gamma) = $ ni;*
- *Let K be a trivalent $\{$in, und, niund$\}$-labelling of $\mathscr{S}$ and from $\{L\}$. For any $\gamma \in \mathscr{S}$:*
  - *if $K(L, O\gamma) = $ in then $K(L, O\bar{\gamma}) = $ niund;*
  - *if $K(L, \neg O\gamma) = $ in then $K(L, O\gamma) = $ niund;*
  - *if $K(L, O\gamma) = $ in then $K(L, \neg O\gamma) = $ niund.*

Proof. Let us provide the proof for the first item only in the case of bivalent $\{$in, ni$\}$-labellings (proofs for the other items follow the same structure). If $K(L, O\gamma) = $ in then there exists an argument $A \in \mathscr{A}_G$ such that $\mathrm{conc}(A) = O\gamma$ and $L(A) = $ IN. Two cases: $O\bar{\gamma}$ is the conclusion of an argument, or not. In the first case, by Lemma 4.1, for any argument $B \in \mathscr{A}_G$ such that $\mathrm{conc}(B) = O\bar{\gamma}$, if $L(A) = $ IN then $L(B) = $ OUT, and thus $K(L, O\bar{\gamma}) = $ ni. In the second case, if $O\bar{\gamma}$ is not the conclusion of any argument in $\mathscr{A}_G$, then $K(L, O\bar{\gamma}) = $ ni. Therefore, if $K(L, O\gamma) = $ in then $K(L, O\bar{\gamma}) = $ ni. $\qquad\square$

We can also remark that if an obligation $O\gamma$ is labelled in then the implied permission $\neg O\bar{\gamma}$ (i.e. $P\gamma$) is also labelled in.

Proposition 4.2. *Let L be a grounded $\{$IN, OUT, UND$\}$-labelling of an argumentation graph constructed from a permissive defeasible theory T, $\mathscr{S}$ a set of literal statements such that $\mathscr{S} = \{p, \neg p \mid p \in \mathrm{Prop}(T)\}$, and K a bivalent $\{$in, ni$\}$-labelling or trivalent $\{$in, und, niund$\}$-labelling of $\mathscr{S}$ and from $\{L\}$. For any $\gamma \in \mathscr{S}$: if $K(L, O\gamma) = $ in then $K(L, \neg O\bar{\gamma}) = $ in.*

Proof. Given an argumentation graph G constructed from a permissive defeasible theory, if $K(L, O\gamma) = $ in, then there exist an argument $A \in \mathscr{A}_G$ and $B \in \mathscr{A}_G$ such that $\text{conc}(A) = O\varphi$ and $B : A \Rightarrow_{d\_\gamma} P\gamma$. Let $\mathscr{A}^\curvearrowright \subseteq \mathscr{A}_G$ be the set of attackers of A, $\mathscr{A}^\curvearrowleft \subseteq \mathscr{A}_G$ the set of arguments attacked by A, and $\mathscr{B}^\curvearrowright \subseteq \mathscr{A}_G$ the set of attackers of B. We have that $\mathscr{B}^\curvearrowright \subseteq \mathscr{A}^\curvearrowright \cup \mathscr{A}^\curvearrowleft$. By Definition 3.12, if $L(A) = $ IN then for any argument $C \in \mathscr{A}^\curvearrowright \cup \mathscr{A}^\curvearrowleft$ $L(C) = $ OUT. Since $\mathscr{B}^\curvearrowright \subseteq \mathscr{A}^\curvearrowright \cup \mathscr{A}^\curvearrowleft$, for any argument $C \in \mathscr{B}^\curvearrowright$ $L(C) = $ OUT, and thus $L(B) = $ IN. Therefore, if $L(A) = $ IN then $L(B) = $ IN, and thus if $K(L, O\gamma) = $ in then $K(L, P\gamma) = $ in, i.e. if $K(L, O\gamma) = $ in then $K(L, \neg O\overline{\gamma}) = $ in. $\square$

Legal scholars, however, may argue that bivalent $\{$in, ni$\}$-labellings are not satisfactory. Such labellings are not legally satisfactory, because, given an argumentation graph from any backgrounded defeasible theory, and though we have arguments supporting weak permissions thanks to background rules, it may the case that all arguments are labelled UND and consequently, using a bivalent $\{$in, ni$\}$-labelling, deontic statements $O\gamma$, $F\gamma$ and $P\gamma$ may be labelled ni, see Example 3. Such labelling outcomes of common sense appear thus inappropriate to address normative completeness. For this reason, we discard bivalent $\{$in, ni$\}$-labelling semantics and put forward trivalent $\{$in, und, niund$\}$-labelling to cater for deontic reasoning and in particular normative completeness.

Example 3. *Let us consider the following arguments, along with the associated argumentation graph and grounded $\{$IN, OUT, UND$\}$-labelling drawn in Figure 3.*

O1:		$\Rightarrow_r O a$	O2:		$\Rightarrow_{r'} O\neg a$
P1:	O1	$\Rightarrow_{d\_a} P a$	P2:	O2	$\Rightarrow_{d\_\neg a} P\neg a$
W1:	$\sim F a$	$\Rightarrow_{k\_a} P a$	W2:	$\sim F\neg a$	$\Rightarrow_{k\_\neg a} P\neg a$

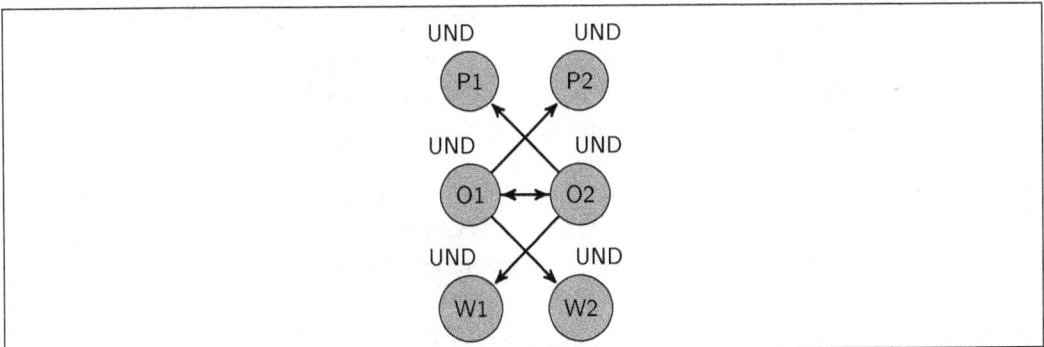

Figure 3: Grounded $\{$IN, OUT, UND$\}$-labelling.

First, a naive common sense bivalent reasoning can be captured by the acceptance bivalent $\{$in, ni$\}$-labelling $\langle \emptyset, \{O a, P a, O\neg a, \neg P a\} \rangle$. However, this bivalent labelling is prob-

lematic from a legal stance because statement a *is here neither obligated, nor permitted nor prohibited. To address this gap, we can employ a trivalent* $\{$in, und, niund$\}$*-labelling* $\langle \emptyset, \{Oa, Pa, O\neg a, \neg Pa\}, \emptyset \rangle$ *according to which the deontic status of* a *is undecided.* □

More formally, the definition of normative gaps, as we may conceive it in terms of statement labellings, depends on whether bivalent $\{$in, ni$\}$-labellings or trivalent $\{$in, und, niund$\}$-labellings are employed.

Definition 4.14. *Let L be a grounded* $\{$IN, OUT, UND$\}$*-labelling of an argumentation graph constructed from a permissive defeasible theory T, and the set of literal statements* $\mathscr{S} = \{p, \neg p \mid p \in \mathrm{Prop}(T)\}$.

- *Let K denote a bivalent* $\{$in, ni$\}$*-labelling of* $\mathscr{S}$ *and from* $\{L\}$. *There is a* $\{$in, ni$\}$*-labelling normative gap iff there exists a literal* $\gamma \in \mathscr{S}$ *such that*

$$K(L, O\gamma) = \mathrm{ni} \text{ and } K(L, O\bar{\gamma}) = \mathrm{ni} \text{ and } K(L, \neg O\bar{\gamma}) = \mathrm{ni}.$$

- *Let K' denote a trivalent* $\{$in, und, niund$\}$*-labelling of* $\mathscr{S}$ *and from* $\{L\}$. *There is a* $\{$in, und, niund$\}$*-labelling normative gap iff there exists a literal* $\gamma \in \mathscr{S}$ *such that*

$$K'(L, O\gamma) = \mathrm{niund} \text{ and } K'(L, O\bar{\gamma}) = \mathrm{niund} \text{ and } K'(L, \neg O\bar{\gamma}) = \mathrm{niund}.$$

As illustrated in Example 3, bivalent $\{$in, ni$\}$-labellings may lead to $\{$in, ni$\}$-labelling normative gaps, whereas trivalent $\{$in, und, niund$\}$-labellings can address normative completeness. To understand why, we can first observe that any backgrounded defeasible theory along with a trivalent labelling semantics lead to a third interpretation of the principle of prohibition in terms of labelling: if something is not prohibited (the prohibition is labelled niund) then it is permitted (the permission is labelled in).

Proposition 4.3. *Let L be a grounded* $\{$IN, OUT, UND$\}$*-labelling of an argumentation graph constructed from a permissive defeasible theory T,* $\mathscr{S}$ *a set of literal statements such that* $\mathscr{S} = \{p, \neg p \mid p \in \mathrm{Prop}(T)\}$, *and K a trivalent* $\{$in, und, niund$\}$*-labelling of* $\mathscr{S}$ *and from* $\{L\}$. *For any* $\gamma \in \mathscr{S}$: *if* $K(L, F\gamma) = \mathrm{niund}$ *then* $K(L, P\gamma) = \mathrm{in}$.

Proof. Given an argumentation graph constructed from a permissive defeasible theory T, for any $\gamma \in \mathscr{S} = \{p, \neg p \mid p \in \mathrm{Prop}(T)\}$, there exists a unique argument $W : \sim F\gamma \Rightarrow_{k\_\gamma} P\gamma$ or $(W : \Rightarrow_{p\_\gamma} P\gamma)$. All attackers of W are arguments whose conclusion is $F\gamma$ (i.e. $O\bar{\gamma}$). If $K(L, O\bar{\gamma}) = \mathrm{niund}$ then all attackers of W are OUT, and thus W is labelled IN, and $P\gamma$ is labelled in. i.e. $K(L, \neg O\bar{\gamma}) = \mathrm{in}$. Therefore, if $K(L, F\gamma) = \mathrm{niund}$ then $K(L, P\gamma) = \mathrm{in}$. □

On the basis of this interpretation in terms of labelling of the principle of prohibition, we now can easily show that trivalent $\{$in, und, niund$\}$-labellings address normative completeness.

Theorem 4.1. *Let L be a grounded* {IN, OUT, UND}*-labelling of an argumentation graph constructed from a permissive defeasible theory T, $\mathscr{S}$ a set of literal statements such that $\mathscr{S} = \{p, \neg p \mid p \in \mathrm{Prop}(T)\}$, and K a trivalent* {in, und, niund}*-labelling of $\mathscr{S}$ and from* {L}*. For any $\gamma \in \mathscr{S}$:*

$$K(L, \mathrm{O}\gamma) \neq \mathsf{niund} \ or \ K(L, \mathrm{O}\overline{\gamma}) \neq \mathsf{niund} \ or \ K(L, \neg\mathrm{O}\overline{\gamma}) \neq \mathsf{niund}.$$

Proof. There are three cases: 1. $K(L, \mathrm{O}\overline{\gamma}) = \mathsf{in}$, 2. $K(L, \mathrm{O}\overline{\gamma}) = \mathsf{und}$, and 3. $K(L, \mathrm{O}\overline{\gamma}) = \mathsf{niund}$. In this last case, by Proposition 4.3, $K(L, \mathrm{P}\gamma) = \mathsf{in}$, i.e. $K(L, \neg\mathrm{O}\overline{\gamma}) = \mathsf{in}$. Therefore in any case, $K(L, \mathrm{O}\gamma) \neq \mathsf{niund}$ or $K(L, \mathrm{O}\overline{\gamma}) \neq \mathsf{niund}$ or $K(L, \neg\mathrm{O}\overline{\gamma}) \neq \mathsf{niund}$. $\square$

Hence, trivalent {in, und, niund}-labellings address normative completeness by means of the status 'undecided' for deontic statements. Eventually, we can remark that such cases of undecidedness can be disentangled in various ways, typically by a competent authority, e.g. a judge.

Above results hold for any permissive theory. Consequently, they hold for backgrounded defeasible theory of any (foreground) defeasible theory with a permissive by default set of background defeasible rule schemata, or with a Kelsenian permissive set of background defeasible rule schemata. In general, it turns out that both sets yield the same trivalent {in, und, niund}-labelling.

Theorem 4.2. *Let*
- *T be a (foreground) defeasible theory;*
- *U be the backgrounded defeasible theory of T with a permissive by default set of background defeasible rule schemata;*
- *V be the backgrounded defeasible theory of T with a Kelsenian permissive set of background defeasible rule schemata;*
- *L_U the grounded* {IN, OUT, UND}*-labelling of the argumentation graph constructed from U;*
- *L_V the grounded* {IN, OUT, UND}*-labelling of the argumentation graph constructed from V;*
- *$\mathscr{S}$ a set of literal statements such that $\mathscr{S} = \{p, \neg p \mid p \in \mathrm{Prop}(T)\}$, and K a trivalent* {in, und, niund}*-labelling of $\mathscr{S}$ and from* {L}*.*

For any $\gamma \in \mathscr{S}$:

$$K(L_U, \gamma) = K(L_V, \gamma) \ and \ K(L_U, \mathrm{O}\gamma) = K(L_V, \mathrm{O}\gamma) \ and \ K(L_U, \neg\mathrm{O}\gamma) = K(L_V, \neg\mathrm{O}\gamma).$$

Proof. Let GU (GV resp.) denote the argumentation graph constructed from U (V resp.). Let f be the bijection such that $f : \mathscr{A}_{GU} \to \mathscr{A}_{GV}$ and $f(A) = B$ iff
- if A is of the form $A : \ \Rightarrow_{\mathrm{p}\_\gamma} \mathrm{P}\gamma$ then B is of the form $B : \ \sim \mathrm{F}\gamma \Rightarrow_{\mathrm{k}\_\gamma} \mathrm{P}\gamma$, and
- if A is of the form $A : A_1, \ldots A_n, \sim \varphi_1, \ldots, \sim \varphi_m \Rightarrow_r \varphi$ ($r \neq \mathrm{p}\_\gamma$) then B is of the form $B : f(A_1), \ldots f(A_n), \sim \varphi_1, \ldots, \sim \varphi_m \Rightarrow_r \varphi$.

In addition, $(A,A') \in \leadsto_{GU}$ iff $(f(A),f(A')) \in \leadsto_{GV}$, and thus (*GU* and *GV* are isomorphic and) $L_U(A) = L_V(f(A))$. Then, as $\mathrm{conc}(A) = \mathrm{conc}(f(A))$, for any $\gamma \in \mathscr{S}$: $K(L_U,\gamma) = K(L_V,\gamma)$, $K(L_U,O\gamma) = K(L_V,O\gamma)$, and $K(L_U,\neg O\gamma) = K(L_V,\neg O\gamma)$. $\qquad\square$

Hence, the adoption of the principle of prohibition 'anything that is not prohibited is permitted' as a schema $\mathsf{k\_\gamma} :\sim O\bar{\gamma} \Rightarrow P\gamma$ or the use of a statement such as 'anything is permitted prima facie' as a schema $\mathsf{p\_\gamma} : \Rightarrow P\gamma$ are two alternatives to cater for normative completeness, and both alternatives actually lead to the same results in terms of statement labellings.

Finally, concerning weak and strong permissions, the reification of doctrinal pieces into defeasible theories blurs somewhat the distinction based on conventional definitions. For example, a permission $P\gamma$ which is the conclusion of an IN-labelled argument $W : \Rightarrow_{\mathsf{p\_\gamma}} P\gamma$ is necessarily labelled in. Consequently we may say that such a permission is derived from a rule and thus, it is a strong permission by definition, whereas it is a weak permission from its conception. For this reason, a third kind of permission may be introduced, which we may call 'doctrinal permission', since such a doctrinal permission for something is derived from the non-existence or rejection of its prohibition and on the basis of reified doctrinal principles.

5 Illustration

To illustrate our system, let us reappraise the policy stating that it is forbidden to enter in a park with a vehicle, unless there is an emergency. This policy and the assumptions may be formalised in different ways. We illustrate our system with one option which is developed in the remainder of the section.

5.1 Backgrounded defeasible theory

We assume that there is a vehicle at the entrance of the park, and that there may be an emergency, maybe not (in the Hart-Fuller debate [19, 25], uncertainty was originally about what can be classified as a vehicle). Let us capture this with the foreground defeasible theory $\langle\{\mathsf{rv},\mathsf{re},\mathsf{r\bar{e}},\mathsf{r}\},\emptyset,\emptyset\rangle$ where

$$
\begin{array}{ll}
\mathsf{rv} & \Rightarrow \mathsf{vehi} \\
\mathsf{re} & \Rightarrow \mathsf{emer} \\
\mathsf{r\bar{e}} & \Rightarrow \neg\mathsf{emer} \\
\mathsf{r} : & \mathsf{vehi}, \sim \mathsf{emer} \Rightarrow \mathsf{Fenter}
\end{array}
$$

Let us adopt a Kelsenian permissive set of background defeasible rule schemata. The foreground theories can be then backgrounded to yield a backgrounded theory featuring, amongst others, background rules as exposed in Example 1.

5.2 Argument and argumentation graph construction

We can construct the following arguments from background rules:

$$
\begin{array}{llll}
W1: & \sim Fvehi & \Rightarrow_{k\_vehi} Pvehi & \qquad W4: & \sim F\neg vehi & \Rightarrow_{k\_\neg vehi} P\neg vehi \\
W2: & \sim Femer & \Rightarrow_{k\_emer} Pemer & \qquad W5: & \sim F\neg emer & \Rightarrow_{k\_\neg emer} P\neg emer \\
W3: & \sim Fenter & \Rightarrow_{k\_enter} Penter & \qquad W6: & \sim F\neg enter & \Rightarrow_{k\_\neg enter} P\neg enter
\end{array}
$$

In addition, we can build the following arguments from the foreground rules and rule d_¬enter:

$$
\begin{array}{lll}
A1: & \Rightarrow_{rv} & vehi & \qquad B1 & \Rightarrow_{re} & emer \\
A2: & A1, \sim emer & \Rightarrow_{r} & Fenter & \qquad C1 & \Rightarrow_{r\bar{e}} & \neg emer \\
A3: & A2 & \Rightarrow_{d\_\neg enter} & P\neg enter &
\end{array}
$$

Consequently, we can form the argumentation graph G such that: $\mathscr{A}_G = \{A1, A2, A3, B1, C1, W1, W2, W3, W4, W5, W6\}$, and $\leadsto_G = \{(B1, C1), (C1, B1), (B1, A2), (B1, A3), (A2, W3)\}$, see Figure 4.

We can note that we have built arguments to support weak/doctrinal permissions, thus we can argue and present full-fledged arguments about such permissions.

5.3 Argument and statement labellings

Let L1 denote the grounded {IN, OUT, UND}-labelling of argumentation graph G (as illustrated in Figure 4). Accordingly, we can lay the bivalent {in, ni}-labelling and trivalent {in, und, niund}-labelling as in Table 1.

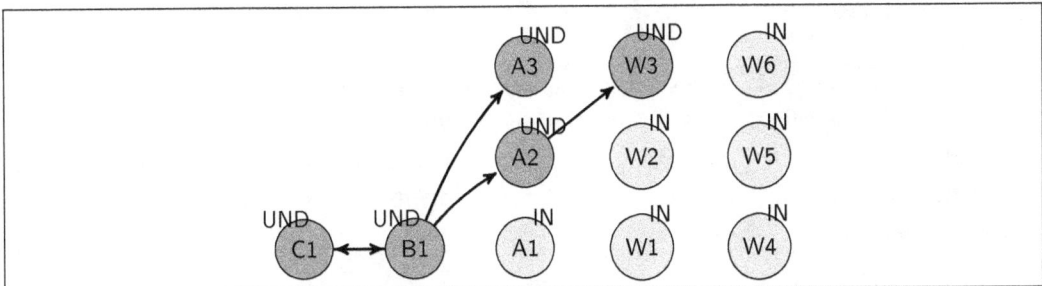

Figure 4: Grounded {IN, OUT, UND}-labelling of argumentation graph G.

	vehi	emer	¬emer	enter	¬enter
$K(L1,\cdot)$	in	ni	ni	ni	ni
$K(L1,\cdot)$	in	und	und	niund	niund

	Ovehi	Oemer	O¬emer	Oenter	O¬enter
$K(L1,\cdot)$	ni	ni	ni	ni	ni
$K(L1,\cdot)$	niund	niund	niund	niund	und

	Pvehi	Pemer	P¬emer	Penter	P¬enter
$K(L1,\cdot)$	in	in	in	ni	in
$K(L1,\cdot)$	in	in	in	und	in

Table 1: Bivalent $\{in, ni\}$-labelling and trivalent $\{in, und, niund\}$-labelling.

The $\{in, no\}$-bivalent labelling results into a normative gap (the statement enter is neither obligatory nor prohibited nor permitted), whereas the trivalent $\{in, und, niund\}$-labelling fills the gap by labelling the permission to enter as undecided.

5.4 Violation and contrary-to-duty obligation

Let us extend the illustration with the formalisation of a violation and a contrary-to-duty obligation. Contrary-to-duty obligations can be a pitfall for deontic formalisms which have a more sophisticated conception of deontic modalities [13, 35], and we would like to illustrate how such obligations can be handled in our argumentation formalism within our KB approach.

Let us suppose that the park policy also states that a violation of the prohibition would be sanctioned by a fine (the amount does not matter for our purposes). To capture such a policy, we can add the following rules.

v :	Fenter, enter $\Rightarrow$ violation	v′ :	$\Rightarrow$ ¬violation
f :	violation $\Rightarrow$ fine	f′ :	$\Rightarrow$ ¬fine

such that $v \succ v'$ and $f \succ f'$. Rules v′ and f′ specify that, by default, we can derive that there is neither violation nor fine, unless the contrary is shown.

Furthermore, a new park management can add a contrary-to-duty obligation: if the prohibition is violated then one should stop driving in the park. We can thus add the following rule.

s : violation $\Rightarrow$ Ostop

A sequence of compensatory obligations can be added along similar lines. Eventually, we also assume that the vehicle enters in the park:

e : $\Rightarrow$ enter

Let L2 denote the grounded {IN, OUT, UND}-labelling of the implied argumentation graph. The acceptance {in, no}-bivalent labelling and trivalent {in, und, niund}-labelling of new statements are exposed in Table 2.

	violation	¬violation	fine	¬fine	stop	¬stop
$K(L2, \cdot)$	ni	ni	ni	ni	ni	ni
$K(L2, \cdot)$	und	und	und	und	niund	niund

	Oviolation	O¬violation	Ofine	O¬fine	Ostop	O¬stop
$K(L2, \cdot)$	ni	ni	ni	ni	ni	ni
$K(L2, \cdot)$	niund	niund	niund	niund	und	niund

	Pviolation	P¬violation	Pfine	P¬fine	Pstop	P¬stop
$K(L2, \cdot)$	in	in	in	in	in	ni
$K(L2, \cdot)$	in	in	in	in	in	und

Table 2: Bivalent {in, ni}-labelling and trivalent {in, und, niund}-labelling.

Again, we can see that the {in, no}-bivalent labelling leads to a normative gap (¬stop is neither obligatory nor prohibited nor permitted), whereas the trivalent {in, und, niund}-labelling fills the gap by labelling the permission to not stop (i.e. the permission to drive through the park) as undecided.

6   Some Design Evaluation

Let us briefly discuss to what extent the framework meets the design considerations given in Section 2, that is, considerations on human interface (explainability, emulation, isomorphism), and inference (parsimony, modularity and computational efficiency), and a legit requirement regarding normative completeness.

Explainability An argumentation framework can inherently ease explanations of acceptance statuses of statements by presenting relevant arguments. The acceptance labelling of arguments is then exposed by the grounded labelling, possibly through a relatively simple algorithm (Algorithm 1) and resulting appealing graphical representations. A distinctive feature of the proposed system is that arguments can be built in support or against permissions and in particular weak/doctrinal permissions.

Emulation The system emulates the way humans argue about norms to the extent that modus ponens is endorsed as a natural reasoning step to build arguments, that attacks amongst arguments are meaningful, and that the sequential multi-labelling model featuring argument and argumentation graph production, argument acceptance/justification, statement acceptance/justification reflects a well-ordered reasoning based on arguments. Beyond that, emulation is further met in that the proposal employs the grounded semantics for which there exist simple dialogical argument games, see e.g. [12, 36, 38], in which full-fledged arguments supporting weak/doctrinal permissions can be put forward.

Isomorphism The system allows an isomorphic representation of normative systems as long as norms can be isomorphically represented by (foreground) defeasible rules. In that regard, we have proposed to have both the 'unless' conjunction ($\sim$) and a superiority relation over rules to get more flexibility to represent norms, thereby yielding a fine-grained expressiveness (as illustrated in Section 5). Furthermore, background rules can be viewed isomorphic to some (possibly very basic) deontic doctrine. The system can be critised too. For example, criminal codes usually do not hold norms with explicit expressions of conditional prohibitions: they usually directly express sanctions with regard to certain conducts, while our system would typically require the expression of prohibitions for automated reasoning.

Parsimony Parsimony can be evaluated at the levels of the inference machiney and knowledge representation. At the level of knowledge representation, parsimony is affected in that core deontic reasoning have been captured by background rules, conflicts and superiority relation to background theories. Deontic reasoning is also captured at the level of statement labellings to address normative gaps, without any impacts on the knowledge representation. At the inference level, the system is parsimonious in that arguments are built with defeasible rules and factual detachment only. Of course, other constructs such as strict rules, 'defeaters' or facts might bring a gain of expressiveness to the framework; the evaluation of such gain is left to further investigations. The system is also parsimonious in that argument acceptance and justification stages coincide, essentially because grounded {IN, OUT, UND}-labelling is used. Importantly, parsimony was required to ease combination with other developments in argumentation. In that respect, the setting can be straightforwardly subject to probabilistic enhancements through, for example, probabilistic labellings, and thus machine learning endeavours, see e.g. [42, 43].

Modularity The modularity of the system is largely based on (the instantiation of) the multi-labelling model for argumentation. At a theoretical level, we can use variants in argument and statement labelling semantics, as evidenced by our use of bivalent and trivalent labellings. At a more practical level, as the multi-labelling model supports a separation of concerns corresponding to different stages of the argumentation process, we can decompose an 'argument-based software system' into well-defined independent modules, for example by developing a module for each labelling stage, thereby easing verification, validation and maintenance of such a system.

Efficiency Concerning computationtal complexity, the reason to focus on the grounded $\{\text{IN}, \text{OUT}, \text{UND}\}$-labelling is that it is unique and it can be computed in a polynomial time, see Algorithm 1. Once the grounded $\{\text{IN}, \text{OUT}, \text{UND}\}$-labelling is computed, one can trivially compute any acceptance trivalent $\{\text{in}, \text{und}, \text{niund}\}$-labellings and other statement labellings. Hence given the argumentation graph built from a permissive defeasible theory, the overall time complexity to compute the labelling over a set of statements is polynomial. However, as previously mentioned, the number of arguments in an argumentation graph constructed from a defeasible theory may not be polynomial in the number of rules of the theory. If we focus on those theories from which argumentation graphs can be constructed efficiently, then argumentation graphs construction, argument labellings and statement labellings can be achieved efficiently, and thus the overall system is efficient in such cases.

Completeness Normative completeness has been primarily addressed by endorsing the principle of prohibition. The principle has been interpreted in three different ways. First, an interpretation has been given as a blunt syntactical equivalence between $\neg O\bar{\gamma}$ and $P\gamma$ (Notation 4.1). Second, the principle has been read as a background rule $k\_\gamma : \sim O\bar{\gamma} \Rightarrow P\gamma$. Finally, the principle has been interpreted in terms of statement labellings (Proposition 4.3). As bivalent statement labellings are not sufficient to obtain normative completeness through the principle, we have supplemented it with a trivalent labelling semantics to deal with cases of undecidedness. Eventually, we have to note that the principle is shared by various legal systems, however the way completeness is resolved depends much on the considered legal system.

This brief evaluation is inherently partial as it is limited to the elicited requirements, see e.g. [18] for some formal issues. A more complete evaluation is left to future investigations, possibly in light of a comparison with related implementations. An overview of related work is given next.

7   Related Work

There exists an increasing amount of work to capture normative reasoning through argumentation or non-monotonic frameworks akin to argumentation, see [10] for a systematic account of legal reasoning and argumentation from a logical, philosophical and legal perspective. Let us focus on some formal works related to our undertaking,

ASPIC$^+$ argumentation framework has first been exploited to express arguments about norms as the application of argument schemes to knowledge bases of facts and norms [39]. However, in [39], norms are expressed without any deontic operators for specifying obligations, permissions and prohibitions. The work was thereupon reappraised by L. van der Torre and S. Villata in [48] to integrate deontic modalities, by adopting input/output logic [34] for the analysis. Despite appearances, there are many differences with our undertaking: the most obvious are briefly exposed here. Conditional obligations and conditional permissions are represented by rules of the form $L_1 \wedge \ldots \wedge L_n \rightsquigarrow OL$ and $L_1, \ldots, L_n \rightsquigarrow \neg OL$ respectively, where L's are literals. Thus, obligations and permissions do not appear in the antecedents of norms. Moreover, the system in [48] does not deal with weak permissions. Furthermore, the conjunction 'unless' ($\sim$) is not used in conditionals. Eventually, neither argument nor statement labelling semantics are specified in [48], though such semantics could be easily integrated in the system.

In another line of research, Beirlaen et al. presents in [6] a formal argumentation system for dealing with the detachment of prioritised conditional obligations and permissions. To do so, Beirlaen et al. devise an argumentation framework where arguments are proof sequences, and they employ Dung's grounded semantics to determine accepted arguments. Again, there are many differences with our work, we expose the most obvious here. A first difference concerns the language. In [6], the language pertains to a modal extension of propositional classical logic, whereas our language is restricted to literals supplemented by deontic operators. Conditional obligations and conditional permissions are denoted by formulas $A \Rightarrow_O B$ and $A \Rightarrow_P B$ respectively, where A and B are propositional formulas. Consequently, neither obligations nor permissions can appear in the antecedents of norms in [6]. Moreover, conditionals in [6] do not cater for the conjunction 'unless' as we do. Then, a major difference holds in that conditional obligations and conditional permissions in [6] are associated with a degree of priority, which is taken into account to define defeats between arguments. In our work, conditionals are directly prioritised through a superiority relation. Beirlaen et al. focus on extension-based grounded semantics, whereas we employ grounded labelling semantics to obtain more refined argument labellings which allow us to implement trivalent statement labellings. Computation complexity is not considered in [6], whereas we have restricted our framework on labellings possibly computed with efficient algorithms.

Another endeavour in defeasible normative reasoning rests on variants of Defeasible Logic (DL), see e.g. [23, 24]. The work includes rich deontic constructs such as temporal deontic modalities and sequence of compensatory obligations. However, DL is not without weaknesses. For example, DL features strict rules, but does not satisfy closure under strict rules [11]. In that regard, our work solves the issue by simply discarding strict rules. DL also features so-called 'defeaters', i.e. rules which cannot be used in arguments supporting a conclusion, and it can be challenging to find succinct counterparts in natural languages for such defeaters. Isomorphism can thus be questioned in DL. Compared to the work reported here, a major difference holds in that DL variants capture legal reasoning patterns in proof theories and thus at the inference level. In that regard, we can say that the loss of parsimony at the knowledge representation level in the framework reported here has been compensated by a gain of parsimony at the inference model. Concerning semantics, a substantial difference with our work holds in that there is no 'undecided' status to tag statements. For example, if a defeasible theory comprises two rules $r: \Rightarrow_O c$ and $s: \Rightarrow_O \neg c$ (capturing that c and $\neg c$ are both obligatory) with no superiority relations and no facts, then the deontic development of DL in [22] does not tag the obligations of c and $\neg c$ as undecided but tags c and $\neg c$ as not obligatory ($-\partial_O c$ and $-\partial_O \neg c$), and eventually it derives c and $\neg c$ as permitted ($+\partial_P c$ and $+\partial_P \neg c$). However, such tagging would be inappropriate in applications where permissions are not entailed from conflicting obligations. Eventually, argumentation semantics exist for DL [21, 30] (with no status for 'undecidedness') but no counterparts have been developed for deontic variants.

In comparison to the above-mentioned works, substantial differences hold in that we have reified doctrinal prices to build our deontic rule-based argumentation system. We have shown that standard bivalent statement labelling semantics fall short to deal with normative completeness, and we have proposed a trivalent statement labelling semantics to address this point.

8 Conclusion

A deontic rule-based argumentation system has been devised to represent and reason upon conditional norms featuring obligations, prohibitions and (strong or weak) permissions. To do so, we have proposed the use of defeasible rule schemata to the greatest extent to capture deontic patterns. By doing so, we could straighforwardly adopt a common model consisting of three stages [4], namely argument and argumentation graph production, argument acceptance/justification and statement acceptance/justification. More specifically, given an argumentation graph, we have proposed to label arguments using grounded {IN, OUT, UND}-labelling semantics, and given the grounded {IN, OUT, UND}-labelling of the graph then we have proposed to simply label (deontic) statements using a trivalent labelling semantics.

We have learnt that it is possible to build deontic argumentation frameworks capturing conditional norms on the basis of common constructs from the literature on argumentation. In particular, the system uses (two possible sets of) inference rule schemata, and only one single (implicit) inference rule, namely (defeasible) modus ponens. In this setting, we have learnt that a standard bivalent labelling may appear insufficient to cover aspects of normative completeness, and that this issue can be addressed by using a trivalent labelling. Through our KB approach, we have retrieved three possible interpretations of the principle of prohibition: as a syntactical equivalence, as a defeasible rule, and as a proposition on the labelling of deontic statement labellings. We have also shown that the adoption of the principle of prohibition 'anything that is not prohibited is permitted' as a schema or the use of a statement such as 'anything is permitted prima facie' as another schema are two alternatives to address normative completeness, and both alternatives actually lead to the same results in terms of statement labellings. Eventually, concerning weak and strong permissions, the reification of doctrinal pieces into defeasible theories blurs somewhat the distinction based on conventional definitions. We have thus introduced a third kind of permission, called doctrinal permission: a doctrinal permission for something is derived from the non-existence or rejection of its prohibition and on the basis of reified doctrinal principles.

Multiple developments are possible. Firstly, the principle of prohibition is not applicable in any normative system. Consequently, the framework can be further developed where the principle cannot/should not be applied. Secondly, in order to enrich the framework, other normative constructs and reasoning patterns may be integrated to build arguments, and labelling variants can be explored. Finally, the system is meant to be combined with other developments in argumentation, ranging from probabilistic argumentation to argumentation in (normative) multi-agent systems.

References

[1] Carlos E. Alchourrón and Eugenio Bulygin. *Normative systems*. Springer-Verlag, 1971.

[2] Pietro Baroni, Martin Caminada, and Massimiliano Giacomin. An introduction to argumentation semantics. *Knowledge Eng. Review*, 26(4):365–410, 2011.

[3] Pietro Baroni, Martin Caminada, and Massimiliano Giacomin. Abstract Argumentation Frameworks and Their Semantics. In P. Baroni, D. Gabbay, M. Giacomin, and L. van der Torre, editors, *Handbook of Formal Argumentation*, pages 159–236. College Publications, 2018.

[4] Pietro Baroni, Guido Governatori, and Régis Riveret. On labelling statements in multi-labelling argumentation. In *Proc. of the 22nd Euro. Conf. on Artificial Intelligence*, pages 489–497. IOS Press, 2016.

[5] William Bechtel and Adele Abrahamsen. Explanation: A mechanist alternative. *Studies in History and Philosophy of Biol and Biomed Science*, 36(2):421–441, 2005.

[6] Mathieu Beirlaen, Jesse Heyninck, and Christian Straßer. Structured argumentation with prioritized conditional obligations and permissions. *Journal of Logic and Computation*, page exy005, 2018.

[7] Mathieu Beirlaen and Christian Straßer. A structured argumentation framework for detaching conditional obligations. *CoRR*, abs/1606.00339, 2016.

[8] Trevor J. M. Bench-Capon and Frans Coenen. Isomorphism and legal knowledge based systems. *Artificial Intelligence and Law*, 1(1):65–86, 1992.

[9] Gustavo Adrian Bodanza, Fernando Tohmé, and Marcelo Auday. Collective argumentation: A survey of aggregation issues around argumentation frameworks. *Argument & Computation*, 8(1):1–34, 2017.

[10] G. Bongiovanni, G. Postema, A. Rotolo, G. Sartor, C. Valentini, and D Walton, editors. *Handbook of Legal Reasoning and Argumentation*. Springer, 2018.

[11] Martin Caminada and Leila Amgoud. On the evaluation of argumentation formalisms. *Artificial Inteligence*, 171(5-6):286–310, 2007.

[12] Martin Caminada and Mikolaj Podlaszewski. Grounded semantics as persuasion dialogue. In *Proc. of 4th Conf. on Computational Models of Argument*, pages 478–485. IOS Press, 2012.

[13] José Carmo and Andrew J. I. Jones. Deontic logic and contrary-to-duties. In *Handbook of Philosophical Logic: Volume 8*, pages 265–343. Springer, 2002.

[14] Federico Cerutti, Sarah A. Gaggl, Matthias Thimm, and Johannes P. Wallner. Foundations of implementations for formal argumentation. *IfCoLog Journal of Logics and their Applications*, 4(8):2623–2706, 2017.

[15] Günther Charwat, Wolfgang Dvořák, Sarah A. Gaggl, Johannes P. Wallner, and Stefan Woltran. Methods for solving reasoning problems in abstract argumentation – a survey. *Artificial Intelligence*, 220:28 – 63, 2015.

[16] Phan Minh Dung. On the acceptability of arguments and its fundamental role in nonmonotonic reasoning, logic programming and n-person games. *Artificial Intelligence*, 77(2):321–358, 1995.

[17] Phan Minh Dung and Phan Minh Thang. Closure and consistency in logic-associated argumentation. *Journal of Artificial Intelligence Research*, 49:79–109, 2014.

[18] Phan Minh Dung, Francesca Toni, and Paolo Mancarella. Some design guidelines for practical argumentation systems. In *In Proc. of the 3rd Conf. on Computational Models of Argument*, pages 183–194. IOS Press, 2010.

[19] Lon L. Fuller. Positivism and fidelity to law: A reply to professor hart. *Harvard Law Review*, 71(4):630–672, 1958.

[20] Dov Gabbay, John Horty, Xavier Parent, Ron van der Meyden, and Leendert van der Torre, editors. *Handbook of Deontic Logic and Normative Systems*. College Publications, 2013.

[21] Guido Governatori, Michael J. Maher, David Billington, and Grigoris Antoniou. Argumentation semantics for defeasible logics. *Journal of Logic and Computation*, 14(5):675–702, 2004.

[22] Guido Governatori, Francesco Olivieri, Antonino Rotolo, and Simone Scannapieco. Computing strong and weak permissions in defeasible logic. *Journal of Philosophical Logic*, 42(6):799–829, 2013.

[23] Guido Governatori and Antonino Rotolo. BIO logical agents: Norms, beliefs, intentions in defeasible logic. *Autonomous Agents and Multi-Agent Systems*, 17(1):36–69, 2008.

[24] Guido Governatori, Antonino Rotolo, Régis Riveret, Monica Palmirani, and Giovanni Sartor. Variants of temporal defeasible logics for modelling norm modifications. In *Proc. of the 11th Int. Conf. on Artificial Intelligence and Law*, pages 155–159. ACM, 2007.

[25] H. L. A. Hart. Positivism and the separation of law and morals. *Harvard Law Review*, 71(4):593Ð629, 1958.

[26] Carl G. Hempel and Paul Oppenheim. Studies in the logic of explanation. *Philosophy of Science*, 15(2):135–175, 1948.

[27] Jörgen Jörgensen. Imperatives and logic. *Erkenntnis*, 7(1):288–296, 1937.

[28] Frank Keil. Explanation and understanding. *Annual review of psychology*, 57:227–254, 2005.

[29] Hans Kelsen. *Pure Theory of Law*. Translated from the 2nd (Revised and Enlarged) German Edition by Knight M. Berkeley: University of California Press, 1967.

[30] Ho-Pun Lam, Guido Governatori, and Régis Riveret. On ASPIC$^+$ and Defeasible Logic. In *Proc. of 6th Int. Conf. on Computational Models of Argument*, Amsterdam, 2016. IOS Press.

[31] Peter Lipton. What good is an explanation? In *Explanation: Theoretical Approaches and Applications*, pages 43–59. Springer, 2001.

[32] Tania Lombrozol. The structure and function of explanations. *Trends in Cognitive Sciences*, 10(10):464–470, 2006.

[33] David Makinson and Leendert van der Torre. Permission from an input/output perspective. *Journal of Philosophical Logic*, 32(4):391–416, 2003.

[34] David Makinson and Leendert W. N. van der Torre. Input/output logics. *Journal of Philosophical Logic*, 29(4):383–408, 2000.

[35] Paul McNamara. Deontic logic. In Dov M. Gabbay and John Woods, editors, *Logic and the Modalities in the Twentieth Century*, volume 7 of *Handbook of the History of Logic*, pages 197 – 288. North-Holland, 2006.

[36] Sanjay Modgil and Martin Caminada. Proof theories and algorithms for abstract argumentation frameworks. In *Argumentation in Artificial Intelligence*, pages 105–129. Springer, 2009.

[37] Sanjay Modgil and Henry Prakken. The *ASPIC*$^+$ framework for structured argumentation: a tutorial. *Argument & Computation*, 5(1):31–62, 2014.

[38] Henry Prakken and Giovanni Sartor. Argument-based extended logic programming with defeasible priorities. *Journal of Applied Non-Classical Logics*, 7(1-2):25–75, 1997.

[39] Henry Prakken and Giovanni Sartor. Formalising arguments about norms. In *Proc. of the 26th Annual Conf. on Legal Knowledge and Information Systems*, pages 121–130. IOS Press, 2013.

[40] Régis Riveret, Alexander Artikis, Jeremy V. Pitt, and Erivelton G. Nepomuceno. Self-governance by transfiguration: From learning to prescription changes. In *Proc. of 8th Int. Conf. on Self-Adaptive and Self-Organizing Systems*, pages 70–79. IEEE Computer Society, 2014.

[41] Régis Riveret, Pietro Baroni, Yang Gao, Guido Governatori, Antonino Rotolo, and Giovanni Sartor. A labelling framework for probabilistic argumentation. *Annals of Mathematics and Artificial Intelligence*, 83(1):21–71, 2018.

[42] Régis Riveret, Yang Gao, Guido Governatori, Antonino Rotolo, Jeremy Pitt, and Giovanni Sartor. A probabilistic argumentation framework for reinforcement learning agents - towards a mentalistic approach to agent profiles. *Autonomous Agents and Multi-Agent Systems*, 33(1-2):216–274, 2019.

[43] Régis Riveret and Guido Governatori. On learning attacks in probabilistic abstract argumentation. In *Proc. of the 15th Int. Conf. on Autonomous Agents & Multiagent Systems*, pages 653–661. ACM, 2016.

[44] Alf Ross. *Directives and Norms*. Humanities Press, 1967.

[45] Giovanni Sartor. *Legal Reasoning: A Cognitive Approach to the Law*. Springer, 2005.

[46] Audun Stolpe. A theory of permission based on the notion of derogation. *Journal of Applied Logic*, 8(1):97–113, 2010.

[47] Christian Straßer and Ofer Arieli. Normative reasoning by sequent-based argumentation. *Journal of Logic and Computation*, page exv050, 2015.

[48] Leendert W. N. van der Torre and Serena Villata. An ASPIC-based legal argumentation framework for deontic reasoning. In *Proc. of 5th Int. Conf. on Computational Models of*

Argument, pages 421–432. IOS Press, 2014.

[49] Georg Henrik von Wright. *Norm and Action: A Logical Enquiry.* Routledge and Kegan Paul, 1963.

Received 4 October 2018

Two Limitations
in Legal Knowledge Base Constructing and Formalizing Law

Réka Markovich[*]

Computer Science and Communications Research Unit, University of Luxembourg
Department of Business Law, Budapest University of Technology and Economics
Department of Logic, Institute of Philosophy, E'otv'os Loránd University
`reka.markovich@uni.lu`

Abstract

Extracting norms from legislative texts confronts us many tasks and requires decisions about approaches, methods, tools, and legal theoretical presuppositions. In this paper I present some examples from the Hungarian legislation showing how challenging sometimes the wording of these texts is from the viewpoint of norm extracting, then I present two limitations we need to face when dealing with this task. On the one hand, I argue that isomorphism cannot be upheld, but it is not necessary to have a faithful formalization. On the other, I argue that however appealing to base on constitutive norms is when formalizing—in order to avoid the necessity of normative reasoning, for instance—the consequential application of their theory in the approach leads us to a very controversial situation.

1 Introduction

In this paper some considerations of information extracting and formalizing law will be presented. The focus of this investigation is on some of the phenomena of law

I am grateful for the comments on my preliminary thoughts expanded in this paper for Trevor Bench-Capon, Marek Sergot, István Szakadát, Gábor Hamp and Csaba Oravecz.

[*]Support provided by the research project K-116191 of the Hungarian Scientific Research Fund is gratefully acknowledged. The research reported in this paper was supported by the Higher Education Excellence Program of the Ministry of Human Capacities in the frame of Artificial Intelligence research area of Budapest University of Technology and Economics (BME FIKP-MI/FM) and the Higher Education Institutional Excellence Grant Autonomous Vehicles, Automation, Normativity: Logical and Ethical Issues at ELTE BTK.

and legislative texts we face and need to deal with if we want to extract norms from the legislative texts and how these influence the methods we need to apply when constructing our knowledge base system and formal setup.[1] The examples used here come from Hungarian law,[2] but we have no reason to suppose that the phenomena their structure and linguistic surface confront us would be particular.

I have no intention to argue for the legal positivist approach in the sense of advertising the absolute primacy of written law, but I restrict this examination to the legislative texts and the linguistic-legal information that can be extracted from them—by human or machine tools (semi-automatized or automatized way). There are several reasons for doing so. One is the Hungarian—and, in general, the continental—legal system's tacit relying on this approach in the sense of requiring clear legislative texts free from ambiguities—as this is what people can access and according to which they are required to behave, and the law has to be understandable by those who are subjects of it. The decree regulating the legislative drafting in the Hungarian legal system (61/2009. (XII. 14.) decree of the Minister of Justice) says: "Legal regulation drafts should be drawn according to the rules of Hungarian language, in a clear, comprehensive, consistent way" (Section 2). The reason because of which the examination in this paper concerns not only the human understanding but the formal representation of laws too in order to prepare "machine understanding" is the amount of laws: at the time of writing these lines there are 7156 statues in force, the (parallel) process of which is way beyond human capacity. We need to consider how (and whether) the (semi-)automated formalization of legislative texts can be managed: how (and whether) the (normative) content of a legislative text is clearly detectable and identifiable.

Obviously, such a topic involves considerations not only from legal theory, but also from computer science, natural language processing (hereafter, NLP) and logic. Nevertheless, in this paper I do not suggest or even consider tools or methodologies for NLP solutions for moving towards a (semi-)automated interpretation of legal texts, as it happened—among many others—in [11], [9], [14], [13] (for project results see Legivoc—[18], Openlaws—[42], EUCases—[8], MIREL, and the most recent DAPRECO—[3] aiming at translating the GDPR into a deontic logic specifically de-

---

[1]This paper is not self/contained in the sense of referring many existing and often discussed issues without introducing or explaining them in detail. Therefore, the intended audience is the audience of this special issue having background in and experience in mining and reasoning with legal texts.

[2]The examples will be presented both in Hungarian and in—as faithfully to the original Hungarian texts' wording as possible—English translation.

signed for natural language semantics in [31]—also related to ProLeMAS), neither will consider specific formalisms—listing of which works would far exceed the limits and scope of this paper;[3] and from the knowledge base constructing issues I only will address those that directly relate to the topics discussed here.

The methodology of this paper is presenting real legislative examples to show that why and how seemingly basic steps of formalizing law and constracting legal knowledge base can be difficult, especially if we consider automated analysis—which, as I argued above, can easily made desirable by the amount of norms of a given legal system. The investigation of the examples will show why the popular and intuitive requirement of isomorphism cannot be upheld, and I will present arguments why should not it even be necessarily pursued. The examples and arguments above will also continuously reflect upon the semantics of legal norms and the differentiation between constitutive and regulative ones, trying to find solutions to the difficulties presented by the examples. This interaction between the examples and the arguments will lead us to another limitation: the limitation of extensively formalizing law in terms of constitutive norms.

2 Approach and Presuppositions

Let's suppose: we are looking for norms in legislative texts. In agreement with the approach of [41] as "a norm must convey information to fulfil its function of communicating standards of behaviour; the way in which one is expected to behave must be clear from the norm", we are looking for the way this can be extracted from the legislative texts (presupposing that they contain this information as we think about them as sets of norms). The paper of [41] refers to [10] when listing five questions to be answered if we want to have a complete norm:
(1) Who is obliged or permitted to do something?
(2) Is there an obligation or a permission to do something or to leave something undone?
(3) What must be done or forborne?
(4) Where must something be done or forborne?
(5) When must something be done or forborne?
These elements more or less correspond to the factors most authors count with when discussing norms. In von [43], the "father of deontic logic" adds the authority to the elements above when talks about the parts or components of norms that are

---

[3]By no means this implies the claim that one or another specific formalism could not be better in dealing with the issues presented in this paper, but here the focus is on preliminary considerations.

prescriptions—as von Wright differentiates between three type of norms: rules (rules of a game), directives (like technical norms of an instruction manual) and prescriptions (commands, permissions, and prohibitions, which are given or issued to agents concerning their conduct. "The laws of states are prescriptions", declares von Wright.

Let's restrict our investigation to the character or deontic status (whether something is obligatory, permitted or prohibited), the action (what is obligatory, permitted or prohibited) and the agent (the subject or addressee of the norm: for whom it is obligatory, permitted or prohibited the given action). The importance of agents is often tacitly underestimated: it is pretty general to discuss deontic logic in an impersonal way, that is, discuss obligations without agents, but in law there is always an addressee of the given norm. And if we take the expectation in [41] above seriously accepting that a norm must convey how *one* is expected to behave, extracting who that 'one' is a fundamental requirement of the whole process. Let's start with the character or deontic status, though, as it is reasonable to suppose that this component is the most salient of the norms—and as we will see, the question of the actions and agents come with it anyway.

3   Language of Legislative Texts: Deontic Character and Linguistic Modalities in Law

The language use in legislation is more formal and bound than other registers of natural language (whether necessarily or not is often discussed, see the Plain English Movement, [7] vs. for instance [12]), even more than other registers in legal language (for instance compared to the language of judgements, contracts or explanations).[4] As we saw above, the legislator itself has formed requirements concerning the legislative language use in a decree on legislative drafting. There are a lot of features of the legal language use in Hungarian (for instance the high proportion of nominal structures) discussed often by linguists and lawyers (this is, of course, is not a specifically Hungarian issue, for English see e.g. [28], [37], for a comparative study to English, French, German, and Latin see [27]), but now we only consider those that are related to the expression of linguistic modality as these are by which we can identify the deontic character of a norm: what is obligatory, what is forbidden and what is permitted.

In the Hungarian language, modalities can be expressed in different ways (not only in Hungarian, of course, for a foundational entry see [23]): the most typical ways

---

[4]Most of the comparative studies of a recent Hungarian project on this topic are only available in Hungarian, see: [35], but one of them will be available in English soon: [44].

are participles, auxiliary verbs with infinitives, and suffixes at the end of the verbs. For instance, *obligatory* can be expressed with saying that something is 'kötelező' (which is the literal translation of 'obligatory'), or 'szükséges' (which is the literal translation of 'necessary')—in its deontic reading, with using the auxiliary verb 'kell' (which means what 'must' or 'shall' means in deontic sense), or with the derivational suffix '-andó, -endő' at the end of the verb denoting the action (for instance, the translation of the word 'fizetendő' is 'is to be paid'). *Permission* can be expressed with the adverb/adjective 'megengedett' (which is the literal translation of 'permitted') or using the adjective 'szabad' (literally meaning 'free to'), but most frequently happens with the inflectional suffix '-hat, -het' attached to the verb which describes the action, or with the derivational suffix '-ható, -hető' attached to the verb (in the predicate place of the sentence) with a—tacit—copula: in this case the translation of the word e.g. 'fizethető' 'it may be paid' or 'it can be paid' with the deontic reading of 'can'. That is, in the Hungarian legal language, these suffixes play the role that in English is played by modal verbs 'may' and 'can' in their deontic reading. These suffixes also have other readings (just like the word 'szükséges', that is, 'necessary', obviously has), but in a legislative text the deontic context is given—or, as [23] calls it, the conversational background is bound: it is *literally* is 'what the law provides' as we are reading *the law itself*. *Forbidden* can be expressed with the adjective 'tilos' (which is the closest version to the translation of 'forbidden'), with the negation of *permitted*, that is, any of the indicated possibilities above with a declarative use of 'no' ('nem' in Hungarian).

3.1 False Friends and Missing Modalities in Legislative Texts

Sometimes, though, other ways of expressing modalities can be faced with: ones that might mislead the reader, both human and computer ones.

In [25] the language of the Hungarian Criminal Code is discussed: as a criminal code lists the actions a society (a legislator) considers the less desired, one would expect to meet the linguistic signs of *forbidden* several times—but would disappoint as the only deontic modality that can be detected in this legislative text is *obligatory*. The dominant form of sections in the Hungarian Criminal Code's special part (listing the felonies) is the following:

Example (1) "Aki mást megöl, bűntett miatt öt évtől tizenöt évig terjedő szabadságvesztéssel büntetendő." (Any person who kills another human is to be punished for five to fifteen years of imprisonment due to having committed a felony.) [Act C of 2012 on the Criminal Code]

As we see, the command to the judge to punish the perpetrator of a forbidden act is not *attached* to the prohibition: it *is* itself the expression of the act being forbidden. Of course, the human interpretator understands that this realizes what we usually refer to as 'forbidden', but the automated processing needs some input in order to properly classify these norms—which, at least in Hungarian—could be the string 'büntetendő' (is to be punished) as these norms are very regular: whenever the computer finds this word, the declarative form verb of the given sentence will provide the action to which we need to assign the deontic status of being forbidden.

A more surprising set of examples can one find in the Hungarian Highway Code. There are several sentences in it like the following ones:

Example(s) (2)
"A fényjelző készülék folyamatos zöld fény jelzésnél kiegészítő hangjelzést is adhat" (The light-signalling device may additionally beep while green.)
"A fény kibocsátására alkalmas jelzőtáblán a jelzőtábla fehér és fekete jelzései egymással felcserélt színekkel is megjelenhetnek." (The white and black markings on the light signal board may appear with commuted colors.)
"A jelzőtábla alatt kiegészítő tábla adhat útmutatást a jelzőtábla jelzésének értelmezésére." Under the traffic sign, an additional sign may give guidance how to interpret the traffic sign.
"Az út mellett vagy közelében lévő egyes létesítményekről kék vagy barna alapszín? jelzőtáblák adhatnak tájékoztatást." (Information about the facilities passed by the road may be given by blue or brown signposts.)
"Az (1) bekezdésben említett jelzőtáblák alatt elhelyezett kiegészítő táblán nyíl jelezheti, hogy a tilalom hatálya a táblától kezdődően vagy a tábláig áll fenn." (Additional signpost with an arrow under the traffic signs mentioned in (1) may indicate whether the prohibition starts or ends with the traffic sign.)

These 'may'-s (the '-hat', '-het' suffixes in Hungarian) might seem permission at first sight. We, of course, know that 'may', just like the given Hungarian suffixes, can be used with other modal tastes—e.g. alethic or metaphysical, but before facing these examples we did not expect to meet with them in legislative texts since, as it was mentioned above, here the context is bound, or, as [23] says, the conversational background is given since the resolving deontic reading can paraphrased as 'in view of what law provides'—which is tacitly the case as we are reading the law itself. But the modalities in examples (2) do not really make sense as permissions: the legislator does not give a permission to the light-signalling device to beep. These rather seem alethic possibilities. Does this mean that we need to introduce a new modality into our formalization? I don't think so. *If* we consider descriptive propositions, these possibilities can be expressed with a disjunction: the light-signalling device beeps or it does not. But what is the deontic character of this rule then? Well, it is an obligation to people using public roads to consider a light-signalling device as such—both if it beeps and if it does not. So we found a conjunction of obligations where the first impression—and the trained machine's result would have—suggested permission. Not an easy tension to resolve with automatic tools.

It also happens that there is no linguistic sign of any modality in a sentence. In [23] we find an example of a sentence missing any linguistic sign of a modality: the explanation Kratzer gives to "The car goes twenty miles an hour" is that "the modality may be inherent in the verb" (p. 639). She does not mention cases, though, where the deontic modality is present inherently in a verb. In legislative texts, actually, we often find seemingly declarative sentences lacking any linguistic sign of modality. The reader probably says promptly: of course, those are constitutive rules. Indeed, in case of constitutive rules, the lack of 'shall' or other phenomenon expressing normativity is not surprising, this is part of the description we got in [33] describing constitutive rules as mostly expressed in non-imperative, 'counts-as' rules. We also can take a step back in history and check what an earlier author whom Searle also leans on: [30] describes practice rules as definitive ones, which definitive feature is so strong that if we don't follow these rules, we don't engage in the given action they define.

If we want to explain the using of indicative mood with this distinction, we need to understood broadly the word 'definition', at least this is how it seems from the Hungarian legislative language use. Let's see the following examples:

Example (3) "Szünetel a biztosítás a fizetés nélküli szabadság ideje alatt." (The insurance intermits during the unpaid leave.) [Act LXXX of 1997 on Social Security]

We might say that the definition of insurance as an institution can be paraphrased in a way it contains this aspects of intermitting. Maybe, we can also say something similar in case of the following rule:

Example (4) "Ha leszármazó és szülő nincs vagy nem örökölhet, az örökhagyó házastársa egyedül örököl." (If there is no descendant or parent, or if they are excluded from succession, the surviving spouse inherits the entire estate.) [Act V of 2013 on the Civil Code]

What does this mean in terms of the norm elements we are looking for? Just to start with the first element we look for: what is the deontic character of these norms? Is there one at all? Constitutive norms are usually something which help us avoid dealing with deontic notions—and therefore deontic logic or normative reasoning. But should we really lean on this option when constructing our knowledge base and formalizing law? Actually, Searle never said that there is no obligation in counts-as norms (what is more, speaks about *deontic power* in their case)—(even if this is typically not the case) they *might* be expressed imperative. (And there are several critiques of the theories of the whole differentiation between constitutive and regulative norms, see for instance [16], [24], [15], [39], [38]). For instance rules listing the required elements of an official document to count as an official document is a paradigmatic case what we usually consider a constitutive rule. Still, the phrasing of the following sentence suggests an obligation:

Example (5) "A meghatalmazás képviseleti jogot létesítő egyoldalú jognyilatkozat. A meghatalmazást a képviselőhöz, az érdekelt hatósághoz, bírósághoz vagy ahhoz a személyhez kell intézni, akihez a meghatalmazás alapján a képviselő jognyilatkozatot jogosult tenni." (A power of attorney is a unilateral act granting the right of representation. The power of attorney shall be addressed to the agent, the competent authority or court, or any person to whom the agent is authorized to make a legal statement.)[Act V of 2013 on the Civil Code]

In von [43] we also find a clear claim about the laws of the state being *prescriptions*, per se, but let's see what the strictly legal approach says.

Legislation as such is subject of a discipline called *legistics* or *legistica* taught at law schools. Hungarian legistics textbooks—obviously—often discuss Hungarian legislative language use, both in descriptive and prescriptive way (discussing how the language should be used when phrasing legal norms). Considering our topic, we can find the following in [40]: "In legal norms indicative mood is dominant. Thus the norm text shuns expressions like 'should', 'ought', 'necessary', since the indicative mood is imperative for the authorities, government agencies (e.g. rules of competence, authorizing rules, rules of procedure); while, less often, to put emphasis, norms expressing obligation literally too may occur—e.g. when the obliged agents are directly citizens, business organizations."[5] Unfortunately, the parsing tools for

---

[5]"A jogi normákban egyeduralkodó a kijelentő mód. Ezért a norma szövege kerüli a "kell", "köteles", "szükséges" kifejezést, mert a hatóságok, állami szervek számára a kijelentő mód imperatív (pl. hatásköri szabályok, felhatalmazó szabályok, eljárási szabályok), míg nyomatékosításként

the Hungarian language does not provide the possibility to check the correlation between the addressees (personal scope) and the linguistic mood of expressing obligations, the claim is still worth to deeper analysis. On the one hand it corresponds nicely with the foundations of the regulative–constitutive rules distinction: as it is often referred, the activities defined by constitutive rules are logically dependent on the rules constituting them, so it is not strange if agents created by the law conducting activities created by the law get their commands from rules having the same linguistic features as the ones created them and their activities. On the other hand, this claim of [40] says this indicative mood is imperative to them, that is, sounds constitutive or not, these rules bear normativity, they convey imperatives. This is confirmed by another textbook of legistics: [36] says that "in Hungarian legislative texts the indicative mood means imperatives or obligation. (...) The predicate in norms' text is always a command, even if it has no linguistic sign."[6] This means that we need to look for and identify deontic character in all the laws' sentences. The correlation indicated above might help is, but there are cases, where this correlation between the official addressee and the phrasing does not hold, like in the following example:

Example (6) "...a vállalkozás egyértelműen és jól látható módon felhívja a fogyasztó figyelmét a 11.§-ban meghatározott információkra" (...the company calls the consumer's attention to the information detailed in section 11 in an univoque and visible way) [45/2014 (II. 26.) Government decree on Detailed Rules of Contracts between Consumer and Company]

Here the obliged agent is a company, that is, not an authority, the mood is still indicative representing no modality. We might be appealed to conclude that all indicative mood predicates cover obligations in legislative texts. For these cases the NLP considerations can be made up with saying that in legislative texts all the verbs in indicative mood—except for the ones in the antecedents of conditional rules[7]—should be detected as obligatory. That is, in the case of example (6) finding 'felhívja' ('calls') should be translated to a formal representation of 'it is obligatory to call'. But this solution does not help in examples above: in example (4) the verb is 'intermits' and the deontic content as 'it is obligatory to intermit' sounds strange, the

---

ritkábban előfordulhat a kötelezést nyelvileg is kifejező norma, (...) például akkor, ha a kötelezettek közvetlenül magánszemélyek, gazdálkodó szervezetek."

[6]"Magyar nyelvű normaszövegekben a kijelentő mód felszólítást vagy kötelezést jelent. (...) A normatív állítmány mindig rendelkezés, akkor is, ha nyelvileg nem az."

[7]About the automated identifiability of the antecedent of conditional sentences in legislative texts in Hungarian see [26]

'it is obligatory to inherit' even more does so. The reason is that inheriting and intermitting are not actions to which oblige someone—or something—would make sense: the spouse has no influence on whether (s)he inherits, the insurance has even less whether it intermits. The principle along which this problem might be solved in NLP and ((semi-)automated) formalization is that instead of acts, we put propositions in the argument of the deontic operators at this level: 'it is obligatory that the insurance intermit' and 'it is obligatory that the spouse inherit' makes much more sense. The satisfaction is temporary, though, as this solution does not provide some crucial information of compliance: whose obligation is that to make things so? In example (6) we see the agent, but in example (3) and (4) the real agent is not in the given sentence or section, and the extraction of this information sometimes is not easy to the human reader either: it requires some legal knowledge where to look for the answer. That is, constructing the knowledge base requires that legal knowledge. Before going further with this direction, note that we already lose something: isomorphism.

4 Isomorphism and its limitation

At the end of the eighties, the term 'isomorphism' has been introduced into the discussion of legal knowledge based systems and formalizing law. In [22] we find five conditions that have been listed in [5] as the following:
(i) Each legal source is presented separately.
(ii) The representation preserves the structure of each legal source.
(iii) The representation preserves the traditional mutual relation, references and connections between the legal sources.
(iv) The representation of the legal sources and their mutual relations (...) is separate from all other parts of the model, notably representation of queries and facts management.
(v) If procedural law is part of the domain of the model then the law module will have representation of material as well as procedural rules and it is demanded that the whole system functions in accordance with and in the order following the procedural rules.

Practically, (ii) is what matters the most, the short or narrow references at least refer to isomorphism mainly covering that: [5] sum up presenting the rules above with saying that "the important demand made by isomorphism is that there is a clear correspondence between items to be found in the source material and items to be found in the knowledge base. (...) Where one to one correspondence is not achieved, however, it is important to relax the constraint only so that one source item corresponds to several knowledge base items and maintain the prohibition on a single knowledge base item capturing the material from several source items." The definition we see in [29] is quite similar: "the situation that one source unit is formalised in one knowledge base unit. By source unit we mean the smallest identifiable unit of the source from which a norm can be extracted. In general this will be a section or a subsection of a code."

[5] convincingly argue that following these rules greatly help to satisfy the concerns of well engineered knowledge systems presented in [19]: verification, validation, and maintenance. They argue that isomorphic formalization also helps the user: "Many of the problems with experts systems come from a mismatch between the rule based conceptualisation of the expert system and the conceptualisation of the user." (Here the authors do not discuss the potential users of expert systems but Bench-Capon does so in [4].) We come back later to the question of users and mismatches between concept structures.

But not only software engineering considerations serve arguments for isomorphism. In [17] and [6] we find legal theoretical considerations on legistics about reasons behind the typical structure of laws providing the general rule first and then a number of specific exceptions. According to the authors, this structure reflects on a need coming from the addressees' side. As [17] says: "This [structure] facilitates the normative function of the law; the law would have little effect on social behaviour if its rules were so convoluted that persons could only with great difficulty, if at all, predict the legal consequences of their actions." According to [6], this presentation structure of general rules and exceptions helps achieve "ease of application, ease of understanding, and the possibility of allocating the burden of proof"—points that can be upheld with keeping on isomorphism when formalizing. In [29] a few years before, the structure above is yet presented as a threat to, or a limitation of isomorphism, which, however, can be overcome (building up formalization upon nonmonotonic reasoning or conflict solving metarules), as it is presented in the paper.

There are other limitations brought up in the literature, though. [5] refer to [34] as latently raising objections against isomophism. It is the critique of isomorphism which is latent in [34] as it is not even mentioned literally since the paper is about following a top down development of logic programming when formalizing the Indian pension rules, but the criticism of the isomorphic approach can be read clearly from it. As [5] rephrase it: "the isomorphism approach is very well if the legislation is itself well stuctured. In such a case, the structure of the problem, the structure of the legis-

951

lation, and the structure of an isomorphic knowledge base would all be in harmony. It is, however, often the case that the legislation is not well structured. Often repeated amendments and 'patching' mean that the legislation is itself a complete mess, and fails to reflect the real structure of the domain. In such a case, basing the structure of the knowledge base on legislation would lead to a poorly structured knowledge base, which fails to correspond to the 'real world' problem." With [34] own words: "in common with many other examples of legislation and regulations, especially those that refer to periods of time, the Pension Rules are imprecise and very casual about many of the key concepts. They are certainly not precise enough to be formulated directly as an executable program, and they are arguably not precise enough to be applied by a human agent either." At their reaction, [5] put a light on the question what is meant by 'poorly structured'—a point to which we come back soon—, and argue by turning the objection into a good argument *for* isomorphism: a poorly structured legislative text very likely becomes subject of amendment, which requires maintenance of our knowledge base—which something that can be efficiently done if we previously followed the isomorphic approach.

At this point, accepting [5]'s reply (and original arguments above), one might be convinced that isomorphism is an obviously advantageous approach when formalizing legal rules, one definitely to pursue, and this whole problem urging us to give it up only came up because we wanted to extract real (regulative) norms from constitutive ones, so we should not do so; examples (3) and (4) refer to actions of official agents/government organizations, so if we would like to help people to comply with law, these can be kept as they are: in descriptive form about what happens in law. In total agreement with the first part of this conclusion, getting rid of the intention to change constitutive into regulative norms won't solve the problem: we encounter it in case of clearly regulative rules, too. Let's have a look at an example looking for a norm in clearly regulative rules.

Rules on advertising (Hungarian Act XLVIII of 2008 on advertising and parts of the Act LVII of 1997 on Fair Competition—which parts earlier were parts of the Act XLVIII of 2008) are full of rules like the following ones:

Example(s) (7)
"Advertising is prohibited for goods whose production or marketing is forbidden."
"Advertisements inciting violent, or personal or public security threatening behavior, are prohibited."
"Comparative advertising may not injure the reputation of another company or the name, merchandise, brand name and other marking of such company."

From the first two, we cannot identify any subject (addressee), from the third, linguistically, we can: the advertising itself. But the advertising or advertisement is clearly not an agent in the sense of being capable of conducting actions, therefore one whose behaviour could be ruled or influenced by imposing a prohibition. Therefore, the advertising (or advertisement) itself cannot be the addressee. But then who is it? Who should see to it that a comparative advertisement does not injure the reputation of other companies? There are several general laws in a society—a phenomenon because of which the harmfulness of the tendency of forgetting about the addressee is not so conspicuous—, the criminal code or the civil code are ones concerning all members of a given society. Is this Act on advertising is one of those? Obviously not, as most people of a given society have absolutely no influence on what happens in advertising. Fortunately, the—overwhelming majority of—legislative texts provide their personal scope in one of the first sections, that is, they denote the agents who shall comply with them. In case of the Act on advertising this section declaring the personal scope does not provide the final solution yet since there are three agents denoted: the advertiser (the company whose product/service the advertisement is about), the advertising service providers (practically the agencies), and the publishers of advertising (the TV and radio companies, the publishers, etc.). From this we could conclude that there are three agents on whom the duties above are imposed. This is not far from the truth, but the situation is a bit more complicated: after checking the beginning of the Act in order to find the agents and put them into the norms we would like to extract, we need to check some sections at the end of the Act, too. After providing the rules in the above form, the legislator put a subtitle (which is one of the possible units of Hungarian legislative texts), the 'Vested responsibility'. This part tells us which addressee will be liable for the violation of a given prohibition above (listed protractedly in the Act)—which practically means that the given agent is the one which is obliged to obey the rule (sometimes the three addressee have joint and several liability—which is a specific problem in terms of formalization, one which we do not discuss here, though—, some rules need to be complied with by one or another of the agents (aligning the reasonable aspect of which phase can be influenced of creating and publishing an advertisement)). If we want to have proper complete norms containing the addressee too (which would be nice considering the information need to be extracted for compliance), we need to get that information from another parts of the Act, not the one where the "main rule" can be found.

We, of course, could say that, in order to keep isomorphism, we formalize them separately when the prohibitions above (let's call (any of) them Rule Nr. N):

Rule Nr. N: *'it is obligatory that action A is executed / that C is the case'*
and then the "rules" in the section on vested responsibility in a way that:
'it is obligatory to subject S to comply with the rule Nr. N', that is,
'it is obligatory to subject S to fulfil the obligation in rule Nr. N'
but this result is rather redundant.

But then what could we do with the laws similar to the Act on advertising (of
which there are a lot) in "ruling" the agents in separate norms? Annotating the
knowledge base slots with the locution in the source (legislative) text could serve
as a solution. Do we lose isomorphism? It depends on how strictly we interpret its
definition. As it is mentioned above, in [29] we see one saying that we need to consider one source unit (whose correspondence with one knowledge base unit should
be upheld) as the smallest identifiable unit of the source from which a norm can
be extracted. Well, in this case it is the whole Act itself—a "structural unit" that
authors would scarcely accept as such: representing the whole Act in one unit of the
knowledge base would completely contradict to what we consider as one unit. We,
therefore, seem to be in the need of loosen up the requirement. A requirement of
an *algorithmizable correspondence* would still be considerable as substantial help in
pursuing the software engineering concerns of verification, validation, and maintenance.

One could say that the Act on advertising is a good example of the poorly structured legislative texts [34] talk about. It is important to emphasize: this is not the
case. Of course, it depends on, as [5] note, what we call 'poor structure', but in the
case of the original example of being poorly structured, the Indian pension rules,
the authors found that the same word was used to refer to slightly different things,
sometimes even in the same sentence, while in other cases the key concepts were
really casually phrased. In the Act on advertising there is no such problem. There
is nothing 'poor' from a legal point of view or, specifically, from the viewpoint of
legistics in this Act's structure. It is, actually, a well structured law. What is behind
the structure of this Act then? From legal theoretical point of view we can call the
attention to the fact that advertising law is an area of law aiming at the protection
of consumers and fair competition. The legislator's view reflects on this aim and the
viewpoint of the consumer/other companies, thus, instead of concentrating on the
obligations of each concerned agent, the legislator ends up in listing the situations
that should not be realized (as for the consumer it does not make a difference which
agent is responsible for a given undesirable situation).

We need to distinguish two phenomena [34] talk about, at least we need to reflect more precisely their relationship. The—clearly problematic—practice of handling concepts and terminology poorly is a sufficient condition of not having matching with the real world problem structure, but far from being necessary. The legal (more precisely, legislative) mapping of the world *is not isomorphic* with the world itself. In [44] we find an—experiment-supported—analysis of the comprehensibility of legal texts. The author's conclusion is that changing the often scolded legal language would not solve the problem that people find it difficult to understand the legal (not just legislative) texts. Offering a classification of three kinds of pragmatic situations where laymen meet legal texts, Ződi points out that the difficulties in understanding legal texts should be investigated together with these situations and not just the syntactical and semantical features of legal texts. He emphasizes: "the texts of legal sources are not organized around everyday problems—they follow a different logic. (...) The texts of legal sources are mainly organized around theoretical legal 'fields', and try to avoid redundancies. Therefore, even for a very simple contractual problem the answer will lie in many places. (...) One is involved in a car accident: there are the rules of the traffic code, rules of the penal law (code), rules of obligatory third-party insurance, and rules for the whole procedure, including the usage of forms, and so on. And even if texts are found, circumstances are fixed, and proper interpretation is in place, the question still remains: What follows from all these rules? What should I do? Where should I go? What should I write down, fill out, submit? Who should I inform, call? And so on."

In an *ideal case*, a knowledge base system can mediate between the two: *faithfully* represents the law (where faithfulness, as we saw, does not necessarily mean isomorphism, especially not verbatim as [20] advocate), and is built up in a way that with normative reasoning can help people find answers to their question in a real life problem. Expecting that the law itself is structured in a way that its verbatim formalizing (if it makes sense at all in terms of feasibility) provides all the answers to real life problems would be reasonable if we thought the legislative texts themselves as doing so. But according to [44]—and the experience—this is not the case. This mediating task is exactly something because of which building legal knowledge base systems would provide a real contribution. This requires legal theoretical knowledge (and of course knowledge of the given legal domain) when constructing. Again, this is the ideal case, presupposing that we want a system for information extraction useful to laymen (too). It can be the case that our target audience is different: specific types of adjudicators. About this case, see [4].

One might feel at this point that the requirement of extracting (regulative) norms demands too many sacrifices, so we should be satisfied with constitutive norms in their "normal, descriptive form", what is more, regulative norms should be rephrased as constitutive ones, too, in order to avoid any complication. Unfortunately, from a legal theoretical or philosophical point of view, consequentially leaning on constitutive rules also has a strong limitation.

5 Dealing with Constitutive Norms in Law and Its Limitation

As it has been mentioned earlier in this paper, there is a view advocating that there is no real difference between these two types of the norms. It might mean that what we call regulative norms, can be rephrased in the way 'constitutive ones' are phrased, that is, exactly in the other way around than we have pursued in this paper so far.

The possibility is appealing: we don't need to deal with normative reasoning or deontic logic, we have simple descriptive propositions. We cannot even argue with calling attention to the presence of deontic mo dalities since Searle only said that constitutive rules are *mostly* phrased in indicative mood, did not say that this happens always: the fact that we find rules using 'ought to' and 'obligatory' in the law of contracts supports this thesis: we all know that if we infringe some of this rule, what comes is not a punishment, but the nullity of the contract—which is actually not a contract then. And the solution of resolving regulative norms into constitutive norms seems to be easy: an obliging norm can be translated into a conditional sentence saying that if the given action is not conducted, then it is a violation.[8] Saying so we already have a paradigmatic constitutive rule: not performing the given action counts as a violation. If it is so easy to lean on constitutive rules when formalizing law, what is the problem?

Let's see a new example. The Hungarian traffic rules are a set of norms—we can call their set as the Highway Code, in Hungarian we use the acronym KRESZ—that look like the classical examples of constitutive rules: the rules of a game. Some rules are general during "playing" (participating in public road use), but the obliging rules are mostly given in a very similar way it happens in board games: if you land on this and that square, you have to do this and that—if you see this and that sign (an octagon with the string 'STOP', down triangle with red edge, circle with red edge and a number in it, etc.), you have to (or should not) do this and that (stop, give priority, go with a specific speed, etc.). This and that (things in the world) count

---

[8]This approach is used when defining specifications for instance in LegalRuleML, see [2] and this thought was the basis of the Andersonian-Kangerian reduction of deontic logic, see [1] and [21].

as this and that (pedestrian way, public road, traffic sign). Even the topographical feature comes, and given situations result in a situation when you miss a (or more) turn(s): your driving license is taken. For a moment we might become uncertain if we thought to 'walking' and 'going by car' as actions that we do anyway, that is, are not defined by these rules. Sure, we can do these: in our home or at our land, we can walk and use our car in any way we would like to, but the whole system of public road use is *defined* by these rules. In Hungary, traffic rules are contained in the 1/1975 (II.5.) joint decree of Minister of Traffic and Minister of Internal Affairs on public road traffic rules, and this statute—not like others—does not contain anything about the consequences of infringing the rules it sets. In Roman law this was called 'lex imperfecta': a law imposing no consequence on its breaching cannot fulfil its role, it is not a proper law. It of course does not mean that in Hungary, breaching traffic rules have no consequences: it has, but these consequences are handled in another law (mostly in the Act on misdemeanors). This structure strengthens the feeling that the traffic rules are just a specific set of rules of a game—the game called public road use.

But we should be cautious with this conclusion. This constitutive rules approach of law has an unwanted consequence: as it has been already referred above, constitutive rules have a definition in [30] according to which if we do not follow these rules, we do not engage with the action itself. But having said that, with breaching the traffic rules, we getting out of the scope of the given law (in case of KRESZ the decree above)—but then what serves as a basis to the policeman to impose a fine? Paying money is something that has nothing to do with the "game rules": we work, earn and spend money independently from the public road use, the word 'money' does not even appear in the text of the decree. It's clearly something that is out of the frame imposed by the traffic rules, we are still exposed to it by breaching them. It is because the system of law is not something discerptible from our life like football or chess is. It provides the structure of our life and—especially—our society in many ways, so thinking about its rules as merely constitutive rules just because there are institutions created by it comes with a controversial consequence. I don't claim that constitutive rules as such make no sense in law: they do—especially the approach we find in [32] about constitutive rules as a manageable and effective technique of presentation of a system of norms as we can connect a set of concrete circumstances to a set of legal consequences by them. But there are always legal consequences.

6   Conclusion

Pursuing extracting norms from legislative text is a reasonable task, just like the presupposition that it can be conducted. As we saw, at least in the Hungarian legislation, there are challenging examples of identifying and extracting the deontic character of the legal norms we would like to have in our knowledge base systems, especially if we would like to use (semi-)automatized methods. This task becomes even more challenging with that, as I argued, it is not reasonable to take isomorphism as a general requirement that should be met, as it not always can be met, even in the case of well structured legislative texts. The strength of this conclusion, of course, might vary according to the goal pursued by the formal representation, but in the case of extracting norms in a way it serves a legal knowledge base which faithfully represents the law and, at the same time, which might be subject of normative reasoning helping for example compliance of agents, being faithful probably won't (cannot be) be equal with insisting isomorphism—even in the case of well structured norms. This good structure, though—having a labelled section providing the agents for instance—gives the opportunity to have another requirement somewhat looser than isomorphism, the algorithmizable correspondence using solutions like annotating the knowledge base slots with the locution in the source (legislative) text. Clusters of different—more specific—technical solutions might be developed according (and corresponding) to different legislative techniques. I also argued in this paper that leaning on constitutive norms—especially in their Rawlsian definition—when formalizing law brings very problematic result as breaching a rule would end up in getting out of the scope of the law, that is, per definitionem not being liable for breaching it.

References

[1] Alan Ross Anderson. A reduction of deontic logic to alethic modal logic. *Mind*, 67(265):100–103, 1958.

[2] Tara Athan, Harold Boley, Guido Governatori, Monica Palmirani, Adrian Paschke, and Adam Wyner. Oasis legalruleml. In *Proceedings of the Fourteenth International Conference on Artificial Intelligence and Law*, ICAIL '13, pages 3–12, New York, NY, USA, 2013. ACM.

[3] Cesare Bartolini, Andra Giurgiu, Gabriele Lenzini, and Livio Robaldo. Towards legal compliance by correlating standards and laws with a semi-automated methodology. In *BNCAI*, volume 765 of *Communications in Computer and Information Science*, pages 47–62. Springer, 2016.

[4] Trevor J. M. Bench-Capon. Deep models, normative reasoning and legal expert systems. In *Proceedings of the Second International Conference on Artificial Intelligence and Law, ICAIL '89, Vancouver, BC, Canada, June 13-16, 1989*, pages 37–45, 1989.

[5] Trevor J. M. Bench-Capon and Frans Coenen. Isomorphism and legal knowledge based systems. *Artif. Intell. Law*, 1(1):65–86, 1992.

[6] Trevor J. M. Bench-Capon and Thomas F. Gordon. Isomorphism and argumentation. In *The 12th International Conference on Artificial Intelligence and Law, Proceedings of the Conference, June 8-12, 2009, Barcelona, Spain*, pages 11–20, 2009.

[7] Robert W. Benson. The end of legalese: The game is over. *New York University Review of Law & Social Change*, 13:519, 1985.

[8] Guido Boella, Luigi Di Caro, Michele Graziadei, Loredana Cupi, Carlo Emilio Salaroglio, Llio Humphreys, Hristo Konstantinov, Kornel Marko, Livio Robaldo, Claudio Ruffini, Kiril Simov, Andrea Violato, and Veli Stroetmann. Linking legal open data: Breaking the accessibility and language barrier in european legislation and case law. In *Proceedings of the 15th International Conference on Artificial Intelligence and Law*, ICAIL '15, pages 171–175, New York, NY, USA, 2015. ACM.

[9] Guido Boella, Luigi Di Caro, and Livio Robaldo. Semantic relation extraction from legislative text using generalized syntactic dependencies and support vector machines. In *Proceedings of the 7th International Conference on Theory, Practice, and Applications of Rules on the Web*, RuleML'13, pages 218–225, Berlin, Heidelberg, 2013. Springer-Verlag.

[10] P.W. Brouwer. *Samenhang in Recht: een analytische studie*. PhD thesis, University of Twente, 6 1990.

[11] John J. Camilleri, Normunds Gruzitis, and Gerardo Schneider. Extracting formal models from normative texts. *CoRR*, abs/1706.04997, 2017.

[12] David Crump. Against plain english: The case for a functional approach to legal document preparation. *Rutgers Law Journal*, 33:713, 2002.

[13] Emile de Maat and Radboud Winkels. A next step towards automated modelling of sources of law. In *Proceedings of the 12th International Conference on Artificial Intelligence and Law*, ICAIL '09, pages 31–39, New York, NY, USA, 2009. ACM.

[14] Emile de Maat, Radboud Winkels, and Tom van Engers. Automated detection of reference structures in law. In *Proceedings of the 2006 Conference on Legal Knowledge and Information Systems: JURIX 2006: The Nineteenth Annual Conference*, pages 41–50, Amsterdam, The Netherlands, The Netherlands, 2006. IOS Press.

[15] J. Garcia. Constitutive rules. *Philosophia*, 17(3):251–270, 1987.

[16] Anthony Giddens. *The constitution of society : outline of the theory of structuration*. Social and political theory from Polity Press. Polity Press, Cambridge, 1984.

[17] T. F. Gordon. Oblog-2: A hybrid knowledge representation system for defeasible reasoning. In *Proceedings of the 1st International Conference on Artificial Intelligence and Law*, ICAIL '87, pages 231–239, New York, NY, USA, 1987. ACM.

[18] Pierre Jouvelot Hughes-Jehan Vibert and Benoit Pin. Legivoc—connectings laws in a

changing world. *Journal of Open Access to Law*, 1(1), 2013.

[19] M. A. Jackson. *Principles of Program Design.* Academic Press, Inc., Orlando, FL, USA, 1975.

[20] Peter Johnson and David Mead. Legislative knowledge base systems for public administration: Some practical issues. In *Proceedings of the Third International Conference on Artificial Intelligence and Law, ICAIL '91, Oxford, England, June 25-28, 1991*, pages 108–117, 1991.

[21] Stig Kanger. New foundations of ethical theory. In Risto Hilpinen, editor, *Deontic Logic: Introductory and Systematic Readings*, pages 36–58. D. Reidel, Dordrecht, 1971.

[22] Jorgen Karpf. Quality assurance of legal expert systems. In *Proceedings of the Third International Conference on "Logica, Informatica, Diritto"*, pages 411–440, 1989.

[23] Angelika Kratzer. Modality. In Arnim Stechow and Dieter Wundelrlich, editors, *Handbook of Semantics*, pages 639–650. de Gruyter, New York, 1991.

[24] Eerik Lagerspetz. *The opposite mirrors : an essay on the conventionalist theory of institutions.* Law and philosophy library ; v. 22. Kluwer, Dordrecht ; Boston, 1995.

[25] Réka Markovich. Order of Norms and Deontic Modality. *South American Journal of Logic*, 1:435?445, 2015.

[26] Réka Markovich, Gábor Hamp, and Syi. A kondicionálisok problémája jogszabályszövegekben [Conditionals in legislative texts]. In *Proceedings of Hungarian Natural Language Processing Conference 2014*, pages 295–302, Szeged, 2014. University of Szeged.

[27] Heikki Mattila. *Comparative Legal Linguistics.* Ashgate Publishing, Aldershot, 2006.

[28] David Mellinkoff. *The Language of the Law.* Resource Publications, Eugene, Oregon, 2004.

[29] Henry Prakken and Joost Schrickx. Isomorphic models for rules and exceptions in legislation. In *Legal knowledge based system JURIX 91: Model-based legal reasoning*, pages 17–27. The Foundation for Legal Knowledge System, 1991.

[30] John Rawls. Two concepts of rules. *The Philosophical Review*, 64, January 1955.

[31] L. Robaldo and X. Sun. Reified input/output logic: Combining input/output logic and reification to represent norms coming from existing legislation. *The Journal of Logic and Computation*, 7, 2017.

[32] Alf Ross. Tu-tu. *Harvard Law Review*, 70(5), March 1957.

[33] John R Searle. *The construction of social reality.* Penguin, London, 1996.

[34] M. J. Sergot, A. S. Kamble, and K. K. Bajaj. Indian central civil service pension rules: A case study in logic programming applied to regulations. In *Proceedings of the 3rd International Conference on Artificial Intelligence and Law*, ICAIL '91, pages 118–127, New York, NY, USA, 1991. ACM.

[35] Miklós Szabó, editor. *A törvény szavai [Words of the Law].* Bíbor Kiadó, Miskolc, 2018.

[36] András Tamás. *Legistica.* Szent Istvn Társulat, Budapest, 2009.

[37] Peter M. Tiersma and Lawrence M. Solan. *The Oxford Handbook of Language and Law.*

Oxford University Press, 2012.

[38] Luca Tummolini and Cristiano Castelfranchi. The cognitive and behavioral mediation of institutions: Towards an account of institutional actions. *Cognitive Systems Research*, 7(2):307–323, 2006.

[39] Raimo Tuomela. *The philosophy of social practices : a collective acceptance view*. Cambridge University Press, Cambridge, 2002.

[40] Judit Tóth. *Jogalkotástan [The Lore of Legislation]*. Corvinus University, Budapest, 2011.

[41] Robert W. van Kralingen, Pepijn R. S. Visser, Trevor J. M. Bench-Capon, and H. Jaap van den Herik. A principled approach to developing legal knowledge systems. *Int. J. Hum.-Comput. Stud.*, 51(6):1127–1154, 1999.

[42] Radboud Winkels, editor. *The OpenLaws project: Big Open Legal Data*. Internationales Rechtsinformatik Symposion IRIS 2015: 26.-28. Februar, Universität Salzburg, Wien: Österreichische Computer Gesellschaft, 2015.

[43] G. H. von (Georg Henrik) Wright. *Norm and Action : a Logical Enquiry*. International library of philosophy and scientific method. Routledge and Kegan Paul, London, 1963.

[44] Zsolt Ződi. The limits of plain legal language – understanding the comprehensible style in law. *International Journal of Law in Context*, 15 (to appear), 2019.

 Received 15 October 2018

Efficient Full Compliance Checking of Concurrent Components for Business Process Models

Silvano Colombo Tosatto
Data61, CSIRO
silvano.colombotosatto@data61.csiro.au

Guido Governatori
Data61, CSIRO
guido.governatori@data61.csiro.au

Nick Van Beest
Data61, CSIRO
nick.vanbeest@data61.csiro.au

Francesco Olivieri
Data61, CSIRO
francesco.olivieri@data61.csiro.au

Abstract

Business process compliance checking is an NP-complete problem, due to concurrency and different mutually exclusive execution paths. Although the complexity of real life process models usually allows for a brute force approach, environments with limited resources or computational power (like for instance blockchain environments) cannot rely on brute force approaches due to the computational complexity of the problem. In this paper, we present an approach to efficiently check a subclass of problems involving concurrent sub-processes. Our approach reduces the computational complexity of concurrent sub-processes from combinatorial to exponential. We prove the correctness of the approach, we experimentally validate the results and evaluate the scalability of the approach. We show that our approach is a significant improvement for highly concurrent processes and easily outperforms existing brute force approaches.

1 Introduction

One of the aspects related to legal reasoning concerns verifying whether a given behaviour complies with a set of given regulations. These so-called *Compliance Checking* procedures can be applied to sets of behaviours, like for instance a model describing the possible behaviours of an agent. The advantage of verifying models describing possible behaviours is that it ensures that all behaviour allowed by the model is proven to be compliant with a given set of regulations. This approach is also referred to as *Compliance by Design* [17], as it ensures that a model contains, by design, only or at least some compliant behaviours. Compliance by design is also known as *forward compliance*, referring to the techniques focused on preventing compliance breaches, which differ from *backward compliance*, as these techniques focus on identifying compliance breaches after they have happened.

Business Process Models are originally designed to formally represent the possible sequences of activities to be executed by an organisation to achieve a certain business goal. In addition to their utility in a legal setting, where they can be used to automatically verify compliance breaches, these models can potentially be used to represent collections of agent's plans and be automatically verified with respect to some given constraints (as as shown by Governatori and Rotolo [11]).

However, proving compliance by design of business process models is in general **NP**-complete, as shown by Colombo Tosatto et al. [6]. Accordingly, no polynomial solutions are possible for the general problem of proving compliance of business process models. Most of the current solutions for the problem, like for instance Regorous [12], adopt a brute force approach over the possible executions of a business process model to prove its compliance. Despite the high theoretical complexity of the problem, solutions like the aforementioned Regorous, still seem to offer practical solutions to some real life instances. Nevertheless, in a number of cases the size of the problem grows enough to not be solvable by brute force approaches, or in others, the environment may be providing only a limited amount of computational resources, such as for instance a blockchain. As a result, a more efficient use of these resources becomes desirable. Part of the complexity of the problem lies in the process model structure, where even compact structures can potentially represent an exponential number of possible executions. Within a business process model, certain sequences of activities can be mutually exclusive, while other activities are concurrent. Concurrency allows for a combinatorial number of possible execution orders of the activities involved, as it considers all possible interleaving of the activities unrestricted by explicit ordering constraints.

To address this issue, we propose an efficient algorithm to prove *full compliance* of business process models with respect to a set of given regulations, by verifying whether a counter-example exists. As such, we provide an approach to resolve compliance over concurrent paths, such that the computational complexity of verifying compliance is reduced from combinatorial to exponential. To keep the theoretical complexity of the solution tractable, we restrict the expressivity of the regulations being checked to literals and with no compensations for eventual violations. Furthermore, we restrict our approach to *structured* process models [26], as they allow to verify their soundness and correctness in polynomial time with respect to the size of the model.

The remainder of the paper is structured as follows: Section 2 discusses the related work. Section 3 introduces the classic regulatory compliance problem, and provides a high-level overview of the proposed solution. Section 4 describes the process models and the decomposition procedure. Section 5 describes the regulations and how Δ-constraints[1] are generated. Section 6 defines full compliance and how it can be proven through decomposed processes and Δ-constraints. Section 7 empirically evaluates the approach against a set of highly complex models. Finally, Section 8 concludes the paper.

2   Related Work

While the area of business process compliance received substantial research interest the past decade (Hashmi et al. [17] identify over 180 research papers between 2000 and 2015 specific to business process regulatory compliance), the study of the computational complexity properties (and solutions to reduce the search space) and whether the proposed techniques offer practical solutions has been largely neglected.

For example, Pulvermüller et al. [27] directly verify temporal logic based specifications on a process, without proper support for different branching options through gateways, and Awad et al. [2] utilise a reduction technique, which results in an incomplete model. Ramezani et al. [28] provide a set of generic compliance patterns, but do not offer the same expressivity as other approaches using defeasible logic or even temporal logic.

Other approaches introduce a large amount of overhead in the state space of the model encoding the process, which is often the result of ignoring the effect of encoding on the internal state machine of the model checker (see e.g. Latvala and Heljanko [23], Bianculli et al. [4], Kherbouche et al. [20], and Kheldoun et al. [19]). In particular, parallel branching constructs may cause a state space explosion, which only few approaches have successfully addressed. Some approaches, like the one proposed by Feja et al. [10] simply disregard parallel information entirely, such that full compliance cannot be ensured. Other approaches, such as Liu et al. [24], do interleave parallel branches correctly, but interleave to such an extent that concurrent executions are linearised entirely, resulting in a state explosion.

Another direction of research has focused on *conformance checking*, such as for instance van der Aalst et al. [30], where they analyse techniques to verify whether executions

---

[1]We discuss the reasoning behind the name used for the alternative representation of the regulation in Section 3.2.

actually belong to the allowed behaviour specified in the business process model. Conformance checking is an orthogonal discipline when compared to *compliance by design*, as the latter focuses on verifying properties of future executions of business process models, while the former focuses on verifying properties of existing executions compared to their corresponding process models. As such, an execution may be conformant (i.e. it matches an allowed execution specified in the model), but not compliant (i.e. it violates a regulation). Similarly, an execution can be non-conformant (i.e. its behaviour is not specified in the model), but it is compliant (i.e. it does not violate any regulation).

Although solutions based on temporal logic benefit from the optimisation and efficiency of modern model checkers (e.g. Awad et al. [3], Elgammal et al. [9], and Pesic et al. [25]), such approaches do not address the reduction of the search space. Moreover Governatori and Hashmi [13] show that given certain circumstances, like requiring compensatory obligations[2] and permissions determining whether an obligation is in force, do not allow temporal logic to soundly model regulatory requirements.

Some solutions have tried to address the complexity of the problem by reducing the required search space. Groefsema et al. [16], for example, propose a solution using Kripke structures to reduce the state explosion derived from concurrency components. However, this approach first creates a full state space, followed by a reduction. In contrast, the approach presented in this paper does not require a subsequent reduction to be feasible for model checking, as it directly allows for efficient compliance checking of concurrent process constructs.

3 Regulatory Compliance Problem and Our Approach

In this section, we first introduce the classic regulatory compliance problem. Subsequently, we introduce the idea behind the proposed approach in this paper. The formal details of both the regulatory compliance problem and the approach are discussed later in the paper.

3.1 The Regulatory Compliance Problem

The regulatory compliance problem is concerned with verifying whether a business process model is compliant with a given set of regulations. The problem contains two separate components: the business process model, describing a set of possible executions capable of achieving the business goal pursued by the model, and the set of regulations defining the constraints that are required to be fulfilled by each execution in the model.

---

[2]A type of obligation that become in force in response to another obligation being violated, and its fulfilment compensates the triggering violation.

Business Process Models

Business process models are formal representations capable of compactly representing multiple executions achieving a particular goal. In order to achieve their compactness, business process models use *coordinators* ([29]) to specify that some of the executions in the model have some parts in common, which the model needs to represent only once. These coordinators are capable of representing e.g. mutually exclusive or concurrent paths of executions. As such, they are based on the semantics of *exclusive* and *parallel* gateways as defined in *BPMN 2.0*, with the restriction of structuredness as described in Definition 2.

The coordinators allow to compactly represent multiple distinct executions within a single business process model. More precisely, a business process model can potentially contain a combinatorial number of executions with respect to the elements composing it. As a result, brute force approaches are potentially impractical as they require the generation and analysis of each possible execution contained within a model.

Regulations

The second component of a regulatory compliance problem concerns the regulations defining the regulatory requirements. These regulations determine which obligations are required to be fulfilled in each possible execution in the business process model.

Given an execution of a business process model and a regulation, the regulation defines a triggering condition, which (if satisfied by the execution) sets an obligation in force that is then required to be fulfilled. Additionally, the regulation also determines a deadline condition, expressing that when the obligation is in force, it must be fulfilled before or until the execution satisfies the deadline. This is illustrated graphically in Figure 1. We can consider an execution to be a sequence of activities each occurring at a distinct point in time. The interval between the point in time satisfying the triggering condition of a regulation and the point in time satisfying the corresponding deadline determines the in force interval of an obligation.

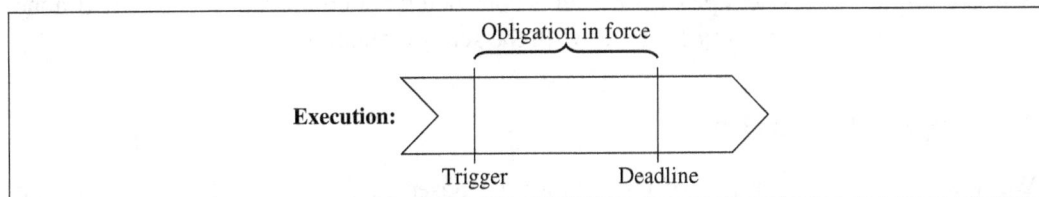

Figure 1: Obligation in force

We consider two types of obligations, *achievement* and *maintenance*, each of which having different properties [6]:

967

Achievement When this type of obligation is in force, the fulfilment condition specified by the regulation must be satisfied by the execution in at least one point in time before the deadline is satisfied. When this is the case, the obligation in force is considered to be satisfied, otherwise it is violated.

Maintenance When this type of obligation is in force, the fulfilment condition must be satisfied *continuously* until the deadline is satisfied. Again, if this is the case, the obligation in force is then satisfied, otherwise it is violated.

Types of Compliance

Given an execution and a regulation, if each in force interval determined by the obligation is satisfied by the regulation, then we can say that this particular execution complies with the regulation. Similarly, when considering a set of regulations instead of a single one, an execution is considered to comply with the set, if and only if it satisfied every in force interval of the obligations determined by the set of regulations.

However, when verifying whether a business process model is compliant with respect to a set of regulations, different types of compliance can be considered. A business process model can be either, *fully compliant*, *partially compliant*, or *not compliant* with respect to a set of regulations. This distinction between the different levels of business process compliance have been already introduced by Governatori and Rotolo [15].

Fully Compliant A business process model is considered fully compliant, if and only if each of its possible executions complies with the set of regulations.

Partially Compliant A business process model is considered partially compliant, if and only if there exists an execution of the business process model that complies with the set of regulations.

Not Compliant A business process model is considered not compliant, if and only if none of its possible execution complies with the set of regulations.

3.2 Approach Overview

We now provide an overview of the solution proposed in this paper. First, the solution focuses on proving whether a business process model is fully compliant with respect to a given set of regulations.

As we formally define in Section 4, mutual exclusion coordinators in a business process model are also referred to as *XOR Blocks*. Our proposed approach verifies whether

968

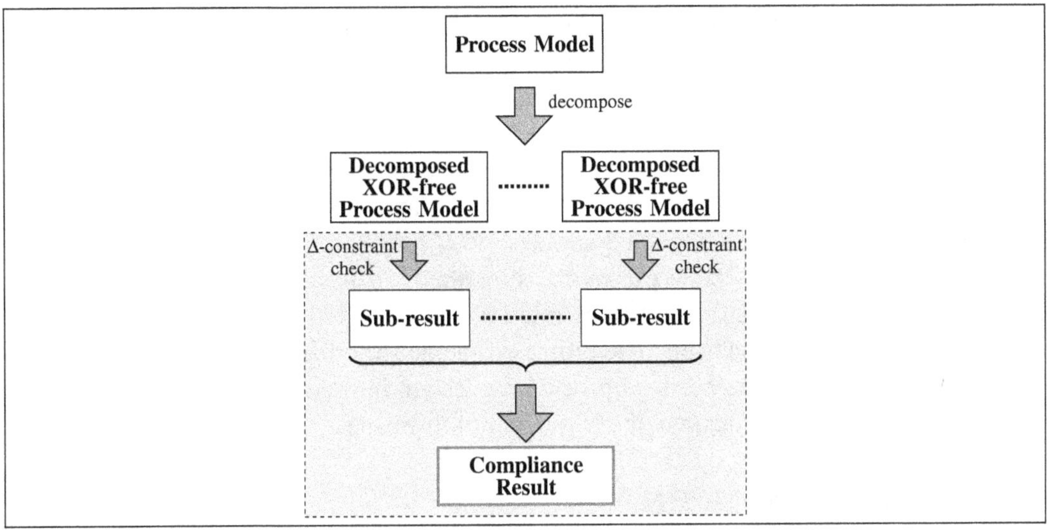

Figure 2: Approach Overview.

a business process model is compliant with a set of regulations as follows: first, it de-composes the full business process model into a set of *XOR-free Process Models*[3], which maintain the expressivity of the original process. That is, the set of executions available in the original process is the same as the sum of the executions of the set of XOR-free Process Models. Second, we represent the set of regulations as Δ-*constraints*, allowing us to verify whether an XOR-free Process Model violates a given regulation. This allows us to prove full compliance for the verified process in case a violation is not found. We adopt the name Δ-constraint to refer to the alternative representation of the regulations, as the new representation focuses on the differences between subsequent process states introduced by the execution of tasks (as described in Definition 8, and updated in accordance to the an-notation of a task being executed as described in Definition 9), which can be understood as the Δ between such process states. Finally, the compliance results relative to the XOR-free Process Models are aggregated, in order to decide whether the original business process model is fully compliant with the given regulations. The procedure is graphically illustrated in Figure 2.

4 Business Process Models and Decomposition

In this section, we first introduce the formal syntax and semantics of business process mod-els. Before introducing the decomposition procedure, we briefly discuss how the structural

---

[3]A set of process models not containing mutual exclusion coordinators.

components of a business process model contribute to the computational complexity of the problem. Finally, we define the decomposition procedure transforming a generic process model in a set of process models free of mutually exclusive choices, such that they can be handled by the solution proposed in this paper.

There exist multiple mainstream business process modelling notations, which include (among others) BPMN, EPC, and Petri nets. The mapping between these different formalisms has been extensively studied and described [31, 8, 22]. Although each of these graph-oriented formalisms allows to model structured business processes, they require a formal definition of the restriction to structured processes on top of the definition of the respective modelling notation itself. Therefore, we adopted a formal description that captures the structured nature of the business process by design, in order to provide a shorter and more intuitive notation throughout the remainder of this paper.

4.1 Business Process Models

We focus on a particular subclass of processes: *Structured Business Processes*, which are a class of processes similar to the structured workflows defined by Kiepuszewski et al. [21]. These processes are defined as a recursive nesting of their components, where each nesting structure is defined as a process block, as well as the process itself and its atomic components, the tasks.

The models used in the sub-problem are both structured and acyclic, such that each execution in the process model is guaranteed to terminate.

Definition 1 (Process Block). *A process block B is a directed graph: the nodes are called* elements *and the directed edges are called* arcs. *The set of elements of a process block are identified by the function $V(B)$ and the set of arcs by the function $E(B)$. The set of elements is composed of tasks and coordinators. The coordinators are of 4 types:* and_split, and_join, xor_split *and* xor_join. *Each process block B has two distinguished nodes called the* initial *and* final *element. The initial element has no incoming arc from other elements in B and is denoted by $b(B)$. Similarly the final element has no outgoing arcs to other elements in B and is denoted by $f(B)$.*

A directed graph composing a process block is defined inductively as follows:

- *A single task constitutes a process block. The task is both initial and final element of the block.*

- *Let $B_1, \ldots, B_n$ be distinct process blocks with $n > 1$:*

  - *SEQ$(B_1, \ldots, B_n)$ denotes the process block with node set $\bigcup_{i=0}^{n} V(B_i)$ and edge set $\bigcup_{i=0}^{n} (E(B_i) \cup \{(f(B_i), b(B_{i+1})) : 1 \leq i < n\})$.*
  The initial element of SEQ$(B_1, \ldots, B_n)$ is $b(B_1)$ and its final element is $f(B_n)$.

- XOR$(B_1, \ldots, B_n)$ *denotes the block with vertex set* $\bigcup_{i=0}^{n} V(B_i) \cup \{\text{xsplit}, \text{xjoin}\}$ *and edge set* $\bigcup_{i=0}^{n} (E(B_i) \cup \{(\text{xsplit}, b(B_i)), (f(B_i), \text{xjoin}) : 1 \leq i \leq n\})$ *where* xsplit *and* xjoin *respectively denote an* xor_split *coordinator and an* xor_join *coordinator, respectively. The initial element of* XOR$(B_1, \ldots, B_n)$ *is* xsplit *and its final element is* xjoin.

- AND$(B_1, \ldots, B_n)$ *denotes the block with vertex set* $\bigcup_{i=0}^{n} V(B_i) \cup \{\text{asplit}, \text{ajoin}\}$ *and edge set* $\bigcup_{i=0}^{n} (E(B_i) \cup \{(\text{asplit}, b(B_i)), (f(B_i), \text{ajoin}) : 1 \leq i \leq n\})$ *where* asplit *and* ajoin *denote an* and_split *and an* and_join *coordinator, respectively. The initial element of* AND$(B_1, \ldots, B_n)$ *is* asplit *and its final element is* ajoin.

By enclosing a process block as defined in Definition 1 along with a start and end task in a sequence block, we obtain a *structured process model*.

Definition 2 (Structured Process Model). *A structured process model P is a directed graph composed of a process block B called the main process block. The process P is represented as a sequence block, as follows:* SEQ$(\text{start}, B, \text{end})$, *where the vertex set of P is* $V(P) = V(B) \cup \{\text{start}; \text{end}\}$ *and its edge set is* $E(P) = E(B) \cup \{(\text{start}, b(B)), (f(B), \text{end})\}$. *The initial element of a structured process model is the pseudo-task* start *and its final element is the pseudo-task* end.

Example 1 (Structured Process Model). *Fig. 3 shows a structured business process containing four tasks labelled* t_1, t_2, t_3, t_4. *The structured process contains an XOR block delimited by the* xor_split *and the* xor_join. *The XOR block contains the tasks* t_1 *and* t_2. *The XOR block is itself nested inside an AND block with the task* t_3. *The AND block is preceded by the* start *and followed by task* t_4 *which in turn is followed by the* end *task.*

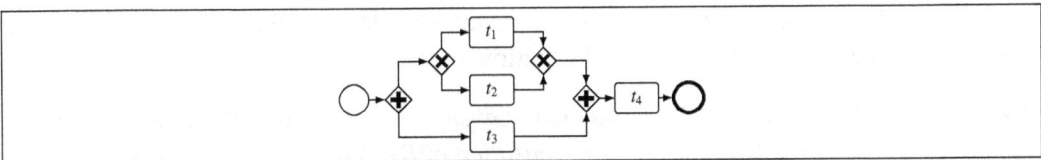

Figure 3: Example of a structured business process.

Considering the structured process in Figure 3 as a sequence block, it can be represented as follows:

$$P = \text{SEQ}(\text{start}, \text{AND}(\text{XOR}(t_1, t_2), t_3), t_4, \text{end})$$

Note that for the process model P, SEQ$(\text{AND}(\text{XOR}(t_1, t_2), t_3), t_4)$ *represents the main process block. The external sequence block of B is absorbed by the sequence block of process model itself, resulting in the final representation.*

971

Business Process Executions

In a structured process model, a valid execution identifies a path from the pseudo-task start to the pseudo-task end, and follows the semantics of the traversed coordinators. An execution is represented as a sequence of a subset of the tasks belonging to the process.

The possible executions allowed by a process model can be represented using partially ordered sets, where the ordering constraints represent the structure of the model. Thus a linear order of tasks following the partially ordered set ordering constraints represents a valid execution.

Definition 3 (Partially Ordered Set). *A partially ordered set* $\mathbb{P} = (\mathscr{S}, \prec_s)$ *is a tuple where* $\mathscr{S}$ *is a set of elements and* $\prec_s$ *is a set of ordering relations between two elements of* $\mathscr{S}$ *such that* $\prec_s \subseteq \mathscr{S} \times \mathscr{S}$ *and for which* transitivity *and* antisymmetry[4] *hold.*

Let $\mathbb{P}_1 = (\mathscr{S}_1, \prec_{s_1})$ *and* $\mathbb{P}_2 = (\mathscr{S}_2, \prec_{s_2})$ *be partially ordered sets, we define the following four operations:*

- *Union:* $\mathbb{P}_1 \cup_{\mathbb{P}} \mathbb{P}_2 = (\mathscr{S}_1 \cup \mathscr{S}_2, \prec_{s_1} \cup \prec_{s_2})$, *where* $\cup$ *is the disjoint union.*

- *Concatenation:* $\mathbb{P}_1 +_{\mathbb{P}} \mathbb{P}_2 = (\mathscr{S}_1 \cup \mathscr{S}_2, \prec_{s_1} \cup \prec_{s_2} \cup \{s_1 \prec s_2 | s_1 \in \mathscr{S}_1 \text{ and } s_2 \in \mathscr{S}_2\})$.

- *Linear Extensions:* $\mathscr{I}(\mathbb{P}_1) = \{(\mathscr{S}, \prec_s) | \mathscr{S} = \mathscr{S}_1, (\mathscr{S}, \prec_s) \text{ is a totally ordered sequence and } \prec_{s_1} \subseteq \prec_s\}$.

The associative *property holds for Union, and Concatenation.*

A serialisation of a process block is a totally ordered sequence of a subset of the tasks. The sequence must follow the semantics of the coordinators contained in the block, and start with the block initial element and finish with its final element. Notice that the definition of serialisation is given as a byproduct of Definition 4.

Definition 4 (Process Block Serialisations). *Given a process block B, the set of serialisations of B, written* $\Sigma(B) = \{\varepsilon | \varepsilon \text{ is a serialisation of } B\}$. *The function* $\Sigma(B)$ *is defined as follows:*

1. *If B is a task t, then* $\Sigma(B) = \{(\{t\}, \emptyset)\}$

2. *if B is a composite block with sub-blocks* $B_1, \ldots, B_n$ *let* ε_i *be the projection of* ε *on block* B_i *(obtained by ignoring all tasks which do not belong to* B_i)

    (a) *If* $B = \mathsf{SEQ}(B_1, \ldots, B_n)$, *then* $\Sigma(B) = \{\varepsilon_1 +_{\mathbb{P}} \cdots +_{\mathbb{P}} \varepsilon_n | \varepsilon_i \in \Sigma(B_i)\}$

    (b) *If* $B = \mathsf{XOR}(B_1, \ldots, B_n)$, *then* $\Sigma(B) = \Sigma(B_1) \cup \cdots \cup \Sigma(B_n)$

---

[4]*Antisymmetry: if* $a \prec_s b$ *and* $b \prec_s a$, *then* $a = b$.

(c) *If $B = \mathsf{AND}(B_1, \ldots, B_n)$, then $\Sigma(B) = \{\varepsilon | \varepsilon \in \mathscr{I}(\varepsilon_1 \cup_\mathbb{P} \cdots \cup_\mathbb{P} \varepsilon_n | \forall \varepsilon_i \in \Sigma(B_i))\}$.*

A serialisation of a process model corresponds to one of its executions, hence the set of serialisations of a business process model corresponds to the set of possible executions of the model itself.

Definition 5 (Execution). *Given a structured process P whose main process block is B, an execution of P corresponds to a serialisation of B including the pseudo-tasks* start *and* end.

$$\Sigma(P) = \{\mathbb{P}_{\mathsf{start}} +_\mathbb{P} \varepsilon +_\mathbb{P} \mathbb{P}_{\mathsf{end}} | \varepsilon \in \Sigma(B)\}$$

Example 2 (Execution). *Consider a business process model P, like the one shown in Example 1:*

$$P = \mathsf{SEQ}(\mathsf{start}, \mathsf{AND}(\mathsf{XOR}(t_1, t_2), t_3), t_4, \mathsf{end})$$

The corresponding possible executions of P, $\Sigma(P)$, are as follows:

ε_1 : start, t_1, t_3, t_4, end,

ε_2 : start, t_3, t_1, t_4, end,

ε_3 : start, t_2, t_3, t_4, end,

ε_4 : start, t_3, t_2, t_4, end

Annotations

The *state* of a process is represented using a set of literals. We assume that executing a task can alter the state of the process, represented using *annotations*, as described by Governatori et al. [14]. The state consists of sets of literals associated to the tasks, where a literal is either an atomic proposition or its negation. A task's annotation describes the changes in the process state when the associated task is executed.

Both the state of a process and the annotations of the tasks are represented by sets of literals, which are required to be consistent.

Definition 6 (Consistent literal set). *A set of literals L is consistent if and only if it does not contain both l and ¬l.*

An annotated process is a process whose tasks are associated with consistent sets of literals.

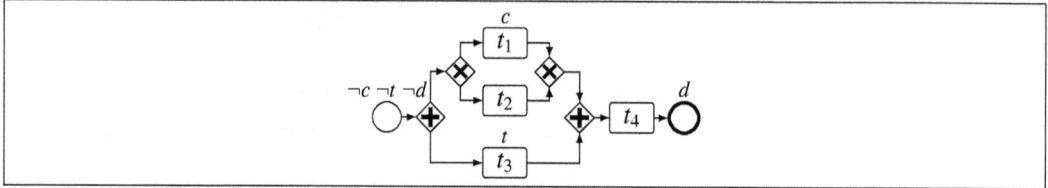

Figure 4: Example of an annotated business process.

Definition 7 (Annotated process). *Let P be a structured process and let T be the set of tasks in P. An annotated process is a pair: (P, ann), where ann is a function associating a consistent set of literals $\mathrm{ann} : T \mapsto L$ to each task in T.*

Definition 8 (State). *Let t_i be a task, and L is a consistent literal set. A state is a tuple $\sigma = (t_i, L)$, and represents the state holding after executing t_i.*

The state of a process is updated by each task's execution through an update operator. This operator is inspired by the AGM belief revision operator [1].

Definition 9 (Literal set update). *Let $\bar{l}$ be the complementary literal as follows:*

- $\bar{l} = \neg\alpha$ *if* $l = \alpha$

- $\bar{l} = \alpha$ *if* $l = \neg\alpha$

Given two consistent sets of literals L_1 and L_2, the update of L_1 with L_2 (denoted by $L_1 \oplus L_2$) is a set of literals defined as follows:

$$L_1 \oplus L_2 = L_1 \setminus \{\bar{l} \mid l \in L_2\} \cup L_2$$

Finally, a *trace* represents the evolution of the state of a process during one of its executions.

Definition 10 (Trace). *Given an annotated process (P, ann) and an execution sequence $\varepsilon = (t_1, \ldots, t_n)$ such that $\varepsilon \in \Sigma(P)$, a trace θ is a finite sequence of states: $(\sigma_1, \ldots, \sigma_n)$. Each state of $\sigma_i \in \theta$ contains a set of literals L_i capturing what holds after the execution of a task t_i. Each L_i is a set of literals such that:*

1. $L_1 = \mathrm{ann}(t_1)$;

2. $L_{i+1} = L_i \oplus \mathrm{ann}(t_{i+1})$, *for* $1 \leq i < n$.

We use $\Theta(B, \mathrm{ann})$ to denote the set of possible traces resulting from an annotated process block (B, ann), where B is a process block and ann is an annotation function.

Example 3 (Trace). *Consider the annotated structured process in Figure 4, containing the following annotations:*

start : $\{\neg c, \neg t, \neg d\}$

$t_1 : \{c\}$

$t_3 : \{t\}$

end : $\{d\}$

The following traces correspond to the executions of P, as illustrated in Example 2, written $\Theta(P)$:

$\theta_1 : (\text{start}; \neg c, \neg t, \neg d), (t_1; c, \neg t, \neg d), (t_3; c, t, \neg d), (t_4; c, t, \neg d), (\text{end}; c, t, d),$

$\theta_2 : (\text{start}; \neg c, \neg t, \neg d), (t_3; \neg c, t, \neg d), (t_1; c, t, \neg d), (t_4; c, t, \neg d), (\text{end}; c, t, d),$

$\theta_3 : (\text{start}; \neg c, \neg t, \neg d), (t_2; \neg c, \neg t, \neg d), (t_3; \neg c, t, \neg d), (t_4; \neg c, t, \neg d), (\text{end}; \neg c, t, d),$

$\theta_4 : (\text{start}; \neg c, \neg t, \neg d), (t_3; \neg c, t, \neg d), (t_2; \neg c, t, \neg d), (t_4; \neg c, t, \neg d), (\text{end}; \neg c, t, d)$

4.2　On the Computational Complexity

Colombo Tosatto et al. [6] have shown that the problem of proving whether a business process model complies with a given set of regulations is an **NP**-complete problem, when structured business process models containing both concurrent components and mutually exclusive ones are used, and the regulations are expressed using literals. The structural components of business process models contribute to the computational complexity of the problem. In particular, two major contributors can be identified: XOR and AND process blocks, each of which will be discussed below.

XOR **Blocks**

XOR blocks contribute to the computational complexity of the problem by allowing the representation of multiple possible executions within a single business process model.

An XOR block allows us to represent possible branches in the executions within a model. The branching factor in an XOR block, represented by the sub-blocks contained, determines the amount of different possible executions obtainable by executing the block, as described in Definition 5.

A single XOR block does not substantially contribute to the computationally complexity of the problem, as it increases the amount of execution available within the model by the number of sub-blocks contained. However, nesting XOR blocks does increase the number of possible executions, as it is no longer polynomial with respect to the business process structure. Let n be the number of sub-blocks in an XOR block and l be the nesting level. Then the amount of possible executions in the worst case scenario, where each branch of an XOR block contains a nested XOR block with n branches, up to a nesting level l, is:

$$n^l$$

In general, when the branching factor of properly nested XOR blocks is not constant and not every branch necessarily nests another XOR block, then the number of possible executions deriving from this structure is calculated by counting the number of branches in the structure not containing a nested XOR block.

Example 4. *Considering the business process model illustrated in Figure 5. The model contains 3 XOR blocks nested within 2 levels, and each block contains 2 branches. Given the structure, we can then calculate the number of possible executions of the model, which is* $2^2 = 4$.

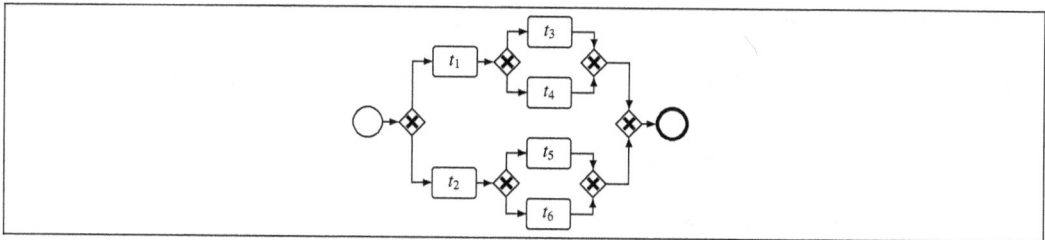

Figure 5: Executions in a model containing nested XOR blocks.

In constrast, when multiple XOR blocks are used sequentially, the amount of possible executions becomes exponential with respect to the number of XOR blocks. Let n be the number of sub blocks in an XOR block and k the number of XOR blocks. Then in the worst case scenario, where each block has the same branching factor, the number of possible executions in a business process model is:

$$n^k$$

Note that in the general case, where the number of branches in the different XOR blocks of a model is not the same, assuming that k is the number of XOR blocks in the model and n_i is the number of branches in the block i for some $1 \leq i \leq k$, the number of executions in the model is calculated as follows:

$$\sum_{i=1}^{k} (n_i)$$

Example 5. *Consider the business process model illustrated in Figure 6. The model contains 2 XOR blocks with 2 branches each. Given the structure, we can then calculate the number of possible executions of the model, which is $2^2 = 4$.*

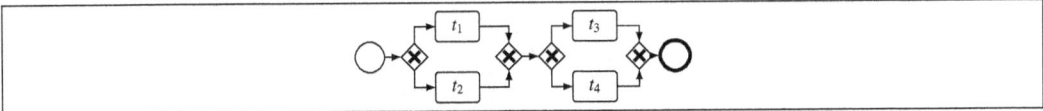

Figure 6: Executions in a model containing sequential XOR blocks.

Given the amount of possible executions with respect to the size of the model in the worst case scenario, a brute force analysis of each execution is theoretically intractable. In other words, it means that the time required to find a solution would exponentially increase with the size of the problem, hence practically making big enough problems unsolvable using brute force approaches.

AND **Blocks**

Similar to XOR blocks, AND blocks contribute to the computational complexity by allowing a compact representation of multiple different possible executions within a single business process model.

Contrary to XOR blocks, however, the positioning of AND blocks in the model does not strongly influence the amount of possible executions available in the business process model. Let k be the number of AND blocks, n the number of branches in an AND block, and m the length of the branches of the block in term of executable tasks. The amount of possible executions in a business process model is combinatorial with respect to the model structure:

$$\left(\frac{(n \times m)!}{(m!)^n} \right)^k$$

In the general case, where n_i is the number of branches of the block i for $1 \leq i \leq k$, and m_{i_j} is the length of the jth branch of block i for $1 \leq j \leq n_i$, the number of executions in a business process model with k AND blocks is calculated as follows:

$$\prod_{i=1}^{k} \left(\frac{(\sum_{j=1}^{n_i}(m_{i_j}))!}{\prod_{j=1}^{n_i}(m_{i_j}!)} \right)$$

Example 6. *Consider the business process model illustrated in Figure 7. The model contains a single AND block with 3 branches of size 2 each. Given the structure, we can then*

977

calculate the number of possible executions of the model, which is $\frac{(2+2+2)!}{(2!)^3} = 90$. However, if we increase the length of each branch by 1 (adding only 3 activities in total), the number of possible executions of the model is $\frac{(3+3+3)!}{(3!)^3} = 1680$.

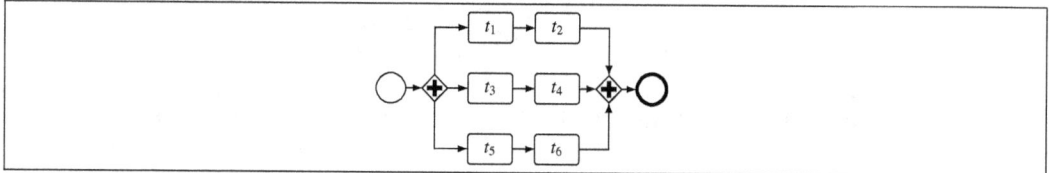

Figure 7: Executions in a model containing AND blocks.

Note that, as this brief analysis suggests, AND blocks contribute more heavily to the computational complexity of the problem than XOR blocks.

Complexity of Real-life Business Process Models

Consider for example the SAP R/3 collection of business process models, used by SAP to customize their R/3 ERP product as documented in [7]. As shown in Table 1, the structural complexity is reasonable with the most complex model having 86 activities and 6 AND splits. However, even when selecting the set of sound workflow models, the complexity in terms of total possible executions can grow as large as $1.76 \cdot 10^{10}$. This shows that the incentive for a more efficient algorithm is not just academic, but a necessity imposed by real-life model complexities.

Metric	Min	Max	Mean
Activities	3	86	15.98
XOR splits	0	6	0.64
Outdegree XOR	2	9	2.73
AND splits	0	6	1.14
Outdegree AND	2	18	3.14

Table 1: Statistics on SAP R/3 model complexity.

4.3   Process Model Decomposition.

A business process model can be decomposed by splitting each of the XOR blocks within the process representation in n different processes, where n corresponds to the branches in the XOR block and each of the new processes contains exactly one of the branches. This procedure is recursively applied until no XOR blocks are left.

The decomposition can potentially lead to an exponential number of decomposed processes, as can be inferred from the contribution of the XOR components to the computational complexity. However, the presence of AND blocks still allows a combinatorial amount of possible executions within the decomposed models. Therefore, a solution is required to prevent analysis of each possible execution to verify the regulatory compliance of the model.

Example 7 (Decomposition). *Consider the business process model described in Example 1:*

$$P = \mathsf{SEQ}(\mathsf{start}, \mathsf{AND}(\mathsf{XOR}(t_1, t_2), t_3), t_4, \mathsf{end})$$

Given that P contains a single XOR block with 2 branches, application of the decomposition procedure leads to the following decomposed processes:

1. $P' = \mathsf{SEQ}(\mathsf{start}, \mathsf{AND}(t_1, t_3), t_4, \mathsf{end})$

2. $P'' = \mathsf{SEQ}(\mathsf{start}, \mathsf{AND}(t_2, t_3), t_4, \mathsf{end})$

A process block serialisation (cf. Definition 4) is constructed through partially ordered sets operations depending on the type of process blocks, as described in Definition 3. Subsequently, a decomposed process can be represented as a partially ordered set by a recursive transformation closely following Definition 4, as reported in Definition 11 below:

Definition 11 (Decomposed Process as Partially Ordered Set). *A decomposed process P can be represented as a partially ordered set by applying the following recursive procedure, $\mathscr{P}(B)$, to each of its process blocks in (B):*

1. If B is a task t, then $\mathscr{P}(B) = \{(\{t\}, \emptyset)\}$

2. if B is a composite block with subblocks $B_1, \ldots, B_n$:

   (a) If $B = \mathsf{SEQ}(B_1, \ldots, B_n)$, then $\mathscr{P}(B) = \{\mathscr{P}(B_1) +_{\mathbb{P}} \cdots +_{\mathbb{P}} \mathscr{P}(B_n)\}$
   (b) If $B = \mathsf{AND}(B_1, \ldots, B_n)$, then $\mathscr{P}(B) = \{\mathscr{P}(B_1) \cup_{\mathbb{P}} \cdots \cup_{\mathbb{P}} \mathscr{P}(B_n)\}$.

Example 8 (Decomposed Partially Ordered Sets). *Given the two decomposed processes from Example 7:*

1. $P' = \mathsf{SEQ}(\mathsf{start}, \mathsf{AND}(t_1, t_3), t_4, \mathsf{end})$

2. $P'' = \mathsf{SEQ}(\mathsf{start}, \mathsf{AND}(t_2, t_3), t_4, \mathsf{end})$

The corresponding partially ordered sets for the two decomposed processes are the following:

1. $\mathscr{P}(P') = (\{\text{start}, t_1, t_3, t_4, \text{end}\}, \{\text{start} \prec t_1, \text{start} \prec t_3, t_1 \prec t_4, t_3 \prec t_4, t_4 \prec \text{end}\})$

2. $\mathscr{P}(P'') = (\{\text{start}, t_2, t_3, t_4, \text{end}\}, \{\text{start} \prec t_2, \text{start} \prec t_3, t_2 \prec t_4, t_3 \prec t_4, t_4 \prec \text{end}\})$

Each decomposed process contains the possible executions given the same choices in the XOR blocks, and the execution set of a decomposed process is disjoint from any other decomposed process obtained. The union of the execution sets of the decomposed processes is exactly the execution set of the original process model.

5 Regulations and Δ-Constraints

In this section, we introduce the regulations that the process model must fulfil, along with an alternative representation of the regulations used by our solution which we refer to as Δ-constraints.

5.1 Regulations

The regulations are defined as conditional obligations. This kind of obligations consists of three conditions: a *trigger* defining when the obligation becomes in force, a *deadline* defining when its in force period terminates, and a *condition* defining the requirement in the in force period.

A conditional obligation can be either of the achievement or maintenance type, as represented in Definition 12 by $o \in \{a, m\}$ respectively. The notion of achievement and maintenance obligations are inspired by the notion of achievement and maintenance goals introduced by Cohen and Levesque [5], while the full semantics of such notions have been discussed by Hashmi et al. [18].

Definition 12 (Conditional Obligation). *A local obligation ω is a tuple $\langle o, c, t, d \rangle$, where $o \in \{a, m\}$ represents the type of the obligation. The elements c, t and d are propositional literals in $\mathscr{L}$. c is the condition of the obligation, t is the trigger of the obligation and d is the deadline of the obligation.*

We use the notation $\omega = \mathscr{O}^o \langle c, t, d \rangle$ to represent a conditional obligation.

Definition 13. *Given an obligation $\mathscr{O}^o \langle c, t, d \rangle$, the annotation of start is assumed to contain the negation of each literal in the obligation tuple. The annotation of end is assumed to contain the literal referring to the deadline.*

Definition 13 provides an initial process state where none of the literals defining an obligation hold. In addition, it ensures that the in force interval of an obligation always terminates when the process execution ends. Note that obligations are evaluated one at a time, hence the annotation of start and end depends on the obligation being evaluated.

Limiting the Expressivity to Literals

Given our goal to reduce the computational complexity of solving sub-problems of verifying the compliance of business process models, we limit the expressivity of the obligations defining the requirements to simple propositional literals instead of propositional formulae. As we discuss later in Section 6, our solution manages to provide a more efficient solution for the sub-problem considered by avoiding to explicitly analyse each possible execution of a business process model, which would otherwise lead to a search state explosion.

The advantage of limiting the expressivity of the obligations by including only propositional literals allows to directly associate the interaction between the obligation's elements to tasks in the business process model. This implies that executing one of such tasks would immediately satisfy one of the elements of the obligation, like the *condition*, *trigger* or *deadline*.

Such direct association would not be possible if these elements in the obligations were to be expressed using propositional formulae. In that case, the satisfaction of an obligation's element can be potentially influenced by a combination of tasks being executed. For instance, assuming that the trigger of an obligation to be the formula $\alpha \wedge \beta$, its satisfaction can be achieved by executing two tasks, one introducing α in the process state, and another introducing β. Moreover, we would also need to track whether no other tasks are executed between those and removing such literals from the process state, which would not lead to the satisfaction of the formula when the second task would be executed. The complication brought by allowing formulae in the obligation's elements would require knowing the exact execution order of the tasks in order to determine whether an obligation is fulfilled. Furthermore, this would be required for each possible execution order of the business process model, which potentially leads to a intractability problem, as the number of executions of a business process model is in general combinatorial with respect to the number of elements composing the model.

Fulfilling Obligations

Before proceeding with the formal introduction of the different obligations, we first introduce two syntactical shorthands to keep the subsequent definitions more compact.

Definition 14 (Syntax Shorthands). *To avoid cluttering, we adopt the following shorthands:*

- *$\sigma \in \theta$ such that $\sigma \models l$ is abbreviated as: σ_l*

- *A task-state pair appearing in a trace: (t, σ) such that $l \in \mathrm{ann}(t)$, is abbreviated as:* contain(l, σ)

Note that an *in force* interval instance of an obligation, having l as trigger, is always started from a state σ, where contain(l, σ) is true. Therefore, multiple in force intervals of an obligation can co-exist at the same time. However, multiple in force intervals can be fulfilled by a single event happening in a trace, as shown in the following definitions.

An achievement obligation is fulfilled by a trace if the fulfilment condition holds at least in one of the trace states when the obligation is in force.

Definition 15 (Comply with Achievement). *Given an achievement obligation $\mathcal{O}^a \langle c, t, d \rangle$ and a trace θ, θ is compliant with $\mathcal{O}^a \langle c, t, d \rangle$ if and only if: $\forall \sigma_t, \exists \sigma_c | \text{contain}(t, \sigma_t)$ and $\sigma_t \preceq \sigma_c$ and $\neg \exists \sigma_d | \sigma_t \preceq \sigma_d \prec \sigma_c$.*

A maintenance obligation, on the other hand, is fulfilled if the condition holds in each of the states where the obligation is in force.

Definition 16 (Comply with Maintenance). *Given a maintenance obligation $\mathcal{O}^m \langle c, t, d \rangle$ and a trace θ, θ is compliant with $\mathcal{O}^m \langle c, t, d \rangle$ if and only if: $\forall \sigma_t, \exists \sigma_d | \text{contain}(t, \sigma_t)$ and $\sigma_t \preceq \sigma_d$ and $\forall \sigma | \sigma_t \preceq \sigma \preceq \sigma_d, c \in \sigma$.*

The two types of obligations considered in this paper allow to represent a variety of obligations existing in real world scenarios, like e.g requirements to achieve a certain condition before a deadline, or maintaining a condition for a period of time. We do not claim that the formalism adopted is sufficient to capture each possible requirement behaviour in real world scenarios, such as for example an obligation whose applicability condition is related to another obligation. In this paper, we do not consider these more complex behaviours, as the additional behavioural complexity would require to explicitly analyse each possible execution of a business process model in order to verify its compliance state.

5.2 On the Computational Complexity

Previously, we have discussed how the structural components of a business process model contribute to the computational complexity of the problem of proving regulatory compliance of business process models. In this subsection, we briefly discuss the impact of the regulatory requirements on the computational complexity of the problem.

Regulatory Complexity

The amount of expressivity allowed into describing the regulatory requirements to be verified directly influences the computational complexity of the problem. The use of logical formulae to represent the components of the regulatory requirements significantly influences the difficulty of finding a solution. As the components of the regulatory requirements need to be checked against the process state, the use of logical formulae to represent such components requires the exact execution history leading to a particular state, as any difference in the execution order can potentially lead to a different state and, hence, a differently evaluated formula.

However, relying on literals to represent the components of a regulations lifts the requirement of having to know the exact execution history. As a consequence, the verification can potentially be performed over the structure of the business process model, instead of the full list of possible executions.

In this paper, we adopt regulatory requirements restricted to be represented using literals instead of formulae, which allows to focus on the structural components contributing to the computational complexity of the problem. Despite the simplified regulatory requirements, however, Colombo Tosatto et al. [6] have shown that proving partial compliance[5] of a business process model against a set of regulatory requirements is still an **NP**-complete problem.

5.3 Translating the Obligations

Instead of checking all subsequent literals over each possible path in the process to ensure full compliance, we propose to verify whether a trace *violates* a given achievement obligation. The main advantage is that the conditions can then be verified directly on the process model, so that it is no longer required to generate and analyse all possible traces. The *failure condition* for achievement obligations is equivalent to the complement of Definition 15, as shown formally below:

Definition 17 (Achievement Failure). *Given an achievement obligation $\mathscr{O}^a\langle c,t,d\rangle$ and a trace θ, θ is not compliant with $\mathscr{O}^a\langle c,t,d\rangle$ if and only if: $\exists \sigma_t, \sigma_d | \text{contain}(t,\sigma_t)$ and $\sigma_t \preceq \sigma_d$ and $\neg \exists \sigma_c | \sigma_t \preceq \sigma_c \preceq \sigma_d$.*

In order to compare the failure conditions and the business process model, we need to standardise the two representations. As such, we transform the failure conditions into so-called Δ-*constraints*, referring to the state update requirements ensuring that a given obligation is violated when such constraints are met. Therefore, Δ-constraints only require to prove the existence of a trace *failing* the fulfilment requirements, instead of proving that each possible trace fulfils them.

Translation for Achievement

The following definition translates Definition 17 into its Δ-constraints representation, describing the required order of state updates proving that a process model contains an execution violating the given obligation. For convenience, we use t_l to denote a task t such that $l \in \text{ann}(t)$.

---

[5]Proving partial compliance requires to prove that a business process model contains at least one execution that is compliant with a set of regulatory requirements.

Definition 18 (Achievement Failure Δ-constraints). *Given an achievement obligation* $\mathscr{O}^a\langle c,t,d\rangle$ *and an execution* ε, ε *is not compliant with* $\mathscr{O}^a\langle c,t,d\rangle$, *if and only if one of the following conditions is satisfied:*

1. $\exists t_t$ *such that:*

$$\exists t_{\neg c}, t_d | t_{\neg c} \preceq t_t \preceq t_d \text{ and}$$
$$\neg \exists t_c | t_{\neg c} \preceq t_c \preceq t_d \text{ and}$$
$$\exists t_{\neg d}, \neg \exists t_d | t_{\neg d} \preceq t_d \preceq t_t$$

2. $\exists t_t$ *such that:*

$$\exists t_{\neg c}, \neg \exists t_c | t_{\neg c} \preceq t_c \preceq t_t \text{ and}$$
$$\exists t_d, \neg \exists t_{\neg d} | t_d \preceq t_{\neg d} \preceq t_t$$

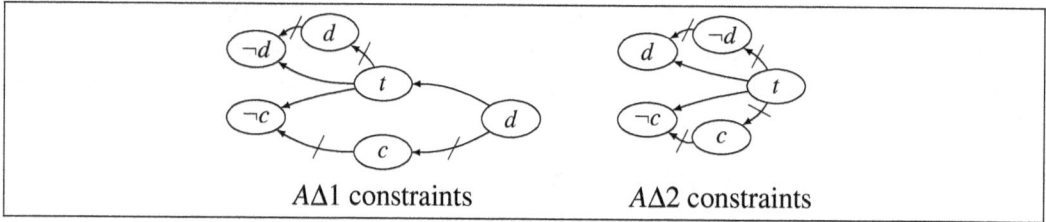

Figure 8: Achievement constraints

Figure 8 provides a graphical representation of the Δ-constraints for achievement obligations. The nodes represent annotations in the tasks and the arrows represent the ordering relations that must be fulfilled by an execution of the process model to fulfil the Δ-constraint. A slashed arrow denotes the required absence of the respective element in the interval identified by the surrounding elements.

Lemma 1. *Given an achievement obligation, executing a task having the obligation's trigger annotated always results in the obligation being in force in the state after the task's execution, where the obligation can be potentially fulfilled or violated.*

Proof. **Sketch**

Either the execution changes the previous state where t was not holding to one where it is, or t had been holding already.

In the first case, the execution brings a new in force period for the obligation.

In the second case, either the obligation is in force and not fulfilled in the previous state, or it becomes fulfilled in the previous state. In both cases, the execution of the task brings the obligation in force again and requires to be fulfilled. □

Theorem 1. *Given an execution ε, represented as a sequence of tasks, if ε satisfies either of the Achievement Failure Δ-Constraints (Definition 18), then ε violates the obligation related to the Achievement Failure Δ-Constraints.*

Proof. **Soundness**:

First case: Achievement Failure Δ-Constraint 1

1. From the hypothesis, and Definition 18, it follows that ε satisfies the following conditions:

    (a) $\exists t_t | \exists t_{\neg c}, t_d | t_{\neg c} \preceq t_t \preceq t_d$

    (b) $\exists t_t | \neg \exists t_c | t_{\neg c} \preceq t_c \preceq t_d$

    (c) $\exists t_t | \exists t_{\neg d}, \neg \exists t_d | t_{\neg d} \preceq t_d \preceq t_t$

2. From 1.(a), and Lemma 1, it follows that: t holds and c does not, in the state holding after the execution of the task t_t.

3. From 2., and 1.(b), it follows that: there is no state included between the state after executing the task t_t and one where d starts holding, where c holds.

4. From 3. and Definition 17, it follows that ε would violate the obligation related to the Achievement Failure Δ-Constraints.

Second case: Achievement Failure Δ-Constraint 2

1. From the hypothesis, and Definition 18, it follows that ε satisfies the following conditions:

    (a) $\exists t_t | \exists t_{\neg c}, \neg \exists t_c | t_{\neg c} \preceq t_c \preceq t_t$

    (b) $\exists t_t | \exists t_d, \neg \exists t_{\neg d} | t_d \preceq t_{\neg d} \preceq t_t$

2. From 1.(a), and Lemma 1, it follows that t holds and c does not, in the state holding after the execution of the task t_t.

3. From 1.(b), it follows that t holds and d holds, in the state holding after the execution of the task t_t.

4. From 2. and 3., it follows that after executing the task t_t, t and d hold and c does not hold.

5. From 4. and Definition 17, it follows that ε would violate the obligation related to the Achievement Failure Δ-Constraints.

Completeness:

1. Given a trace violating a given achievement obligation.

2. From 1. and Definition 17 it follows that $\exists \sigma_t, \sigma_d | \text{contain}(t, \sigma_t)$ and $\sigma_t \preceq \sigma_d$ and $\neg\exists \sigma_c | \sigma_t \preceq \sigma_c \preceq \sigma_d$.

3. From 2., it follows that a task with t annotated is executed.

4. From 2., it follows that a task with d annotated is executed.

5. From 2., it follows that $\neg c$ holds and no task with c annotated is executed between the one with t and the one with d.

6. Following from 3., 4., and 5. two cases are possible:

   $t_d \preceq t_t$ this case is covered by the second set of conditions in Definition 18.

   $t_t \preceq t_d$ this case covered by the first set of conditions in Definition 18.

7. Thus, all cases are covered and a violating trace is always identified by the Achievement Failure Δ-constraints. □

Translation for Maintenance

The translation is also applied to maintenance obligations. Definition 19 describes the failure condition for maintenance obligations, which is the complement of Definition 16. Definition 20 describes the corresponding Δ-constraints.

Definition 19 (Maintenance Failure). *Given a maintenance obligation $\mathcal{O}^m \langle c, t, d \rangle$ and a trace θ, θ is not compliant with $\mathcal{O}^m \langle c, t, d \rangle$ if and only if:*

$$\exists \sigma_t \forall \sigma_d | \text{contain}(t, \sigma_t) \text{ and } \sigma_t \preceq \sigma_d \text{ and } \exists \sigma_{\neg c} | \sigma_t \preceq \sigma_{\neg c} \preceq \sigma_d$$

Definition 20 (Maintenance Failure Δ-constraints). *Given an achievement obligation $\mathcal{O}^m \langle c, t, d \rangle$ and an execution ε, ε is not compliant with $\mathcal{O}^m \langle c, t, d \rangle$ if and only if one of the following conditions is satisfied:*

1. *$\exists t_t$ such that:*

   $$\exists t_{\neg c}, \neg\exists t_c | t_{\neg c} \preceq t_c \preceq t_t$$

2. *$\exists t_t$ such that:*

   $$\exists t_c, \neg\exists t_{\neg c} | t_c \preceq t_{\neg c} \preceq t_t \text{ and}$$
   $$\forall t_d (\exists t_{\neg c} | t_t \preceq t_{\neg c} \prec t_d)$$

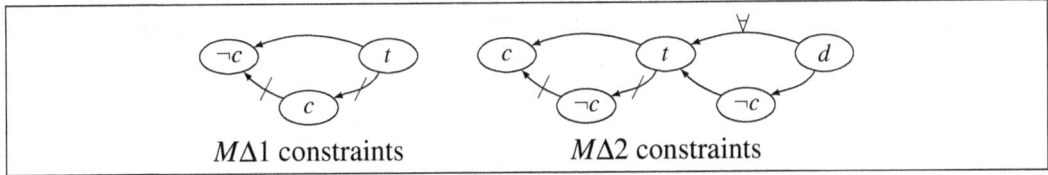

Figure 9: Maintenance constraints

Theorem 2. *Given an execution ε, represented as a sequence of tasks, if ε satisfies either of the Maintenance Failure Δ-Constraints (Definition 20), then ε violates the obligation related to the Maintenance Failure Δ-Constraints.*

Proof. **Soundness**:

First case: Maintenance Failure Δ-Constraint 1

1. From the hypothesis, and Definition 20, it follows that ε satisfies the following condition:

    (a) $\exists t_t | \exists t_{\neg c}, \neg \exists t_c | t_{\neg c} \preceq t_c \preceq t_t$

2. From 1.(a), and Lemma 1, it follows that: t holds and c does not, after the execution of the task t_t annotated.

3. From 2. and Definition 19, it follows that ε would violate the obligation related to the Maintenance Failure Δ-Constraints.

Second case: Maintenance Failure Δ-Constraint 2

1. From the hypothesis, and Definition 20, it follows that ε satisfies the following condition:

    (a) $\exists t_t | \exists t_c, \neg \exists t_{\neg c} | t_c \preceq t_{\neg c} \preceq t_t$ and

    (b) $\exists t_t | \forall t_d (\exists t_{\neg c} | t_t \preceq t_{\neg c} \prec t_d)$

2. From 1.(a), and Lemma 1, it follows that t holds and c holds, in the state holding after executing t_t.

3. From 1.(b), and Lemma 1, it follows that after executing the t_t, t it is always the case that in the following states c stops holding before d starts holding.

4. From 2. it follows that after the execution of the task t_t, c holds due to the constraint preventing a task having $\neg c$ annotated to be executed and cancelling c from the process state, which is already holding due to the execution of a task with c annotated.

5. From 3. it follows before the execution of any task having d in its annotation, a task having $\neg c$ annotated is executed, leading to the removal of c from the process state.

6. From 4. and 5. and Definition 19, it follows that ε would violate the obligation related to the Maintenance Failure Δ-Constraints.

Completeness:

1. Given a trace violating a given maintenance obligation.

2. From 1. and Definition 19 it follows that $\exists \sigma_t \ \forall \sigma_d \mid \text{contain}(t, \sigma)$ and $\sigma_t \preceq \sigma_d$ and $\exists \sigma_{\neg c} \mid \sigma_t \preceq \sigma_{\neg c} \preceq \sigma_d$.

3. From 2., it follows that a task with t annotated is executed.

4. From 2., it follows that a task with $\neg c$ annotated is executed.

5. Following from 3., and 4. two cases are possible:

   $t_{\neg c} \preceq t_t$ case covered by the first set of conditions in Definition 20.

   $t_t \preceq t_{\neg c}$ case covered by the second set of conditions in Definition 20.

6. Thus, all cases are covered and a violating trace is always identified by the Maintenance Failure Δ-constraints.

$\square$

Verifying Δ-constraints

Verifying whether a business process model satisfies a Δ-constraint instead of the original regulation is equivalent to looking for a counter-example falsifying whether a model is fully compliance. Thus, if such an example cannot be found in any possible execution, then the business process model is proven to be fully compliant. Note that every possible execution is implicitly checked by analysing the decomposed business processes as partially ordered sets, as discussed in Section 4.3.

Translation Complexity

Translating a given obligation in the corresponding set of Δ-constraints can be done in constant time, since depending on the type of obligations, the Δ-constraints need to be instantiated with the parameters of the obligation.

6   Verifying Full Compliance

In this section, we show how full compliance is verified for a process model with respect to a given regulation, by proving that the Δ-constraints are consistent with the set of partially ordered set representations of a decomposed process.

In addition, we argue that extending the procedure to prove full compliance against a *set of* regulations only requires to iterate the process over each obligation in the set, thus increasing the complexity of proving compliance for a single regulation by a polynomial factor.

6.1   Full Compliance

A trace is compliant with a set of obligations if it fulfils all obligations belonging to the set. Note that according to Definitions 15 and 16 an obligation that is never activated by a trace is considered to be fulfilled by such a trace.

Definition 21 (Set Fulfilment). *Given a trace θ and a set of obligations $\mathbb{O} = \{\omega_1, \ldots, \omega_n\}$, $\theta \vdash \mathbb{O}$, iff:*

$$\forall \omega_i \in \mathbb{O}, (\theta \vdash \omega_i)$$

Otherwise $\theta \nvdash \mathbb{O}$.

Finally, a process model is said to be fully compliant with a set of obligations, if and only if each of its executions fulfils each of the obligations belonging to the given set.

Definition 22 (Process Full Compliance). *Given a process (P, ann) and a set of obligations $\mathbb{O}$.*

- *Full Compliance: $(P, \mathsf{ann}) \vdash^F \mathbb{O}$*
  iff $\forall \theta \in \Theta(P, \mathsf{ann}), \theta \vdash \mathbb{O}$.

6.2   Δ-constraints Verification

Given a partially ordered set representation of a decomposed business process model and the Δ-constraints representation of a given obligation, we illustrate how the constraints can be verified in the partially ordered sets, signifying that the original business process model contains at least one execution failing the original obligation.

Relevant Tasks

The first step of the verification consists of populating the sets of *relevant tasks*. Each set of relevant tasks contains the tasks having annotated a parameter of the obligation being checked. For instance, the set of relevant tasks for d contains every task in the business process model having d annotated.

From Definition 18 and Definition 20, it follows that viable task sets are required for: t, c, $\neg c$, d, and $\neg d$. These viable task sets are respectively represented as follows: $\mathbb{T}$, $\mathbb{C}$, $anti\mathbb{C}$, $\mathbb{D}$, $anti\mathbb{D}$. It follows from Definition 13 that the end task always belongs to $\mathbb{D}$.

Algorithm

Intuitively, to prove whether a process model is not fully compliant with a given regulation, we have to show that there exists a task in the process model having t annotated, and that it is possible to find instantiations of the relevant tasks for the existential quantifiers[6] in the Δ-constraints satisfying their ordering conditions.

```
input  : Relevant tasks: T, antiC, C, D, antiD and a partially ordered set representation of a decomposed business
         process model 𝒫(P′)
output: Whether a partially ordered representation P contains an execution violating ω by a trigger of t

 1 for tₜ ∈ T do
 2 |   for t₋c ∈ antiC do
 3 |   |   for t_d ∈ D do
 4 |   |   |   if t₋c ⪯ tₜ ⪯ t_d compatible with 𝒫(P′) then
 5 |   |   |   |   good = true;
 6 |   |   |   |   for t_c ∈ C do
 7 |   |   |   |   |   if t₋c ⪯ t_c ⪯ t_d compatible with 𝒫(P′) then
 8 |   |   |   |   |   |   good = false;
           |   |   |   |   end
           |   |   |   end
 9 |   |   |   if good then
10 |   |   |   |   for t₋d ∈ antiD do
11 |   |   |   |   |   if t₋d ⪯ tₜ then
12 |   |   |   |   |   |   good = true;
13 |   |   |   |   |   |   for t2_d ∈ D do
14 |   |   |   |   |   |   |   if t₋d ⪯ t2_d ⪯ tₜ compatible with 𝒫(P′) then
15 |   |   |   |   |   |   |   |   good = false;
           |   |   |   |   |   |   end
           |   |   |   |   |   end
16 |   |   |   |   |   if good then
17 |   |   |   |   |   |   return true;
           |   |   |   |   end
           |   |   |   |   end
           |   |   |   end
           |   |   |   end
           |   |   end
           |   end
           end
18 return false;
```

Algorithm 1: Achievement Δ-constraint 1 ($A\Delta1$)

Algorithm 1 illustrates how to verify whether a partially ordered set representation of a

---

[6]In Definition 20, the universal quantifier can be understood as it is referring to the earliest happening element quantified.

decomposed business process model satisfies the first Δ-constraint for achievement obligations. Note that checking whether a given ordering of tasks is compatible with the ordering of a partially ordered set, appearing in Algorithm 1 in lines 4, 7, and 14, can be done in polynomial time.

Complexity of Algorithm 1

Let $\mathbb{X}$ be the number of tasks in P, and assuming the worst case scenario, where the cardinality of each relevant set is $\mathbb{X}$. The computational complexity of Algorithm 1 is the following:

$$\mathbf{O}(|\mathbb{X}|^2 \times (|\mathbb{X}| + |\mathbb{X}|^2))$$

Thus the computational complexity of Algorithm 1 is polynomial in time with respect to the size of the process model. Algorithms to verify the other Δ-constraints closely resemble Algorithm 1[7], hence we do not explicitly illustrate and discuss them in this paper. Moreover, as Achievement Failure Δ-Constraint 1 is the one containing more ordering conditions (as shown in Definition 18 and Definition 20), the computational complexity of the other algorithms not explicitly discussed is at most as high as Algorithm 1.

6.3 Proving Full Compliance

Algorithm 2 shows how the algorithms verifying whether the Δ-constraints are satisfied in a decomposed business process model can be used to prove whether a business process model is fully compliant with a given obligation.

---

input : An obligation ω, a set of decomposed processes $\mathbb{P}$ and its viable task sets with respect to the obligation
output: Whether the process model corresponding to $\mathbb{P}$ is fully compliant with ω

foreach *decomposed process $P' \in \mathbb{P}$* **do**
    if *ω is achievement* **then**
        if $A\Delta1(\mathbb{T}, anti\mathbb{C}, \mathbb{C}, \mathbb{D}, anti\mathbb{D}, \mathscr{P}(P'))$ **then** return false;
        if $A\Delta2(\mathbb{T}, anti\mathbb{C}, \mathbb{C}, \mathbb{D}, anti\mathbb{D}, \mathscr{P}(P'))$ **then** return false;
    else
        if $M\Delta1(\mathbb{T}, anti\mathbb{C}, \mathbb{C}, \mathbb{D}, anti\mathbb{D}, \mathscr{P}(P'))$ **then** return false;
        if $M\Delta2(\mathbb{T}, anti\mathbb{C}, \mathbb{C}, \mathbb{D}, anti\mathbb{D}, \mathscr{P}(P'))$ **then** return false;
    end
end
return true;

Algorithm 2: Proving Full Compliance

---

[7]Other Δ-constraints checking algorithms would be structurally equivalent to Algorithm 1, with the only difference that the instantiations of the relevant tasks would be done on the ordering conditions of the other Δ-constraints.

Complexity of Algorithm 2

Let $\mathbb{P}$ be the set of the decomposed representations of the original process. The computational complexity of proving whether the process is compliant with a given regulation is the following:

$$\mathbf{O}(|\mathbb{P}| \times (\mathbf{O}(A\Delta1) + \mathbf{O}(A\Delta2) + \mathbf{O}(M\Delta1) + \mathbf{O}(M\Delta2)))$$

As $\mathbb{P}$ can be potentially exponential in size with respect to the size of the original process, we cannot claim that the complexity of Algorithm 2 is polynomial. However, as the computational complexity of any brute force approach to solve regulatory compliance is combinatorial with respect to the size of the problem (see Example 6), the proposed solution represents a more efficient approach as its computational complexity is exponential in time with respect to the size of the problem.

Illustrating the Verification

Examples 9 and 10 show how the Δ-constraints allow to identify fully compliant processes by analysing their executions. However, note that the thorough analysis of the executions is given only for illustration purposes, reminding that the proposed approach in the paper verifies the Δ-constraints directly on the partially ordered sets, as show by Algorithm 1.

Example 9 (Non-compliance). *Consider the annotated business process depicted in Figure 3 and its corresponding partially ordered sets of Example 8. P' denotes a concurrent execution of t_1 and t_3 after* start, *followed by t_4.*

P' allows two valid executions: ε_1: start, t_1, t_3, t_4, end *and ε_2:* start, t_3, t_1, t_4, end*. We substitute the task's labels with their annotations, making it easier to observe whether the Δ-constraints are fulfilled.*
ε_1: $(\neg c, \neg t, \neg d)$, (c), (t), $()$, (d)
ε_2: $(\neg c, \neg t, \neg d)$, (t), (c), $()$, (d)

Similarly, $\mathbb{P}''$ denotes a concurrent execution of t_2 and t_3 after start, *followed by t_4. As such, $\mathbb{P}''$ allows two valid executions:*
ε_3: $(\neg c, \neg t, \neg d)$, $()$, (t), $()$, (d)
ε_4: $(\neg c, \neg t, \neg d)$, (t), $()$, $()$, (d)

Given an obligation $\mathcal{O}^a\langle c, t, d \rangle$, applying the achievement patterns to ε_1, it is easy to see that c exists before d. As such, Achievement Failure Δ-Constraint 1 fails, as it requires the absence of c before d. Equivalently, there exists no d before t and Achievement Failure Δ-Constraint 2 fails immediately as well. Similarly for ε_2.

For ε_3, there exists no d before t, as in ε_1 and ε_2. From start *$\neg c$ holds, continues to hold through t_2, and still holds when t occurs in t_3. As t_1 is not part of the trace, c does not occur before d. Therefore, ε_3 fulfils the Achievement Failure Δ-Constraint 1 pattern and is, as a* *result, not compliant. This is similar for ε_4. Consequently, the process of Figure 3 is not fully compliant.*

Example 10 (Full compliance). *Consider the following annotated business process model:*

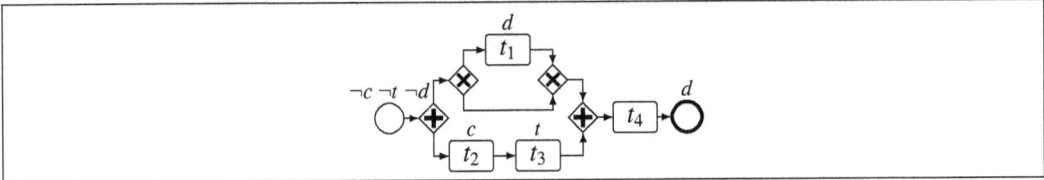

Figure 10: Example of a compliant process

This process can be decomposed as follows:

1. $P' = \mathsf{SEQ}(\mathsf{start}, \mathsf{AND}(t_1, \mathsf{SEQ}(t_2, t_3)), t_4, \mathsf{end})$

2. $P'' = \mathsf{SEQ}(\mathsf{start}, \mathsf{AND}(\emptyset, \mathsf{SEQ}(t_2, t_3)), t_4, \mathsf{end})$

The corresponding partially ordered sets for the two decomposed processes are the following:

1. $\mathbb{P}' = (\{\mathsf{start}, t_1, t_2, t_3, t_4, \mathsf{end}\}, \{\mathsf{start} \prec t_1, \mathsf{start} \prec t_2, t_1 \prec t_4, t_2 \prec t_3, t_3 \prec t_4, t_4 \prec \mathsf{end}\})$

2. $\mathbb{P}'' = (\{\mathsf{start}, t_2, t_3, t_4, \mathsf{end}\}, \{\mathsf{start} \prec t_2, t_2 \prec t_3, t_3 \prec t_4, t_4 \prec \mathsf{end}\})$

$\mathbb{P}'$ allows three valid executions[8]:
ε_1: $(\neg c, \neg t, \neg d), (d), (c), (t), (), (d)$
ε_2: $(\neg c, \neg t, \neg d), (c), (d), (t), (), (d)$
ε_3: $(\neg c, \neg t, \neg d), (c), (t), (d), (), (d)$
$\mathbb{P}''$ allows only one valid execution:
ε_4: $(\neg c, \neg t, \neg d), (c), (t), (), (d)$.

All four executions have c before t. Given two obligations: $\mathcal{O}^a \langle c, t, d \rangle$ and $\mathcal{O}^m \langle c, t, d \rangle$, the respective Δ-constraints Achievement Failure Δ-Constraint 1, Achievement Failure Δ-Constraint 2 and Maintenance Failure Δ-Constraint 1 fail, as they require the absence of c before t. Maintenance Failure Δ-Constraint 2 does have c before t, but also requires $\neg c$ between t and d for every d and fails, therefore, as well for all executions. As none of the patterns apply to the process, we can conclude that the process is fully compliant.

	Nesting			Executions	Dec.	Time	Check T.	Total Time
	Level 1	Level 2	Level 3					
1	$1 \cdot AND_{2\times5}$	–	–	252	1	0.08 ms	0.04 ms	0.12 ms
2	$1 \cdot AND_{4\times5}$	–	–	1.17E+10	1	0.08 ms	0.05 ms	0.12 ms
3	$2 \cdot AND_{2\times5}$	–	–	63 504	1	0.07 ms	0.04 ms	0.11 ms
4	$2 \cdot AND_{4\times5}$	–	–	1.38E+20	1	0.08 ms	0.20 ms	0.29 ms
5	$1 \cdot AND_{2\times5}$	$1 \cdot XOR_{2\times5}$	–	194 480	4	0.06 ms	0.30 ms	0.36 ms
6	$1 \cdot AND_{4\times5}$	$1 \cdot XOR_{4\times5}$	–	3.43E+20	256	3.13 ms	2.21 ms	5.34 ms
7	$1 \cdot AND_{2\times5}$	$1 \cdot AND_{2\times5}$	–	2.55E+12	1	0.05 ms	0.14 ms	0.19 ms
8	$1 \cdot AND_{4\times5}$	$1 \cdot AND_{4\times5}$	–	1.27E+95	1	0.42 ms	2.85 ms	3.27 ms
9	$1 \cdot AND_{2\times5}$	$1 \cdot XOR_{2\times5}$	$1 \cdot AND_{2\times5}$	2.30E+15	4	0.18 ms	0.85 ms	1.02 ms
10	$1 \cdot AND_{4\times5}$	$1 \cdot XOR_{4\times5}$	$1 \cdot AND_{4\times5}$	1.11E+107	256	11.88 ms	120.50 ms	132.38 ms

Table 2: Evaluation models and performance.

7 Evaluation

We implemented the proposed method as a standalone Java tool. We tested our approach over a set of synthetic process models of increasing complexity, up to the point where the amount of concurrency is well beyond realistic business scenarios. The models consist of a set of nested process constructs, which are either a structured AND-block or structured XOR-block with n branches of m activities long. Each of the synthetic models is randomly annotated, which ensures that one every three tasks in the model is annotated with a randomly selected set of literals. Figure 11 shows the basic structure of the models.

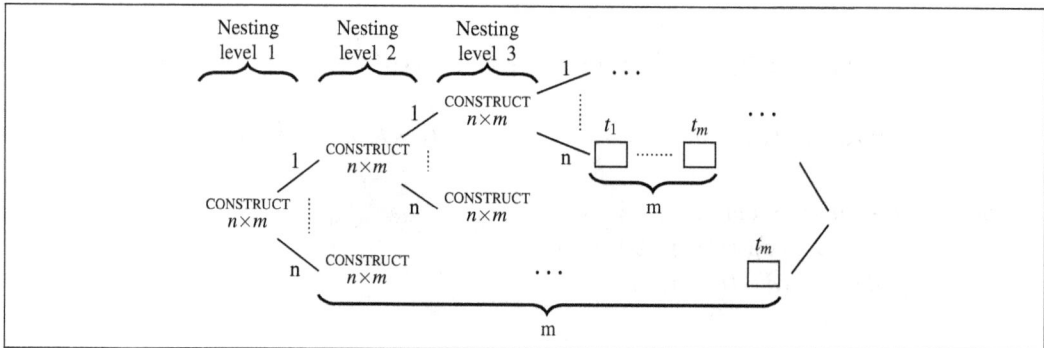

Figure 11: Synthetic process structure.

All tests were performed on a computer equipped with a quad core Intel® Core™ i7-7700HQ CPU @ 3.80GHz, 16GB RAM, running Ubuntu 16.10 and Java 1.8.0_131. To eliminate load times, each test was executed five times, where the average time of three executions was recorded while removing the fastest and the slowest.

---

[8]We use again the tasks' annotation instead of their labels to clearly show whether a Δ-constraint is satisfied.

The results are shown in Table 2. The first three columns of the table contain details about the process structure. The columns under *Nesting* contain information on the process structure, by depicting for each nesting level the number of blocks, their type, the number of branches, and the length of each branch. As such, the column *Level 2* describes the blocks nested in each branch of each block described in *Level 1*. Similarly, *Level 3* contains information of the blocks nested in each branch of each block in *Level 2*. The fourth column, *Executions*, describes how many different executions are hypothetically possible in the given model when linearising the concurrent paths to indicate the theoretical complexity of the models when adopting a brute force approach. The simplest model (1) has 252 possible executions, while the most complex model (10) has 1.11E+107 possible executions.

The column *Dec.* contains the number of decomposed processes generated from the original process and *Time* contains the time required for their generation. Finally, *Check T.* contains the time required to solve all decomposed processes, and *Total Time* contains the total time required by the procedure to obtain a result concerning the compliance of the process. That is, it records the total time required for decomposition and evaluation of the decomposed processes.

Existing approaches, such as Regorous [12], use brute force, thereby evaluating all possible executions. Regorous is able to solve the first process and third process from Table 2 in 23 seconds and 47 seconds, respectively. For the other processes, we stopped Regorous after 10min, without being able to decide on the solution within the given time.

8 Conclusion

In this paper, we proposed an approach capable of efficiently verifying whether process models comprising concurrency are fully compliant with a set of obligations. Some of the key contributions of the proposed approach are the introduction of Δ-constraints, an alternative representation of the obligations used to specify the compliance requirements, and the ability to verify whether a business process model is fully compliant directly analysing its structure, without explicitly generating its executions. Compared to other approaches trying to solve the compliance problem through *brute force* or using *heuristics*, our proposed approach reduces the overall computational complexity of solving a sub-problem of the compliance problem by using a divide an conquer approach, while still steering clear from approximate solutions.

Although theoretically exponential in complexity (due to exclusive paths), we have empirically shown that the combined approach is capable of solving highly complex processes that are otherwise infeasible using existing brute force approaches. Even processes with more than 250 possible paths and 1.11E+107 possible executions were checked within 132ms.

However, the demonstrated performance gain does come at a tradeoff, as *brute force based* approaches are capable of solving more expressive instances of the problem than our approach, as we allow only literals and not formulae. While this limitation prevents us from compliance checking with full regulatory specifications, it can be successfully used for many aspects of structural compliance (i.e. conditions about the tasks appearing and their mutual relationships). Despite the structural limitations over the process model, we show how our solution can be combined with additional procedures in order to solve more generic problems.

As future work, we plan to improve the current solution in order to be able to resolve some of the current limitations. We reckon that a possibility to improve the current approach is to investigate how the Δ-constraints introduced in the solution can be generalised, and potentially reused in a more modular fashion to create efficient solutions for more generic sub-problems of business process compliance.

Acknowledgments

This research is supported by the science and industry endowment fund.

References

[1] Carlos E. Alchourrón, Peter Gärdenfors, and David Makinson. On the logic of theory change: Partial meet contraction and revision functions. *J. of Symbolic Logic*, 50(2):510–530, 1985.

[2] Ahmed Awad, Gero Decker, and Mathias Weske. Efficient compliance checking using bpmn-q and temporal logic. In *BPM 2008*, pages 326–341. Springer, 2008.

[3] Ahmed Awad, Matthias Weidlich, and Mathias Weske. Visually specifying compliance rules and explaining their violations for business processes. *J. Vis. Lang. Comput.*, 22(1):30–55, 2011.

[4] Domenico Bianculli, Carlo Ghezzi, and Paola Spoletini. A model checking approach to verify bpel4ws workflows. In *Service-Oriented Computing and Applications, 2007. SOCA'07. IEEE International Conference on*, pages 13–20, 2007.

[5] Philip R. Cohen and Hector J. Levesque. Intention is choice with commitment. *Artif. Intell.*, 42(2-3):213–261, 1990.

[6] Silvano Colombo Tosatto, Guido Governatori, and Pierre Kelsen. Business process regulatory compliance is hard. *IEEE Transactions on Services Computing*, 8(6):958–970, 2015.

[7] Thomas Curran, Gerhard Keller, and Andrew Ladd. *SAP R/3 Business Blueprint: Understanding the Business Process Reference Model*. Prentice-Hall, Inc., Upper Saddle River, NJ, USA, 1998.

[8] Remco M. Dijkman, Marlon Dumas, and Chun Ouyang. Semantics and analysis of business process models in bpmn. *Information and Software Technology*, 50(12):1281–1294, 2008.

[9] Amal Elgammal, Oktay Turetken, Willem-Jan van den Heuvel, and Mike Papazoglou. Formalizing and Applying Compliance Patterns for Business Process Compliance. *Software & Systems Modeling*, pages 1–28, 2014.

[10] Sven Feja, Andreas Speck, and Elke Pulvermüller. Business process verification. In *GI Jahrestagung*, pages 4037–4051, 2009.

[11] G. Governatori and A. Rotolo. How do agents comply with norms? In *2009 IEEE/WIC/ACM International Joint Conference on Web Intelligence and Intelligent Agent Technology*, volume 3, pages 488–491, Sept 2009.

[12] Guido Governatori. The Regorous approach to process compliance. In *2015 IEEE 19th International EDOC Workshop*, pages 33–40. IEEE Press, 2015.

[13] Guido Governatori and Mustafa Hashmi. No time for compliance. In Sylvain HallÃl' and Wolfgang Mayer, editors, *2015 IEEE 19th Enterprise Distibuted Object Computing Conference*, pages 9–18. IEEE, 2015.

[14] Guido Governatori, Jörg Hoffmann, Shazia Wasim Sadiq, and Ingo Weber. Detecting regulatory compliance for business process models through semantic annotations. In *Business Process Management Workshops*, volume 17 of *LNBIP*, pages 5–17. Springer, 2008.

[15] Guido Governatori and Antonino Rotolo. An algorithm for business process compliance. In Enrico Francesconi, Giovanni Sartor, and Daniela Tiscornia, editors, *The Twenty-First Annual Conference on Legal Knowledge and Information Systems*, volume 189 of *Frontieres in Artificial Intelligence and Applications*, pages 186–191. IOS Press, 2008.

[16] Heerko Groefsema, Nick R T P van Beest, and Marco Aiello. A formal model for compliance verification of service compositions. *IEEE Transactions on Services Computing*, 11:466 – 479, 2018.

[17] Mustafa Hashmi, Guido Governatori, Ho-Pun Lam, and Moe Thandar Wynn. Are we done with business process compliance: State-of-the-art and challenges ahead. *Know. and Inf. Sys.*, 2018.

[18] Mustafa Hashmi, Guido Governatori, and Moe Thandar Wynn. Normative requirements for regulatory compliance: An abstract formal framework. *Information Systems Frontiers*, 18(3):429–455, 2016.

[19] Ahmed Kheldoun, Kamel Barkaoui, and Malika Ioualalen. Specification and verification of complex business processes - a high-level petri net-based approach. In *Business Process Management*, volume 9253 of *LNCS*, pages 55–71. Springer International Publishing, 2015.

[20] Oussama M. Kherbouche, Adeel Ahmad, and Henri Basson. Using model checking to control the structural errors in bpmn models. In *Research Challenges in Information Science (RCIS), 2013 IEEE Seventh International Conference on*, pages 1–12, 2013.

[21] Bartek Kiepuszewski, Arthur H. M. ter Hofstede, and Christoph Bussler. On structured workflow modelling. In *Proceedings of the 12th International Conference on Advanced Information Systems Engineering*, pages 431–445, London, UK, UK, 2000. Springer-Verlag.

[22] Marcello La Rosa, Hajo A Reijers, Wil M P Van Der Aalst, Remco M Dijkman, Jan Mendling, Marlon Dumas, and Luciano García-Bañuelos. Apromore: An advanced process model repository. *Expert Systems with Applications*, 38(6):7029–7040, 2011.

[23] Timo Latvala and Keijo Heljanko. Coping with strong fairness. *Fundamenta Informaticae*, 43(1-4):175–193, 2000.

[24] Ying Liu, Samuel Müller, and Ke Xu. A static compliance-checking framework for business process models. *IBM Systems Journal*, 46:335–361, 2007.

[25] M. Pesic, H. Schonenberg, and Wil van der Aalst. DECLARE: Full Support for Loosely-Structured Processes. In *Procedings of 11th IEEE International Conference on Enterprise Distributed Object Computing (EDOC'07)*, pages 287–287, 2007.

[26] Artem Polyvyanyy, Luciano García-Bañuelos, and Marlon Dumas. Structuring acyclic process models. In *International Conference on Business Process Management*, pages 276–293. Springer, 2010.

[27] Elke Pulvermüller, Sven Feja, and Andreas Speck. Developer-friendly verification of process-based systems. *Knowl.-Based Syst.*, 23(7):667–676, 2010.

[28] Elham Ramezani, Dirk Fahland, and Wil MP van der Aalst. Supporting domain experts to select and configure precise compliance rules. In *International Conference on Business Process Management*, pages 498–512. Springer, 2013.

[29] Shazia Sadiq, Guido Governatori, and Kioumars Namiri. Modeling control objectives for business process compliance. In *International conference on business process management*, pages 149–164. Springer, 2007.

[30] van der Aalst Wil, Adriansyah Arya, and van Dongen Boudewijn. Replaying history on process models for conformance checking and performance analysis. *Wiley Interdisciplinary Reviews: Data Mining and Knowledge Discovery*, 2(2):182–192, 2012.

[31] Boudewijn F van Dongen, Wil M P Van der Aalst, and Henricus M W Verbeek. Verification of epcs: Using reduction rules and petri nets. In *International Conference on Advanced Information Systems Engineering*, pages 372–386. Springer, 2005.

Received 27 September 2018

www.ingramcontent.com/pod-product-compliance
Lightning Source LLC
Chambersburg PA
CBHW051206200326
41519CB00025B/7023